KEW INDEX
FOR 1987

KEW INDEX FOR 1987

NAMES OF SEED-BEARING PLANTS, FERNS, AND FERN ALLIES AT THE RANK OF FAMILY AND BELOW PUBLISHED DURING 1987 WITH SOME OMISSIONS FROM EARLIER YEARS

COMPILED BY

R. A. DAVIES

and

K. M. LLOYD

AT THE HERBARIUM OF THE ROYAL BOTANIC GARDENS

KEW

CLARENDON PRESS · OXFORD

1988

Oxford University Press, Walton Street, Oxford OX2 6DP

Oxford New York Toronto
Delhi Bombay Calcutta Madras Karachi
Petaling Jaya Singapore Hong Kong Tokyo
Nairobi Dar es Salaam Cape Town
Melbourne Auckland

and associated companies in
Berlin Ibadan

Oxford is a trade mark of Oxford University Press

Published in the United States
by Oxford University Press, New York

British Library Cataloguing in Publication Data
Kew index for . . . : names of seed-bearing plants, ferns, and fern allies at the rank of family and below published during . . . with some omissions from earlier years.—1987–
1. Phanerogams 2. Botany—Nomenclators
I. Index Kewensis
582'.0014 QK11
ISBN 0–19–854245–3

ISSN 0951–7502

Typeset by Promenade Graphics Ltd., Cheltenham
Printed and bound in
Great Britain by Biddles Ltd.
Guildford and King's Lynn

FOREWORD

In this, the second annual *Kew Index*, we continue to use the same journal abbreviations as *Kew Record*, which publishes bibliographic details as additional journals are scanned.

The entries for ferns and fern-allies continue to be listed in an appendix, as in 1986. These entries will form the basis of *Index Filicum*, Supplement 7.

Cyrillic characters have been transliterated in accordance with British Standard 2979 (1958): *Transliteration of Cyrillic and Greek Characters*.

Names at all taxonomic ranks from family downwards are included. Names at ranks above genus are arranged alphabetically under families in a separate sequence at the beginning of the volume. We have adopted the term 'isonym' to denote a 'new combination' which has already been made earlier by another author.

A dagger (†) indicates the corrected re-entry of a name cited in an earlier volume.

Books and journals received at Kew or seen at the British Museum (Natural History) during 1987 have been scanned. A few extra entries missed in previous years are also included.

We are grateful to Dr R. K. Brummitt for much nomenclatural advice and to the Kew Computer Section, especially Mr M. D. Jackson and Mr D. F. Martindale, for enabling the material to be processed.

Royal Botanic Gardens
Kew
January 1988

E. ARTHUR BELL
Director

CONTENTS

KEW INDEX FOR 1987
NAMES OF ALL RANKS ABOVE GENUS UP TO AND INCLUDING FAMILY

ANACARDIACEAE

subfam. **Dobineoideae** (Engler) A. Takhtadzhyan, Sistema Magnoliofitov (Systema Magnoliophytorum): 180 (1987): tribus Dobineeae.

subfam. **Julianioideae** (Hemsley) A. Takhtadzhyan, Sistema Magnoliofitov (Systema Magnoliophytorum): 180 (1987): fam. Julianiaceae.

subfam. **Pistacioideae** (Caruel) A. Takhtadzhyan, Sistema Magnoliofitov (Systema Magnoliophytorum): 180 (1987): fam. Pistaciaceae.

subfam. **Spondioideae** (Kunth) A. Takhtadzhyan, Sistema Magnoliofitov (Systema Magnoliophytorum): 180 (1987): fam. Spondiaceae.

APOCYNACEAE

tribus **Anodendreae** Trân Đínhh Lý in Feddes Repert., 97(5–6): 272 (1986), without latin descr.

subfam. **Carissoideae** Trân Đính Lý in Feddes Repert., 97(5–6): 266 (1986), without latin descr.

APODANTHACEAE

fam. **Apodanthaceae** (R. Br.) Van Tieghem ex A. Takhtadzhyan, Sistema Magnoliofitov (Systema Magnoliophytorum): 42 (1987): tribus Apodantheae.

ARGOPHYLLACEAE

fam. **Argophyllaceae** (Engler) A. Takhtadzhyan, Sistema Magnoliofitov (Systema Magnoliophytorum): 208 (1987): Saxifragaceae tribus Argophylleae.

BRUNIACEAE

subfam. **Berzelioideae** (Nakai) A. Takhtadzhyan, Sistema Magnoliofitov (Systema Magnoliophytorum): 145 (1987): fam. Berzeliaceae.

CACTACEAE

subtribus **Pfeifferinae** S.A. Volgin in Feddes Repert., 97(9–10): 563 (1986).

subtribus **Schlumbergerinae** S.A. Volgin in Feddes Repert., 97(9–10): 563 (1986).

CONOSTYLIDACEAE

fam. **Conostylidaceae** (Pax) A. Takhtadzhyan, Sistema Magnoliofitov (Systema Magnoliophytorum): 313 (1987): Amaryllidaceae tribus Conostylideae.

CURTISIACEAE

fam. **Curtisiaceae** (Harms) A. Takhtadzhyan, Sistema Magnoliofitov (Systema Magnoliophytorum): 214 (1987): Cornaceae subfam. Curtisioideae.

DACTYLANTHACEAE

fam. **Dactylanthaceae** (Engler) A. Takhtadzhyan, Sistema Magnoliofitov (Systema Magnoliophytorum): 43 (1987): Balanophoraceae subfam. Dactylanthoideae.

GRAMINEAE

tribus **Atractocarpeae** (E.-G. Camus) Jacq.-Félix ex N.N. Tsvelev in Komarov. Chten., 37: 22 (1987): subtribus Atractocarpeae.

tribus **Dendrocalameae** (Benth.) Keng ex N.N. Tsvelev in Komarov. Chten., 37: 20 (1987): subtribus Dendrocalameae.

tribus **Leptaspideae** N.N. Tsvelev in Komarov. Chten., 37: 25 (1987).

subtribus **Monerminae** (C.E. Hubb.) N.N. Tsvelev in Komarov. Chten., 37: 33 (1987): tribus Monermeae.

subtribus **Neurachninae** (S.T. Blake) W.D. Clayton & S.A. Renvoize, Genera Graminum (Kew Bull. Addit. Ser. 13): 377 (1986): tribus Neurachneae.

subfam. **Oryzoideae** H. Duistermaat in Blumea, 32(1): 158 (1987).

tribus **Oxytenanthereae** N.N. Tsvelev in Komarov. Chten., 37: 22 (1987).

subtribus **Scolochloinae** (Tzvel.) N.N. Tsvelev in Komarov. Chten., 37: 33 (1987): tribus Scolochloeae.

GRISELINIACEAE

fam. **Griseliniaceae** (Wangerin) A. Takhtadzhyan, Sistema Magnoliofitov (Systema Magnoliophytorum): 209 (1987): Cornaceae tribus Griselinieae.

HAEMODORACEAE

subfam. **Conostylidoideae** (Lindley) T.D. Macfarlane & S.D. Hopper in Fl. Australia, 45: 454 (1987): Haemodoraceae tribus Conostylideae.

HAEMODORACEAE :–
tribus **Tribonantheae** T.D. Macfarlane & S.D. Hopper in Fl. Australia, 45: 454, 131 (1987).

LEGUMINOSAE
subtribus **Trigonellinae** (Schulz) E. Small in C.H. Stirton (ed.), Advances in Legume System., 3: 177 (1987): tribus Papilionatae-Trigonelleae.

MEDEOLACEAE
fam. **Medeolaceae** (S. Watson) A. Takhtadzhyan, Sistema Magnoliofitov (Systema Magnoliophytorum): 291 (1987): Liliaceae subtribus Medeoleae.

MELIACEAE
tribus **Nymanieae** F. White in Bothalia, 16(2): 145 (1986).

MONIMIACEAE
tribus **Hennecartieae** W.R. Philipson in Nordic J. Bot., 7(1): 28 (1987), without latin descr.
tribus **Palmerieae** W.R. Philipson in Nordic J. Bot., 7(1): 28 (1987), without latin descr.
tribus **Peumieae** (Schodde) W.R. Philipson in Nordic J. Bot., 7(1): 28 (1987), without basionym ref.: Monimiaceae subfam. Peumoidieae.

MONOTACEAE
fam. **Monotaceae** (Gilg) Maury ex A. Takhtadzhyan, Sistema Magnoliofitov (Systema Magnoliophytorum): 129 (1987): Dipterocarpaceae subfam. Monotoideae.

NYMPHAEACEAE
subfam. **Nupharioideae** (Endl.) M. Ito in Bot. Mag. Tokyo, 100(1057): 33 (1987): Nymphaeaceae tribus Nuphaeae.

ORCHIDACEAE
subtribus **Galeolinae** L.A. Garay in Bot. Mus. Leafl. Harvard Univ., 30(4): 233 (1986).

PEGANACEAE
fam. **Peganaceae** (Engler) Van Tieghem ex A. Takhtadzhyan, Sistema Magnoliofitov (Systema Magnoliophytorum): 178 (1987): Zygophyllaceae subfam. Peganoideae.

POLYGALACEAE
subfam. **Moutabeoideae** (Endlicher) A. Takhtadzhyan, Sistema Magnoliofitov (Systema Magnoliophytorum): 192 (1987): fam. Moutabeaceae.

POLYGALACEAE :–
subfam. **Xanthophylloideae** (Gagnepain) A. Takhtadzhyan, Sistema Magnoliofitov (Systema Magnoliophytorum): 192 (1987): fam. Xanthophyllaceae.

POTTINGERIACEAE
fam. **Pottingeriaceae** (Engler) A. Takhtadzhyan, Sistema Magnoliofitov (Systema Magnoliophytorum): 212 (1987): Saxifragaceae tribus Pottingerieae.

RHOPALOCARPACEAE
fam. **Rhopalocarpaceae** Hemsley ex A. Takhtadzhyan, Sistema Magnoliofitov (Systema Magnoliophytorum): 130 (1987), nom. nov.: Bixaceae tribus Sphaerosepaleae Warburg.

RIPOGONACEAE
fam. **Ripogonaceae** Huber ex A. Takhtadzhyan, Sistema Magnoliofitov (Systema Magnoliophytorum): 306 (1987), isonym; see J.G. Conran & H.T. Clifford in Nordic J. Bot., 5(3): 219 (1985).

ROSACEAE
subtribus **Coluriinae** (Rydb.) E.I. Golubkova in Bot. Zhurn., 72(6): 777 (1987): Rosaceae tribus Colurieae.

RUBIACEAE
tribus **Aulacocalyceae** E. Robbrecht & C. Puff in Bot. Jahrb., 108(1): 123 (1986).
tribus **Brachytomeae** D.D. Tirvengadum in Bull. Mus. Nation. Hist. Nat., B, Adansonia, Ser. 4, 8(3): 292 (1986).

SARCOPHYTACEAE
fam. **Sarcophytaceae** (Engler) Van Tieghem ex A. Takhtadzhyan, Sistema Magnoliofitov (Systema Magnoliophytorum): 43 (1987): Balanophoraceae subfam. Sarcophytoideae.

TAPISCIACEAE
fam. **Tapisciaceae** (F. Pax) A. Takhtadzhyan, Sistema Magnoliofitov (Systema Magnoliophytorum): 171 (1987): Staphyleaceae subfam. Tapiscioideae.

VIOLACEAE
subfam. **Fusispermoideae** W.H.A. Hekking in Proc. Kon. Nederl. Akad. Wetensch., C, 87(2): 128 (1984).

KEW INDEX FOR 1987
NAMES AT GENERIC RANK AND BELOW

A

ABILDGAARDIA (Cyperac.).
acutespicata *K.A. Lye in Nordic J. Bot.*, 7(1): 42 (1987)—Zambia.
angustespicata *K.A. Lye in Nordic J. Bot.*, 7(1): 43 (1987)—Tanzania.
densiflora *K.A. Lye in Nordic J. Bot.*, 7(1): 41 (1987)—Zambia.
disticha *K.A. Lye in Nordic J. Bot.*, 6(6): 765 (1986)—Brazil.
lacunosa *K.A. Lye in Nordic J. Bot.*, 7(1): 44 (1987)—Zambia.
microcephala *K.A. Lye in Nordic J. Bot.*, 7(1): 46 (1987)—Zambia.
microrotundata *K.A. Lye in Nordic J. Bot.*, 7(1): 47 (1987)—Tanzania.
pluricephala *K.A. Lye in Nordic J. Bot.*, 7(1): 49 (1987)—Zambia.
squarrosa *K.A. Lye in Nordic J. Bot.*, 7(1): 39 (1987)—Kenya.

ABUTILON (Malvac.).
burandtii *P.A. Fryxell in Syst. Bot.*, 12(2): 274 (1987)—Venezuela.

ACACIA (Legumin.).
janzenii *J.E. Ebinger & D.S. Seigler in Southwest. Nat.*, 32(2): 245 (1987)—Mexico.
sousae *L. Rico in Brittonia*, 39(1): 130 (1987)—Mexico.

ACALYPHA (Euphorbiac.).
marissima *M.G. Gilbert in Kew Bull.*, 42(2): 360 (1987)—Ethiopia.

ACANTHOLIMON (Plumbaginac.).
aegaeum *F.K. Meyer in Haussknechtia, Mitt. Thüring. Bot. Ges.*, 3: 38 (1987)—East Aegean Is.
albanicum *O. Schwarz & F.K. Meyer in Haussknechtia, Mitt. Thüring. Bot. Ges.*, 3: 31 (1987)—Albania.
diversifolium *O. Schwarz & F.K. Meyer in Haussknechtia, Mitt. Thüring. Bot. Ges.*, 3: 43 (1987)—Turkey.
gabrieljaniae *N.V. Mirzoeva in Biol. Zhurn. Arm.*, 34(9): 973 (1981), as 'gabrieljanii'—U.S.S.R. (Caucasus).
graecum *F.K. Meyer in Haussknechtia, Mitt. Thüring. Bot. Ges.*, 3: 20 (1987)—Greece, Albania.

ACANTHOLIMON :-
muradicum *O. Schwarz & F.K. Meyer in Haussknechtia, Mitt. Thüring. Bot. Ges.*, 3: 35 (1987)—Turkey.
olympicum *(Boiss.) F.K. Meyer in Haussknechtia, Mitt. Thüring. Bot. Ges.*, 3: 24 (1987): A. androsaceum β olympicum.
trojanum *F.K. Meyer in Haussknechtia, Mitt. Thüring. Bot. Ges.*, 3: 27 (1987)—Turkey.

ACANTHOPANAX (Araliac.).
baoxinensis *X.P. Fang & C.K. Hsieh in Bull. Bot. Res. North-East. Forest. Inst.*, 7(4): 89 (1987)—China.
nanpingensis *X.P. Fang & C.K. Hsieh in Bull. Bot. Res. North-East. Forest. Inst.*, 7(4): 87 (1987)—China.
wilsonii
var. pilosulus *(Rehd.) X.P. Fang & C.K. Hsieh in Bull. Bot. Res. North-East. Forest. Inst.*, 7(4): 90 (1987): A. giraldii var. pilosulus.

ACER (Acerac.).
guizhouense *Y.K. Li in Guihaia*, 7(3): 211 (1987)—China.
legonsanicum *Y.K. Li in Guihaia*, 7(3): 212 (1987)—China.

ACETOSELLA (Polygonac.).
multifida *(L.) Á. Löve, Fl. Iceland:* 168 (1983): Rumex multifidus.

ACHILLEA (Compos.).
ser. Lingulatae *Botsch. ex M.L. Samutina in Vestn. Leningrad. Univ., Biol.*, 1983(21): 96 (1983).
ser. Ptarmicae *Botsch. ex M.L. Samutina in Vestn. Leningrad. Univ., Biol.*, 1983(21): 96 (1983).
ser. Salicifoliae *Botsch. ex M.L. Samutina in Vestn. Leningrad. Univ., Biol.*, 1983(21): 96 (1983).

ACHIMENES (Gesneriac.).
nayaritensis *L.E. Skog in Op. Bot.*, 92: 225 (1987)—Mexico.

ACHMANDRA G.A.W. Arnott in Madras J. Lit. Sci., 12: 49 (1840). *CUCURBITACEAE.*

ACIDOSASA (Gramin.).
purpurea *(Hsueh & Yi) P.C. Keng in J. Wuhan Bot. Res.*, 4(4): 335 (1986): Indosasa purpurea.

ACINOS (Labiat.).
troodi
subsp. grandiflorus *P. Hartvig & A. Strid in Bot. Jahrb.,* 108(2–3): 331 (1987)—Turkey.

ACONITUM (Ranunculac.).
abietetorum *W.T. Wang & L.Q. Li in Acta Phytotax. Sin.,* 25(1): 32 (1987)—China.
chasmanthum
subsp. baltistanicum *R.A. Qureshi & M.N. Chaudhri in Pakistan Syst.,* 3(1): 4 (1987)—Pakistan.
gezaense *W.T. Wang & L.Q. Li in Acta Phytotax. Sin.,* 25(1): 31 (1987)—China.
hemsleyanum
var. lasianthum *W.T. Wang & L.Q. Li in Acta Phytotax. Sin.,* 25(1): 29 (1987)—China.
var. pilopetalum *W.T. Wang & L.Q. Li in Acta Phytotax. Sin.,* 25(1): 29 (1987)—China.
henryi
var. pilocarpum *W.T. Wang & L.Q. Li in Acta Phytotax. Sin.,* 25(1): 29 (1987)—China.
heterophyllum
subsp. parciflorum *R.A. Qureshi & M.N. Chaudhri in Pakistan Syst.,* 3(1): 6 (1987)—Pakistan.
jaluense
subsp. iwatekense *(Nakai) Y. Kadota in Ann. Tsukuba Bot. Gard.,* 2: 101 (1983): A. iwatekense.
japonicum
subsp. maritimum *(Nakai ex Tamura & Namba) Y. Kadota in Ann. Tsukuba Bot. Gard.,* 2: 100 (1983): A. deflexum var. maritimum.
kiyomiense *Y. Kadota in Bull. Nation. Sci. Mus., B (Tokyo),* 13(1): 26 (1987)—Japan.
kongboense
var. polycarpum *W.T. Wang & L.Q. Li in Acta Phytotax. Sin.,* 25(1): 33 (1987): China.
luningense *W.T. Wang in Acta Phytotax. Sin.,* 25(1): 24 (1987)—China.
lycoctonifolium *W.T. Wang & L.Q. Li in Acta Phytotax. Sin.,* 25(1): 27 (1987)—Xizang.
napiforme
var. albiflorum *Y.N. Lee in Korean J. Bot.,* 25(4): 177 (1982)—Korea.
ouvrardianum
var. pilopes *W.T. Wang & L.Q. Li in Acta Phytotax. Sin.,* 25(1): 32 (1987)—China.
pilopetalum *W.T. Wang & L.Q. Li in Acta Phytotax. Sin.,* 25(1): 30 (1987)—China.
swatense
var. bracteatum *R.A. Qureshi & M.N. Chaudhri in Pakistan Syst.,* 3(1): 5 (1987)—Pakistan.
var. kashmeriense *R.A. Qureshi & M.N. Chaudhri in Pakistan Syst.,* 3(1): 5 (1987)—Pakistan.

ACONITUM :–
tabatae *M. Tamura in Acta Phytotax. Geobot.,* 37(4–6): 152 (1986)—Nepal.
wardii
var. trisectum *W.T. Wang & L.Q. Li in Acta Phytotax. Sin.,* 25(1): 24 (1987)—China.
yangii *W.T. Wang & L.Q. Li in Acta Phytotax. Sin.,* 25(1): 25 (1987)—China.
var. villosulum *W.T. Wang & L.Q. Li in Acta Phytotax. Sin.,* 25(1): 26 (1987)—China.
yanyuanense *W.T. Wang & L.Q. Li in Acta Phytotax. Sin.,* 25(1): 28 (1987)—China.
zigzag
subsp. ryohakuense *Y. Kadota in Mem. Nation. Sci. Mus. (Tokyo),* 19: 137 (1986)—Japan.

ACRODON (Aizoac.).
duplessiae *(L. Bol.) H.F. Glen in Bothalia,* 16(2): 217 (1986): Ruschia duplessiae.
leptophyllus *(L. Bol.) H.F. Glen in Bothalia,* 16(2): 217 (1986): Ruschia leptophylla.

ACROPOGON (Sterculiac.).
austrocaledonicus *(J.D. Hook.) P. Morat in Bull. Mus. Nation. Hist. Nat., B, Adansonia,* Ser. 4, 8(4): 361 (1986 publ. 1987): Sterculia austro-caledonica.
bullatus *(Pancher & Sebert) P. Morat in Bull. Mus. Nation. Hist. Nat., B, Adansonia,* Ser. 4, 8(4): 361 (1986 publ. 1987): Sterculia bullata.
dzumacensis *(Guillaumin) P. Morat in Bull. Mus. Nation. Hist. Nat., B, Adansonia,* Ser. 4, 8(4): 361 (1986 publ. 1987): Sterculia dzumacensis.
francii *(Guillaumin) P. Morat in Bull. Mus. Nation. Hist. Nat., B, Adansonia,* Ser. 4, 8(4): 361 (1986 publ. 1987): Sterculia francii.
megaphyllus *(Bureau & Poisson ex Guillaumin) P. Morat in Bull. Mus. Nation. Hist. Nat., B, Adansonia,* Ser. 4, 8(4): 361 (1986 publ. 1987): Sterculia megaphylla.
schefferaefolius *(Guillaumin) P. Morat in Bull. Mus. Nation. Hist. Nat., B, Adansonia,* Ser. 4, 8(4): 361 (1986 publ. 1987): Sterculia scheffleraefolia.

ACTEPHILA (Euphorbiac.).
daii *N.N. Thìn in J. Biol. (Vietnam),* 9(2): 39 (1987)—Vietnam.

ACTINIDIA (Actinidiac.).
arguta
var. erythrocarpa *S. Ohtani in Sci. Rep. Yokosuka City Mus.,* 26: 7 (1979)—Japan.

ACTINODAPHNE (Laurac.).
monantha *(Yang & P.H. Huang) H.P. Tsui in Bull. Bot. Res. North-East. Forest. Inst.,* 7(4): 4 (1987): Litsea monantha.
sessilifructa *C.J. Qi & K.W. Liu in Guihaia,* 7(3): 215 (1987)—China.

ADELOBOTRYS (Melastomatac.).
ayangannensis *J.J. Wurdack in Brittonia,* 39(2): 159 (1987)—Guyana.

ADENOPHORA (Campanulac.).
pereskiifolia
subsp. alternifolia *(Fuh ex Y.Z. Zhao) C.X. Fu & M.Y. Liu in Acta Phytotax. Sin.,* 25(3): 186 (1987): A. pereskiifolia var. alternifolia.

ADENOPHYLLUM (Compos.).
anomalum *(Canby & Rose) J.L. Strother in Sida,* 11(4): 376 (1986): Hymenatherum anomalum.
appendiculatum *(Lagasca) J.L. Strother in Sida,* 11(4): 376 (1986): Dyssodia appendiculata.
aurantium *(L.) J.L. Strother in Sida,* 11(4): 376 (1986): Aster aurantius.
cooperi *(A. Gray) J.L. Strother in Sida,* 11(4): 376 (1986): Dyssodia cooperi.
glandulosum *(Cav.) J.L. Strother in Sida,* 11(4): 377 (1986): Willdenowa glandulosa.
porophylloides *(A. Gray) J.L. Strother in Sida,* 11(4): 377 (1986): Dyssodia porophylloides.
porophyllum
var. cancellatum *(Cass.) J.L. Strother in Sida,* 11(4): 377 (1986): Lebetina cancellata.
var. radiatum *(DC.) J.L. Strother in Sida,* 11(4): 377 (1986): Dyssodia porophylla var. radiata.
speciosum *(A. Gray) J.L. Strother in Sida,* 11(4): 378 (1986): Dyssodia speciosa.
squamosum *(A. Gray) J.L. Strother in Sida,* 11(4): 378 (1986): Dyssodia squamosa.
wrightii
var. pulcherrimum *(Strother) J.L. Strother in Sida,* 11(4): 377 (1986): Dyssodia neomexicana var. pulcherrima.

ADESMIA (Legumin.).
confusa *E.A. Ulibarri in Darwiniana,* 27(1–4): 371 (1986), nom. nov.: A. arborea *Bertero ex Clos.*
dessaueri *(Reiche) E.A. Ulibarri in Darwiniana,* 27(1–4): 335 (1986): Patagonium dessaueri.
mendozana *E.A. Ulibarri in Darwiniana,* 27(1–4): 354 (1986)—Argentina.
pumahuasiana *E.A. Ulibarri in Darwiniana,* 27(1–4): 380 (1986)—Argentina, Bolivia.
renjifoana *(Reiche) E.A. Ulibarri in Darwiniana,* 27(1–4): 365 (1986): Patagonium renjifoanum.
volckmannii *R.A. Philippi in An. Univ. Chile,* 21: 444 (1862)—Chile. †

ADOXA (Adoxac.).
ser. Inodorae *O.A. Nepomnyashchaya in Bot. Zhurn.,* 72(1): 89 (1987).
ser. Omeienses *O.A. Nepomnyashchaya in Bot. Zhurn.,* 72(1): 90 (1987).
ser. Orientales *O.A. Nepomnyashchaya in Bot. Zhurn.,* 72(1): 89 (1987).

ADOXA :-
sect. Orientales *(Nepomn.) S.Y. Li & Z.H. Ning in Bull. Bot. Res. North-East. Forest. Inst.,* 7(4): 98 (1987): A. ser. Orientales.
subgen. Tetradoxa *(C.Y. Wu) S.Y. Li & Z.H. Ning in Bull. Bot. Res. North-East. Forest. Inst.,* 7(4): 96 (1987): gen. Tetradoxa.
inodora *(Falc. ex C.B. Clarke) O.A. Nepomnyashchaya in Bot. Zhurn.,* 72(1): 89 (1987): A. moschatellina var. inodora.
moschatellina
var. insularis *(Nepomn.) S.Y. Li & Z.H. Ning in Bull. Bot. Res. North-East. Forest. Inst.,* 7(4): 100 (1987): A. insularis.

AECHMEA (Bromeliac.).
carvalhoi *E. Pereira & E.M.C. Leme in Bradea,* 4(34): 267 (1986)—Brazil.
discordiae *E.M.C. Leme in Bradea,* 4(33): 255 (1986)—Brazil.
esseri *Gross & W. Rauh in Abh. Akad. Wiss. Lit. Mainz, Math.-Naturwiss. Klasse, Trop. Subtrop. Pflanzenwelt,* 58: 53 (1986)—Paraguay.
orlandiana
subsp. belloi *E. Pereira & E.M.C. Leme in Bradea,* 4(34): 266 (1986)—Brazil.
sergipana *E. Pereira & E.M.C. Leme in Bradea,* 4(34): 268 (1987)—Brazil.
weilbachii
f. leodiensis *(André) E. Pereira & E.M.C. Leme in J. Bromeliad Soc.,* 37(3): 124 (1987): A. weilbachii var. leodiensis.
f. viridisepala *E. Pereira & E.M.C. Leme in J. Bromeliad Soc.,* 37(3): 124 (1987)—Brazil.

AERANGIS (Orchidac.).
distincta *J. Stewart & I.F. la Croix in Kew Bull.,* 42(1): 217 (1987)—Malawi.
splendida *J. Stewart & I.F. la Croix in Kew Bull.,* 42(1): 215 (1987)—Malawi, Zambia.

AERANTHES (Orchidac.).
subgen. Denticulatae *K. Senghas in Orchidee,* 38(1): 4 (1987).
erectiflora *K. Senghas in Orchidee,* 38(1): 7 (1987)—Madagascar.
robusta *K. Senghas in Orchidee,* 38(1): 5 (1987)—Madagascar.

AETANTHUS (Loranthac.).
dichotomus *(Ruiz & Pavón) J. Kuijt in Fl. Ecuador,* 24(32C): 167 (1986): Loranthus dichotomus.
macranthus *(Hook.) J. Kuijt in Fl. Ecuador,* 24(32C): 169 (1986): Loranthus macranthus.

AETHIONEMA (Crucif.).
cordatum
subsp. pichleri *(Boiss.) P. Hartvig & A. Strid in Bot. Jahrb.,* 108(2–3): 314 (1987), without basionym ref.: A. cordatum var. pichleri.

AETHIONEMA :-
speciosum
subsp. compactum *P. Hartvig & A. Strid in Bot. Jahrb.*, 108(2–3): 314 (1987)—Turkey.

AGALINIS (Scrophulariac.).
bandeirensis *K. Barringer in Brittonia,* 39(3): 353 (1987)—Brazil.
bangii *(Kuntze) K. Barringer in Brittonia,* 39(3): 353 (1987): Gerardia bangii.
digitalis *(Benth.) K. Barringer in Brittonia,* 39(3): 355 (1987): Gerardia digitalis.
linarioides
subsp. rojasii *K. Barringer in Brittonia,* 39(3): 357 (1987)—Paraguay.
meyeniana *(Benth.) K. Barringer in Brittonia,* 39(3): 355 (1987): Gerardia rigida β meyeniana.
ramulifera *K. Barringer in Brittonia,* 39(3): 355 (1987)—Brazil.

AGAPETES (Ericac.).
camelliifolia *S.H. Huang in Acta Phytotax. Sin.*, 25(4): 318 (1987)—Xizang.
ciliata *S.H. Huang in Acta Phytotax. Sin.*, 25(4): 314 (1987)—Xizang.
leiocarpa *S.H. Huang in Acta Phytotax. Sin.*, 25(4): 317 (1987)—Xizang.
leucocarpa *S.H. Huang in Acta Phytotax. Sin.*, 25(4): 317 (1987)—Xizang.
medogensis *S.H. Huang in Acta Phytotax. Sin.*, 25(4): 315 (1987)—Xizang.
rugosus *(Hook. f.) K. Haridasan & R.R. Rao, Forest Fl. Meghalaya,* 2: 542 (1987): Pentapterygium rugosum.
subansirica *G.D. Pal & K. Thothathri in Bull. Bot. Surv. India,* 26(3–4): 239 (1984 publ. 1986)—India.
xizangensis *S.H. Huang in Acta Phytotax. Sin.*, 25(4): 315 (1987)—Xizang.

AGASTACHE (Labiat.).
ser. Brittonastrum *(Briquet) R.W. Sanders, Taxon. Agastache sect. Brittonastrum (Lamiac.-Nepeteae) (Syst. Bot. Monog., 15):* 30 (1987): gen. Brittonastrum.
ser. Canae *R.W. Sanders, Taxon. Agastache sect. Brittonastrum (Lamiac.-Nepeteae) (Syst. Bot. Monog., 15):* 68 (1987).
ser. Mearnsianae *R.W. Sanders, Taxon. Agastache sect. Brittonastrum (Lamiac.-Nepeteae) (Syst. Bot. Monog., 15):* 47 (1987).
ser. Micranthae *R.W. Sanders, Taxon. Agastache sect. Brittonastrum (Lamiac.-Nepeteae) (Syst. Bot. Monog., 15):* 78 (1987).
ser. Pallidae *R.W. Sanders, Taxon. Agastache sect. Brittonastrum (Lamiac.-Nepeteae) (Syst. Bot. Monog., 15):* 41 (1987).
mexicana
subsp. xolocotziana *R. Bye, E. Linares & T.P. Ramamoorthy in Phytologia,* 62(3): 157 (1987)—Mexico.

AGATHIS (Araucariac.).
silbae *D.J. De Laubenfels in Phytologia,* 61(7): 448 (1987), as 'silbai'—Vanuatu.

AGERATINA (Compos.).
aromatica
var. incisa *(A. Gray) C.F. Reed in Phytologia,* 63(5): 412 (1987): Eupatorium aromaticum var. incisum.
astellera *(B.L. Turner) B.L. Turner in Phytologia,* 63(1): 5 (1987): Eupatorium astellerum.
oppositifolia *(A. Gray) B.L. Turner in Phytologia,* 63(1): 5 (1987): Bigelovia oppositifolia.
seleri *B.L. Turner in Phytologia,* 63(1): 4 (1987)—Mexico.

AGERATINASTRUM (Compos.).
lejolyanum *(Adamska & Lisowski) K. Kalanda in Bull. Jard. Bot. Nation. Belg.*, 56(3–4): 384 (1986): Vernonia lejolyana.

AGLAIA (Meliac.).
koordersii *S.S. Jain & S.S.R. Bennet in Indian J. Forest.*, 9(3): 271 (1986 publ. 1987), nom. nov.: A. hiernii *Koord. & Valet. ex Koord.*
spectabilis *(Miq.) S.S. Jain & S.S.R. Bennet in Indian J. Forest.*, 9(3): 271 (1986 publ. 1987): Amoora spectabilis.

AGOSERIS (Compos.).
aurantiaca
subsp. purpurea *(Gray) G.W. Douglas in Canad. J. Bot.*, 64(11): 2726 (1986): Macrorhynchus purpureus.

AGRIMONIA (Rosac.).
gorovoii *S.D. Rumyantsev in Bot. Zhurn.*, 72(9): 1258 (1987)—U.S.S.R. (Soviet Far East).
pilosa
subsp. dahurica *(Willd. ex Sér.) R.V. Kamelin in Byull. Mosk. Obshch. Ispȳt. Prir., Biol.*, 91(6): 92 (1986): A. dahurica.

×AGROPOGON (Gramin.).
robinsonii *(Druce) Melderis & D. McClintock in D. McClintock, Suppl. Wild Flow. Guernsey:* 49 (1987), as 'Agropyron': Agrostis × robinsonii.

AGROPYRON (Gramin.).
fragile
var. longe-aristatum *A. Ataeva in Izv. Akad. Nauk Turkm. SSR, Biol. Nauk,* 1987(3): 54 (1987)—U.S.S.R. (Middle Asia).
intermedium *(Host) Beauvois, Ess. Agrostogr.:* 102, 146 (1812): Triticum intermedium. †
pectinatum
var. gluma-villosa *A. Ataeva in Izv. Akad. Nauk Turkm. SSR, Biol. Nauk,* 1987(3): 54 (1987)—U.S.S.R. (Middle Asia).

×AGROSITANION (Gramin.).
pinalenoensis *G.L. Pyrah in Great Basin Nat.*, 43(1): 135 (1983)—U.S.A. (Arizona). (Agropyron arizonicum × Sitanion hystrix var. brevifolium).

AGROSTIS (Gramin.).
calderoniae *S. Acosta Castellanos in Phytologia*, 62(6): 449 (1987)—Mexico.
canina
subsp. granatensis *A.T. Romero García, G. Blanca López & C. Morales Torres in An. Jard. Bot. Madrid*, 43(1): 52 (1986)—Spain.
hesperica *A.T. Romero García, G. Blanca López & C. Morales Torres in An. Jard. Bot. Madrid*, 43(1): 52 (1986)—Spain, Portugal.
koelerioides *A.M. Romo in Treb. Inst. Bot. Barcelona*, 11: 37 (1987)—Morocco.
mertensii
f. macrantha *(Eames) J. Cayouette in Nat. Canad.*, 113(4): 331 (1986): A. borealis var. macrantha.
rubra *F. Dujardin, Fl. Indre et Loire:* 278 (1833), non L. (1753)—France.

AIDIA (Rubiac.).
chantonea *D.D. Tirvengadum in Bull. Mus. Nation. Hist. Nat., B, Adansonia*, Ser. 4, 8(3): 272 (1986)—Vietnam, Cambodia, Thailand.
coriacea *(Miq.) D.D. Tirvengadum in Bull. Mus. Nation. Hist. Nat., B, Adansonia*, Ser. 4, 8(3): 268 (1986): Urophyllum coriaceum.
graeffei *(Reinecke) D.D. Tirvengadum in Bull. Mus. Nation. Hist. Nat., B, Adansonia*, Ser. 4, 8(3): 270 (1986): Randia graeffei.
merrillii *(Chun) D.D. Tirvengadum in Bull. Mus. Nation. Hist. Nat., B, Adansonia*, Ser. 4, 8(3): 271 (1986): Randia merrillii.
oblongata *(Miq.) D.D. Tirvengadum in Bull. Mus. Nation. Hist. Nat., B, Adansonia*, Ser. 4, 8(3): 271 (1986): Gynopachis oblongata.
spicata *(Valeton) D.D. Tirvengadum in Bull. Mus. Nation. Hist. Nat., B, Adansonia*, Ser. 4, 8(3): 271 (1986): Randia spicata.
thozetiana *(F. Muell.) D.D. Tirvengadum in Bull. Mus. Nation. Hist. Nat., B, Adansonia*, Ser. 4, 8(3): 272 (1986): Ixora thozetiana.
wallichiana *D.D. Tirvengadum in Bull. Mus. Nation. Hist. Nat., B, Adansonia*, Ser. 4, 8(3): 263 (1986), nom. nov.: Webera densiflora *Wall.*

AIDIOPSIS D.D. Tirvengadum in Bull. Mus. Nation. Hist. Nat., B, Adansonia, Ser. 4, 8(3): 287 (1986). *RUBIACEAE.*
forbesii *(King & Gamble) D.D. Tirvengadum in Bull. Mus. Nation. Hist. Nat., B, Adansonia*, Ser. 4, 8(3): 288 (1986): Randia forbesii.

AIRA (Gramin.).
collina *(Dumort.) B.D. Jackson in Index Kewensis*, 1: 67 (1753–1885 publ. 1893): A. flexuosa * (subsp.) collina.

AIRA :–
hartmaniana *(Nyman) B.D. Jackson in Index Kewensis*, 1: 68 (1753–1885 publ. 1893): A. caespitosa * hartmaniana.
insularis *(Parl.) J. Woods, Tourist's Fl.*: 403 (1850): Antinoria insularis.
littoralis *(Gaudin) C.-H. Godet, Fl. Jura*, 2: 803 (1853): A. caespitosa β littoralis.

AJANIA (Compos.).
abolinii *S.S. Kovalevskaya in Novosti Sist. Vÿssh. Rast.*, 23: 247 (1986)—U.S.S.R. (Middle Asia).
korovinii *S.S. Kovalevskaya in Novosti Sist. Vÿssh. Rast.*, 23: 246 (1986)—U.S.S.R. (Middle Asia).

AJUGA (Labiat.).
decumbens
var. rosa *Y.N. Lee in Korean J. Bot.*, 25(4): 177 (1982)—Korea.

ALBIZIA (Legumin.).
blanchetii *(Benth.) G.P. Lewis, Legumes of Bahia:* 162 (1987): Enterolobium blanchetii.
inopinata *(Harms) G.P. Lewis, Legumes of Bahia:* 162 (1987): Serianthes inopinata.
lucidior *(Steud.) I. Nielson ex H. Hara in Enum. Fl. Pl. Nepal*, 2: 104 (11 June 1979): Inga lucidior.
polyantha *(Spreng. f.) G.P. Lewis, Legumes of Bahia:* 164 (1987): Acacia polyantha.

ALBUCA (Hyacinthac.).
tenuis *S.H. Knudtzon in Nordic J. Bot.*, 6(6): 785 (1986)—Kenya.

ALCHEMILLA (Rosac.).
nothosect. Coriaceae *S. Fröhner in Gleditschia*, 14(1): 41 (1986), contrary to Arts. 32.1(b) and H7 ICBN 1983. (A. sect. Erectae × sect. Pentaphylleae × sect. Ultravulgares).
nothosect. Decumbentes *S. Fröhner in Gleditschia*, 14(1): 36 (1986), contrary to Arts. 32.1(b) and H7 ICBN 1983. (A. sect. Pentaphylleae × sect. Ultravulgares).
sect. Erectae *S. Fröhner in Gleditschia*, 14(1): 30 (1986).
nothosect. Flabellatae *S. Fröhner in Gleditschia*, 14(1): 44 (1986), contrary to Arts. 32.1(b) and H7 ICBN 1983. (A. sect. Alpinae × sect. Erectae × sect. Pentaphylleae).
nothosect. Glaciales *S. Fröhner in Gleditschia*, 14(1): 39 (1986), contrary to Arts. 32.1(b) and H7 ICBN 1983. (A. sect. Alpinae × sect. Pentaphylleae).
nothosect. Plicatae *S. Fröhner in Gleditschia*, 14(1): 45 (1986), contrary to Arts. 32.1(b) and H7 ICBN 1983. (A. sect. Alpinae × sect. Pentaphylleae × sect. Ultravulgares).
sect. Ultravulgares *S. Fröhner in Gleditschia*, 14(1): 31 (1986).
contractilis *(Plocek) S. Fröhner in Gleditschia*, 14(1): 18 (1986): A. monticola var. contractilis.

ALCHEMILLA :–
crassa *(Plocek) A. Plocek in Folia Geobot. Phytotax.,* 21(4): 424 (1986): A. monticola var. crassa.
delitescens *A. Plocek in Folia Geobot. Phytotax.,* 21(4): 425 (1986)—Czechoslovakia.
dostalii *A. Plocek in Folia Geobot. Phytotax.,* 21(4): 426 (1986)—Czechoslovakia.
galkinae *S. Fröhner in Candollea,* 41(2): 255 (1986)—Switzerland.
gingolphiana *S. Fröhner in Candollea,* 41(2): 256 (1986)—Switzerland.
moncophila *A. Plocek in Folia Geobot. Phytotax.,* 21(4): 426 (1986)—Czechoslovakia.
obtusa
subsp. trapezialis *A. Plocek in Folia Geobot. Phytotax.,* 21(4): 423 (1986)—Czechoslovakia.
pungentiflora *(Plocek) A. Plocek in Folia Geobot. Phytotax.,* 21(4): 425 (1986): A. glabra var. pungentiflora.
reniformis
subsp. ursina *(Fröhner) A. Plocek in Folia Geobot. Phytotax.,* 21(4): 424 (1986): A. ursina.
rhodobasis *A. Plocek in Folia Geobot. Phytotax.,* 21(4): 426 (1986)—Czechoslovakia.
sojakii *A. Plocek in Folia Geobot. Phytotax.,* 21(4): 427 (1986)—Czechoslovakia.

ALECTRYON (Sapindac.).
cardiocarpus *P.W. Leenhouts in Blumea,* 32(1): 221 (1987)—New Guinea (W. Irian, Papua New Guinea).
diversifolius *(F. Muell.) S.T. Reynolds in Austrobaileya,* 2(4): 335 (1987): Heterodendrum diversifolium.
kangeanensis *P.W. Leenhouts in Blumea,* 32(1): 222 (1987)—Java.
myrmecophilus *P.W. Leenhouts in Blumea,* 32(1): 223 (1987)—New Guinea (Papua New Guinea).
oleifolius *(Desf.) S.T. Reynolds in Austrobaileya,* 2(4): 335 (1987): Heterodendrum oleifolium.
subsp. canescens *S.T. Reynolds in Austrobaileya,* 2(4): 337 (1987), nom. nov.: Heterodendrum macrocalyx *Radlk.*
subsp. elongatus *S.T. Reynolds in Austrobaileya,* 2(4): 337 (1987), nom. nov.: Heterodendrum microcalyx *Radlk.*
pubescens *(S. Reyn.) S.T. Reynolds in Austrobaileya,* 2(4): 335 (1987): Heterodendrum pubescens.
ramiflorus *S.T. Reynolds in Austrobaileya,* 2(4): 332 (1987)—Australia (Queensland).
tropicus *(S. Reyn.) S.T. Reynolds in Austrobaileya,* 2(4): 335 (1987): Heterodendrum tropicum.

ALETRIS (Melanthiac.).
yaanica *G.H. Yang in Acta Phytotax. Sin.,* 25(3): 237 (1987)—China.

ALEXGEORGEA (Restionac.).
nitens *(Nees) L.A.S. Johnson & B. Briggs in Telopea,* 2(6): 781 (24 April 1986); *also published by S. Carlquist* in Aliso, 11(2): 158 (1985 publ. 29 April 1986): Restio nitens. †

ALFREDIA (Compos.).
integrifolia *(Iljin) M. Tulyaganova in Bot. Zhurn.,* 72(6): 827 (1987): A. nivea f. integrifolia.

ALGERNONIA (Euphorbiac.).
gibbosa *(Pax & K. Hoffm.) M. Emmerich in Arq. Mus. Nac. (Rio de Janeiro),* 56: 93 (1981): Tetraplandra gibbosa.
glazioui *M. Emmerich in Arq. Mus. Nac. (Rio de Janeiro),* 56: 94 (1981)—Brazil.

ALLIUM (Alliac.).
altyncolicum *N.V. Frizen in Bot. Zhurn.,* 72(6): 816 (1987)—U.S.S.R. (Middle Asia).
eivissanum *F. Garbari & P. Miceli in Willdenowia,* 16(2): 383 (1987)—Balearic Is.
nemrutdaghense *Kit Tan & F. Sorger in Pl. Syst. Evol.,* 154(1–2): 126 (1986)—Turkey.
nudicaule *J.G.C. Lehmann in Sem. Hort. Bot. Hamb.:* 17 (1826)—Locality not stated. †
pangasicum *I. Turakulov in Dokl. Akad. Nauk Tadzh. SSR,* 29(3): 181 (1986)—U.S.S.R. (Middle Asia).
paniculatum
subsp. salinum *(Debeaux) F. Botté & M. Kerguélen in Lejeunia,* 120: 41 (1987): A. paniculatum var. salinum.
paradoxum
var. normale *W.T. Stearn in Kew Mag.,* 4(4): 196 (1987)—Iran.
prokhanovii *(Worosch.) V.Yu. Barkalov in Sosud. Rast. Sovet. Dal'nego Vostoka,* 2: 385 (1987): A. splendens subsp. prokhanovii.
rudolfii *I. Turakulov in Dokl. Akad. Nauk Tadzh. SSR,* 29(3): 180 (1986)—U.S.S.R. (Middle Asia).
shevockii *D.W. McNeal in Madroño,* 34(2): 150 (1987)—U.S.A. (California).
vuralii *Kit Tan in Pl. Syst. Evol.,* 155(1–4): 102 (1987)—Turkey.

ALLOISPERMUM (Compos.).
insuetum *C.F. Fernandez, L.E. Urbatsch & G. Sullivan in Madroño,* 34(2): 162 (1987)—Colombia.

ALLOPLECTUS (Gesneriac.).
baguensis *L.E. Skog in Op. Bot.,* 92: 225 (1987)—Peru.
spectabilis *Wiehler ex L.E. Skog & J. Steyermark in Op. Bot.,* 92: 228 (1987)—Venezuela.

ALOE (Aloeac.).
cooperi
subsp. pulchra *H.F. Glen & D.S. Hardy in Flow. Pl. Afr.,* 49(3–4): t. 1944 (1987)—South Africa (Natal), Swaziland.

ALOE :–
fouriei *D.S. Hardy & H.F. Glen in Flow. Pl. Afr.,* 49(3–4): t. 1941 (1987)—South Africa (Transvaal).
hardyi *H.F. Glen in Flow. Pl. Afr.,* 49(3–4): t. 1942 (1987)—South Africa (Transvaal).
marlothii
subsp. orientalis *H.F. Glen & D.S. Hardy in Flow. Pl. Afr.,* 49(3–4): t. 1943 (1987)—South Africa (Natal), Mozambique.

ALOPECURUS (Gramin.).
aquaticus *(Dumort.) F.A. Tinant, Fl. Luxembourgeoise:* 75 (1836): A. pratensis * (subsp.) aquaticus.
aristulosus *N.A. Desvaux, Obs. Pl. Environs d'Angers:* 46 (1818)—France.

ALSTROEMERIA (Alstroemeriac.).
psittacina *J.G.C. Lehmann in Sem. Hort. Bot. Hamb.:* 17 (1826)—Locality not stated. †

ALTERNANTHERA (Amaranthac.).
albotomentosa
var. ecuadoriensis *U. Eliasson in Fl. Ecuador,* 28(44): 69 (1987)—Ecuador.
corymbiformis *U. Eliasson in Fl. Ecuador,* 28(44): 81 (1987)—Ecuador.
grandis *U. Eliasson in Fl. Ecuador,* 28(44): 99 (1987)—Ecuador.
porrigens
var. mearsii *U. Eliasson in Fl. Ecuador,* 28(44): 89 (1987)—Ecuador.
var. piurensis *(Standley) U. Eliasson in Fl. Ecuador,* 28(44): 87 (1987): A. piurensis.

ALTHAEA (Malvac.).
australis *W.R. Barker in Fl. S. Australia,* 4: 2146 (1986)—Australia (S. Australia).

ALTHENIA (Zannichelliac.).
filiformis
var. barrandonii *(Duval-Jouve) P. García Murillo & S. Talavera in Lagascalia,* 14(1): 107 (1986): A. barrandonii.
orientalis *(Tzvelev) P. García Murillo & S. Talavera in Lagascalia,* 14(1): 108 (1986): A. filiformis subsp. orientalis.
subsp. betpakdalensis *(Tzvelev) P. García Murillo & S. Talavera in Lagascalia,* 14(1): 112 (1986): A. filiformis subsp. betpakdalensis.

ALYOGYNE (Malvac.).
cravenii *P.A. Fryxell in Syst. Bot.,* 12(2): 277 (1987)—Australia (N. Terr.).

AMARANTHUS (Amaranthac.).
emarginatus
subsp. pseudogracilis *(Thell.) G. Hügin in Willdenowia,* 16(2): 463 (1987): A. lividus f. pseudogracilis.

AMHERSTIA Wall. in Taylor & Phillips, Phil. Mag., n.s., 3: 233 (1826); cf. D.J. Mabberley in Taxon, 31(1): 67 (1982). *LEGUMINOSAE.* †

AMOMUM (Zingiberac.).
kinabaluense *R.M. Smith in Notes Roy. Bot. Gard. Edinburgh,* 44(2): 233 (1987)—Borneo (Sabah).

AMORIA (Legumin.).
glomerata *(L.) R.M. Burton, Fl. London Area:* 225 (1983): Trifolium glomeratum.
suffocata *(L.) R.M. Burton, Fl. London Area:* 225 (1983): Trifolium suffocatum.

AMORPHA (Legumin.).
herbacea
var. crenulata *(Rydberg) D. Isely in Sida,* 11(4): 433 (1986): A. crenulata.

AMPELOCALAMUS (Gramin.).
naibunensis *(Hayata) T.H. Wen in J. Bamboo Res.,* 6(3): 34 (1987): Arundinaria naibunensis.
yongshanensis *J.R. Xue & D.Z. Li in J. Bamboo Res.,* 6(2): 10 (1987)—China.

AMPHIBLEMMA (Melastomatac.).
monticola *H. Jacques-Félix in Bull. Mus. Nation. Hist. Nat., B, Adansonia,* Ser. 4, 9(2): 125 (1987), as 'monticole'—Cameroun.

AMPHICARPAEA (Legumin.).
bracteata
var. comosa *(L.) C.F. Reed in Phytologia,* 63(5): 410 (1987): Glycine comosa.

AMYRIS (Rutac.).
metopioides *T.A. Zanoni & M.M. Mejía P. in Moscosoa,* 4: 125 (1986)—Hispaniola (Dominican Republic).

×**ANACAMPTORCHIS** (Orchidac.).
vanlookenii *C. Bernard & G. Fabre in Orchidophile,* 78: 1345 (1987), as 'van lookenii'—France. (Anacamptis pyramidalis × Orchis papilionacea).

ANACARDIUM (Anacardiac.).
fruticosum *J.D. Mitchell & S.A. Mori in Mem. New York Bot. Gard.,* 42: 54 (1987)—Guyana.

ANAGALLIS (Primulac.).
×hadidii *J. Chrtek & J. Osbornová-Kosinová in Folia Geobot. Phytotax.,* 21(4): 413 (1986)—Egypt. (A. arvensis var. arvensis × latifolia).

ANAPHALIS (Compos.).
sinica
var. yakusimensis *(Masamune) T. Yahara in J. Fac. Sci. Univ. Tokyo, Bot.,* 14(2): 112 (1987): A. yakusimensis.

ANAXAGOREA (Annonac.).
angustifolia *A. Timmerman in Proc. Kon. Nederl. Akad. Wetensch., C,* 87(3): 298 (1984)—Brazil.
brevipedicellata *A. Timmerman in Proc. Kon. Nederl. Akad. Wetensch., C,* 87(3): 299 (1984)—Brazil, French Guiana.

ANAXAGOREA :–
floribunda *A. Timmerman in Proc. Kon. Nederl. Akad. Wetensch., C,* 87(3): 300 (1984)—Peru.
manausensis *A. Timmerman in Proc. Kon. Nederl. Akad. Wetensch., C,* 87(3): 301 (1984)—Brazil.
rufa *A. Timmerman in Proc. Kon. Nederl. Akad. Wetensch., C,* 87(3): 301 (1984)—Colombia, Venezuela, Brazil.

ANDRACHNE (Euphorbiac.).
ephemera *M.G. Gilbert in Kew Bull.,* 42(2): 351 (1987)—Ethiopia.

ANDROSACE (Primulac.).
subsect. Dasyphyllae *Ovcz. ex V.B. Kuvaev in Bot. Zhurn.,* 72(7): 955 (1987).
sect. Filiformes *V.B. Kuvaev & N.M. Pirozhkova in Bot. Zhurn.,* 72(7): 955 (1987).
subsect. Oreades *Schischk. & Bobr. ex V.B. Kuvaev in Bot. Zhurn.,* 72(7): 956 (1987), without type.
subsect. Perennes *V.B. Kuvaev & N.M. Pirozhkova in Bot. Zhurn.,* 72(7): 956 (1987).
sect. Pseudodouglasia *Ovcz. ex V.B. Kuvaev in Bot. Zhurn.,* 72(7): 955 (1987).
amurensis *N.S. Probatova in Sosud. Rast. Sovet. Dal'nego Vostoka,* 2: 155 (1987), nom. nov.: A. lactiflora *Fisch. ex Duby.*
arctisibirica *(Korobkov) N.S. Probatova in Sosud. Rast. Sovet. Dal'nego Vostoka,* 2: 152 (1987): A. chamaejasme subsp. arctisibirica.
cylindrica
subsp. hirtella *(Dufour) W. Greuter & Burdet in Willdenowia,* 16(2): 452 (1987): A. hirtella.

ANEMONE (Ranunculac.).
sect. Fuscopurpurea *V.F. Tarasevich & R.P. Chaudkhari in Bot. Zhurn.,* 72(7): 895 (1987).
mirajabii *R.A. Qureshi & M.N. Chaudhri in Pakistan Syst.,* 3(1): 16 (1987)—Pakistan.
obtusiloba
var. leiocarpa *M. Tamura in Acta Phytotax. Geobot.,* 37(4–6): 153 (1986)—Nepal.
rupicola
subsp. laceratoincisa *(W.T. Wang) R.P. Chaudkhari in Bot. Zhurn.,* 72(6): 824 (1987): A. laceratoincisa.
subsp. reniformis *R.P. Chaudkhari in Bot. Zhurn.,* 72(6): 825 (1987)—India.
subsp. sericea *(Hook. f. & Thoms.) R.P. Chaudkhari in Bot. Zhurn.,* 72(6): 823 (1987): A. rupicola var. sericea.
wendelboi *R.P. Chaudkhari in Bot. Zhurn.,* 72(6): 825 (1987), as 'wendelboii'—Pakistan.

ANERINCLEISTUS (Melastomatac.).
phyllagathoides
var. inaequalis *(Stapf) J.F. Maxwell in Gard. Bull. Singapore,* 35(2): 215 (1982 publ. 1983): Pomatostoma inaequalis.

ANGELICA (Umbellif.).
archangelica
subsp. himalaica *(Clarke) G. Singh in G. Singh & P. Kachroo, Forest Fl. Srinagar:* 183 (1987): A. officinalis var. himalaica.
cyclocarpa *(C. Norman) J.F.M. Cannon in Enum. Fl. Pl. Nepal,* 2: 184 (11 June 1979), isonym; see M. Hiroe, *Umbellif. of World:* 1430 (28 Feb. 1979): Archangelica cyclocarpa.
czernaevia
var. flaccida *(Kom.) M.G. Pimenov in Sosud. Rast. Sovet. Dal'nego Vostoka,* 2: 258 (1987): A. flaccida.
harae *M.G. Pimenov in Feddes Repert.,* 98(7–8): 379 (1987)—Nepal.
tichomirovii *V.M. Vinogradova in Novosti Sist. Vyssh. Rast.,* 23: 93 (1986)—Xinjiang.

×**ANGREONIELLA** hort. in Orchid Rev., 95(1119): centre page pull-out p. 8 (1987). *ORCHIDACEAE.* (Angraecum × Oeoniella).

ANIGOZANTHOS (Haemodorac.).
sect. Concatenati *S.D. Hopper in Fl. Australia,* 45: 455, 118 (1987).
subgen. Haplanthesis *(Benth.) S.D. Hopper in Fl. Australia,* 45: 455 (1987): A. sect. Haplanthesis.
bicolor
subsp. decrescens *S.D. Hopper in Fl. Australia,* 45: 455, 123 (1987)—Australia (W. Australia).
subsp. exstans *S.D. Hopper in Fl. Australia,* 45: 455, 122 (1987)—Australia (W. Australia).
subsp. minor *(Benth.) S.D. Hopper in Fl. Australia,* 45: 455 (1987): A. bicolor var. minor.
humilis
subsp. chrysanthus *S.D. Hopper in Fl. Australia,* 45: 456, 120 (1987)—Australia (W. Australia).
manglesii
subsp. quadrans *S.D. Hopper in Fl. Australia,* 45: 456, 125 (1987)—Australia (W. Australia).
viridis
subsp. terraspectans *S.D. Hopper in Fl. Australia,* 45: 456, 124 (1987)—Australia (W. Australia).

ANISOPAPPUS (Compos.).
lejolyanus *S. Lisowski in Bull. Jard. Bot. Nation. Belg.,* 56(3–4): 370 (1986)—Zaïre.
robynsianus *S. Lisowski in Bull. Jard. Bot. Nation. Belg.,* 56(3–4): 369 (1986)—Zaïre.
upembensis *S. Lisowski in Bull. Jard. Bot. Nation. Belg.,* 56(3–4): 371 (1986)—Zaïre.

ANISOPHYLLEA (Anisophylleac.).
myriosticta *J.-J. Floret in Bull. Mus. Nation. Hist. Nat., B, Adansonia,* Ser. 4, 8(4): 373 (1986 publ. 1987)—Nigeria, Cameroun, Congo - Brazzaville, Gabon.

ANISOPHYLLEA :–
polyneura *J.-J. Floret in Bull. Mus. Nation. Hist. Nat., B, Adansonia,* Ser. 4, 8(4): 377 (1986 publ. 1987)—Gabon, Cameroun, Congo - Brazzaville, Nigeria, Zaïre.

ANODA (Malvac.).
sect. Clausanoda *P.A. Fryxell in Aliso,* 11(4): 490 (1987).
sect. Liberanoda *P.A. Fryxell in Aliso,* 11(4): 489 (1987).
sect. Pseudanoda *P.A. Fryxell in Aliso,* 11(4): 489 (1987).
albiflora *P.A. Fryxell in Aliso,* 11(4): 493 (1987)—Cultivation ex Mexico.
guatemalensis *P.A. Fryxell in Aliso,* 11(4): 498 (1987)—Guatemala.
hintoniorum *P.A. Fryxell in Aliso,* 11(4): 500 (1987)—Mexico.
hirta *P.A. Fryxell in Aliso,* 11(4): 502 (1987)—Mexico.
leonensis *P.A. Fryxell in Aliso,* 11(4): 504 (1987)—Mexico.
maculata *P.A. Fryxell in Aliso,* 11(4): 506 (1987)—Mexico.
palmata *P.A. Fryxell in Aliso,* 11(4): 507 (1987)—Mexico.
polygyna *P.A. Fryxell in Aliso,* 11(4): 512 (1987)—Mexico.
speciosa *P.A. Fryxell in Aliso,* 11(4): 516 (1987)—Mexico.
succulenta *P.A. Fryxell in Aliso,* 11(4): 518 (1987)—Mexico.
zuccagnii *(Sprengel) P.A. Fryxell in Aliso,* 11(4): 519 (1987): Sida zuccagnii.

ANOMANTHODIA (Rubiac.).
alveolata *(Elmer) D.D. Tirvengadum in Bull. Mus. Nation. Hist. Nat., B, Adansonia,* Ser. 4, 8(3): 279 (1986): Randia alveolata.
bakeri *(Merr.) D.D. Tirvengadum in Bull. Mus. Nation. Hist. Nat., B, Adansonia,* Ser. 4, 8(3): 279 (1986): Randia bakeri.
dilleniacea *(Baillon) D.D. Tirvengadum in Bull. Mus. Nation. Hist. Nat., B, Adansonia,* Ser. 4, 8(3): 277 (1986): Randia dilleniacea.
lancifolia *(Wong) D.D. Tirvengadum in Bull. Mus. Nation. Hist. Nat., B, Adansonia,* Ser. 4, 8(3): 279 (1986): Aidia lancifolia.

ANTENNARIA (Compos.).
friesiana
subsp. beringensis *V.V. Petrovskiĭ in Novosti Sist. Vȳssh. Rast.,* 23: 193 (1986)—U.S.S.R. (Soviet Far East).
subsp. pseudoisolepis *V.V. Petrovskiĭ in Novosti Sist. Vȳssh. Rast.,* 23: 194 (1986)—U.S.S.R. (Soviet Far East).
pseudoarenicola *V.V. Petrovskiĭ in Novosti Sist. Vȳssh. Rast.,* 23: 195 (1986)—U.S.S.R. (Soviet Far East).
pulcherrima
var. anaphaloides *(Rydb.) G.W. Douglas in Canad. J. Bot.,* 64(11): 2726 (1986): A. anaphaloides.

ANTHAENANTIOPSIS (Gramin.).
racemosa *S.A. Renvoize in Kew Bull.,* 42(4): 924 (1987)—Brazil.

ANTHOPTERUS (Ericac.).
ecuadorensis *J.L. Luteyn in Op. Bot.,* 92: 113 (1987)—Ecuador.
verticillatus *J.L. Luteyn in Op. Bot.,* 92: 110 (1987)—Ecuador.

ANTHURIUM (Arac.).
betanianum *T.B. Croat in Aroideana,* 9(1–4): 9 (1986)—Venezuela.
bonplandii
subsp. cuatrecasasii *T.B. Croat in Aroideana,* 9(1–4): 11 (1986), as 'cuatrecasii'—Venezuela, Colombia.
caraboboense *T.B. Croat in Aroideana,* 9(1–4): 14 (1986)—Venezuela.
cataniapoense *T.B. Croat in Aroideana,* 9(1–4): 18 (1986)—Venezuela.
davidsei *T.B. Croat in Aroideana,* 9(1–4): 22 (1986)—Venezuela.
ginesii *T.B. Croat in Aroideana,* 9(1–4): 30 (1986)—Venezuela.
humboldtianum
subsp. viridispadix *T.B. Croat in Aroideana,* 9(1–4): 36 (1986)—Venezuela.
humoense *T.B. Croat in Aroideana,* 9(1–4): 36 (1986)—Venezuela.
liesneri *T.B. Croat in Aroideana,* 9(1–4): 43 (1986)—Venezuela.
marinoanum *T.B. Croat in Aroideana,* 9(1–4): 44 (1986)—Venezuela.
puberulinervium *T.B. Croat in Aroideana,* 9(1–4): 51 (1986)—Venezuela.
tachiranum *T.B. Croat in Aroideana,* 9(1–4): 56 (1986)—Venezuela.

ANTHYLLIS (Legumin.).
vulneraria
subsp. boscii *M. Kerguélen in Lejeunia,* 120: 45 (1987), nom. nov.: A. vulneraria var. pyrenaica *Rouy.*

ANTIDAPHNE (Eremolepidac.).
andina *J. Kuijt in Fl. Ecuador,* 24(32A): 7 (1986)—Ecuador.

ANTIRRHINUM (Scrophulariac.).
graniticum
subsp. boissieri *(Rothm.) B. Valdés in Lagascalia,* 14(1): 91 (1986): A. boissieri.
subsp. onubensis *(Fernández Casas) B. Valdés in Lagascalia,* 14(1): 91 (1986): A. meonanthum subsp. onubensis.

AOTUS (Legumin.).
subspinescens *(Benth.) M.D. Crisp in Fl. S. Australia,* 2: 658 (1986): A. villosa var. subspinescens.

APHELANDRA (Acanthac.).
gunnarii *D.C. Wasshausen in Op. Bot.,* 92: 267 (1987), as 'gunnari'—Ecuador.

APHELANDRA :–
harlingii *D.C. Wasshausen in Op. Bot.*, 92: 267 (1987)—Ecuador.
lilacina *C. Ezcurra in Darwiniana,* 27(1–4): 399 (1986)—Argentina.

APLUDA (Gramin.).
aristata
var. ciliata *(Anderss.) R.K. Jain in Indian J. Forest.*, 9(4): 345 (1986 publ. 1987): A. ciliata.
var. jainii *R.K. Jain in Indian J. Forest.*, 9(4): 345 (1986 publ. 1987)—India.
mutica
subsp. aristata *(L.) R.D. Gaur in J. Econ. Taxon. Bot.*, 9(1): 59 (1987): A. aristata.
var. major *(Hack.) R.K. Jain in Indian J. Forest.*, 9(4): 347 (1986 publ. 1987): A. varia var. major.

APODOSTIGMA (Celastrac.).
pallens
var. buchholzii *(Loesener) N. Hallé in Fl. Gabon,* 29: 197 (1986): Hippocratea ? buchholzii.
var. dummeri *N. Hallé in Fl. Gabon,* 29: 196 (1986)—Uganda, Zaïre, Cameroun.

APURIMACIA (Legumin.).
boliviana *(Britton) M. Lavin in C.H. Stirton (ed.), Advances in Legume System.*, 3: 64 (1987): Coursetia boliviana.

AQUILARIA (Thymelaeac.).
banaense *Pham-hoang Ho in Bull. Mus. Nation. Hist. Nat., B, Adansonia,* Ser. 4, 8(3): 241 (1986)—Vietnam.

AQUILEGIA (Ranunculac.).
pubiflora
var. hazarica *R.A. Qureshi & M.N. Chaudhri in Pakistan Syst.*, 3(1): 6 (1987)—Pakistan.

ARABIDOPSIS (Crucif.).
erysimoides *I.C. Hedge & Kit Tan in Pl. Syst. Evol.*, 156(3–4): 202 (1987)—Saudi Arabia.

ARALIA (Araliac.).
bipinnata
var. inermis *(Yanagita) T. Yamazaki in J. Jap. Bot.*, 62(6): 190 (1987): A. hypoleuca var. inermis.
elata
f. canescens *(Fr. & Sav.) T. Yamazaki in J. Jap. Bot.*, 62(6): 190 (1987): A. elata β canescens.
guilfoylei *W. Bull, Cat.*, 83: 4 (1873)—South Sea Is.

ARBUTUS (Ericac.).
tessellata *P.D. Sørensen in Brittonia,* 39(2): 263 (1987)—Mexico.

ARCHIATRIPLEX G.L. Chu in J. Arnold Arbor., 68(4): 461 (1987). *CHENOPODIACEAE.*

ARCHIATRIPLEX :–
nanpinensis *G.L. Chu in J. Arnold Arbor.*, 68(4): 462 (1987)—China.

ARCTOSTAPHYLOS (Ericac.).
glandulosa
f. adamsii *(Munz) P.V. Wells in Four Seasons,* 7(4): 18 (1987): A. glandulosa var. adamsii.
f. glaucoides *P.V. Wells in Four Seasons,* 7(4): 17 (1987)—U.S.A. (California).
subsp. glaucomollis *P.V. Wells in Four Seasons,* 7(4): 21 (1987)—U.S.A. (California).
f. howellii *(Eastwood) P.V. Wells in Four Seasons,* 7(4): 17 (1987), without basionym ref.: Basionym not stated.
f. wieriana *P.V. Wells in Four Seasons,* 7(4): 19 (1987)—Mexico, U.S.A. (California).
f. zacaensis *(Eastwood) P.V. Wells in Four Seasons,* 7(4): 16 (1987), without basionym ref.: Basionym not stated.
tomentosa
f. hebeclada *(DC.) P.V. Wells in Four Seasons,* 7(4): 12 (1987): Andromeda bracteosa β hebeclada.

ARDISIA (Myrsinac.).
dawnensis *M.P. Nayar & G.S. Giri in Bull. Bot. Surv. India,* 27(1–4): 222 (1985 publ. 1987)—Burma.

ARENARIA (Caryophyllac.).
angustifolioides *Kit Tan & F. Sorger in Pl. Syst. Evol.*, 155(1–4): 96 (1987)—Turkey.
eliasiana *Kit Tan & F. Sorger in Pl. Syst. Evol.*, 155(1–4): 93 (1987)—Turkey.
gothica
var. fugax *(Gay ex Gren.) M.B. Wyse Jackson & J.A.N. Parnell in Watsonia,* 16(4): 382 (1987): A. ciliata β fugax.
subsp. moehringioides *(J. Murr) M.B. Wyse Jackson & J.A.N. Parnell in Watsonia,* 16(4): 382 (1987): A. moehringioides.
ledebouriana
var. grandiflora *P. Hartvig & A. Strid in Bot. Jahrb.*, 108(2–3): 318 (1987)—Turkey.
leptoclados
subsp. viscidula *(Rouy & Fouc.) F. Dvořák in Biológia, A (Czechoslovakia),* 39(9): 835 (1984): A. serpyllifolia β. viscidula.
mons-cragus *Kit Tan & F. Sorger in Pl. Syst. Evol.*, 155(1–4): 94 (1987)—Turkey.
redowskii
subsp. ajanensis *V.N. Voroshilov in Byull. Mosk. Obshch. Ispўt. Prir., Biol.*, 91(3): 109 (1986)—U.S.S.R. (Soviet Far East).
subsp. baburinii *V.N. Voroshilov in Byull. Mosk. Obshch. Ispўt. Prir., Biol.*, 91(3): 108 (1986)—U.S.S.R. (Soviet Far East).
sivasica *Kit Tan & F. Sorger in Pl. Syst. Evol.*, 155(1–4): 94 (1987)—Turkey.

ARISAEMA (Arac.).
griffithii
var. pradhanii *(C.E.C. Fischer) U.C. Pradhan in Himalayan Pl. J.*, 4(9–10): 16 (1986): A. pradhanii.
minus *(Serizawa) J. Murata in Acta Phytotax. Geobot.*, 37(1–3): 37 (1986): A. kishidae var. minus.

ARISARUM (Arac.).
simorrhinum
var. clusii *(Schott) S. Talavera in Lagascalia,* 14(1): 116 (1986): A. clusii.
var. subexertum *(Webb & Berthelot) S. Talavera in Lagascalia,* 14(1): 115 (1986), without exact basionym date: A. subexertum.

ARISTEA (Iridac.).
angolensis
var. robusta *W. Marais in Kew Bull.*, 42(4): 932 (1987)—Tanzania.
cognata
subsp. abyssinica *(Pax) W. Marais in Kew Bull.*, 42(4): 932 (1987): A. abyssinica.
lukwangulensis *W. Marais in Kew Bull.*, 42(4): 932 (1987)—Tanzania.

ARISTOLOCHIA (Aristolochiac.).
austroszechuanica *Chien & Cheng ex C.Y. Cheng & J.L. Wu in J. Wuhan Bot. Res.*, 5(3): 221 (1987)—China.
cinnabarina *C.Y. Cheng & J.L. Wu in J. Wuhan Bot. Res.*, 5(3): 219 (1987)—China.
jinshanensis *Z.L. Yang & S.X. Tan in Bull. Bot. Res. North-East. Forest. Inst.*, 7(2): 129 (1987)—China.
neolongifolia *J.L. Wu & Z.L. Yang in J. Wuhan Bot. Res.*, 5(3): 223 (1987)—China.
rechingeriana *Kit Tan & F. Sorger in Pl. Syst. Evol.*, 155(1–4): 101 (1987)—Turkey.

ARMENIACA (Rosac.).
vulgaris
var. rushanica *Korsh. ex E.A. Sokolova in Byull. Vses. Ord. Lenina Inst. Rast. N.I. Vavilova,* 166: 54 (1986)—U.S.S.R. (Middle Asia).

ARMERIA (Plumbaginac.).
arenaria
subsp. praecox *(Jordan) M. Kerguélen in Lejeunia,* 120: 49 (1987): Statice praecox.
belgenciensis *Donadille ex M. Kerguélen in Lejeunia,* 120: 49 (1987)—France.

ARMORACIA (Crucif.).
lacustris *(A. Gray) I.A. Al-Shehbaz & V. Bates in J. Arnold Arbor.*, 68(3): 357 (1987): Nasturtium lacustre.

ARNOCRINUM (Anthericac.).
gracillimum *G.J. Keighery in Fl. Australia,* 45: 466, 246 (1987)—Australia (W. Australia).

ARRHENATHERUM (Gramin.).
rupestre *(Dumort.) B.D. Jackson in Index Kewensis,* 1: 193 (1753–1885 publ. 1893): A. avenaceum * (subsp.) rupestre.

ARROJADOA (Cactac.).
polyantha *(Werdermann) D. Hunt in Bradleya,* 5: 92 (1987): Cephalocereus polyanthus.

ARTEMISIA (Compos.).
subsect. Aralocaspicae *N.S. Filatova in Novosti Sist. Vyssh. Rast.*, 23: 231 (1986).
sect. Calciphilum *N.S. Filatova in Novosti Sist. Vyssh. Rast.*, 23: 217 (1986).
subsect. Europaeae *N.S. Filatova in Novosti Sist. Vyssh. Rast.*, 23: 229 (1986).
sect. Halophilum *N.S. Filatova in Novosti Sist. Vyssh. Rast.*, 23: 227 (1986).
subsect. Herbae-albae *N.S. Filatova in Novosti Sist. Vyssh. Rast.*, 23: 222 (1986).
subsect. Heterophyllae *N.S. Filatova in Novosti Sist. Vyssh. Rast.*, 23: 231 (1986).
subsect. Junceae *N.S. Filatova in Novosti Sist. Vyssh. Rast.*, 23: 219 (1986).
sect. Junceum *Poljak. ex N.S. Filatova in Novosti Sist. Vyssh. Rast.*, 23: 219 (1986).
subsect. Kazachstanicae *N.S. Filatova in Novosti Sist. Vyssh. Rast.*, 23: 227 (1986).
subsect. Knorringianae *N.S. Filatova in Novosti Sist. Vyssh. Rast.*, 23: 236 (1986).
subsect. Lehmannianae *N.S. Filatova in Novosti Sist. Vyssh. Rast.*, 23: 239 (1986).
sect. Leucophyton *N.S. Filatova in Novosti Sist. Vyssh. Rast.*, 23: 222 (1986).
subsect. Mongolicae *N.S. Filatova in Novosti Sist. Vyssh. Rast.*, 23: 234 (1986).
subsect. Montanae *N.S. Filatova in Novosti Sist. Vyssh. Rast.*, 23: 225 (1986).
subsect. Orientales *N.S. Filatova in Novosti Sist. Vyssh. Rast.*, 23: 220 (1986).
subsect. Prasinae *N.S. Filatova in Novosti Sist. Vyssh. Rast.*, 23: 225 (1986).
sect. Pycnanthum *N.S. Filatova in Novosti Sist. Vyssh. Rast.*, 23: 236 (1986).
subsect. Rhodanthae *N.S. Filatova in Novosti Sist. Vyssh. Rast.*, 23: 236 (1986).
subsect. Robustae *N.S. Filatova in Novosti Sist. Vyssh. Rast.*, 23: 220 (1986).
sect. Sclerophyllum *N.S. Filatova in Novosti Sist. Vyssh. Rast.*, 23: 224 (1986).
subsect. Tauricae *N.S. Filatova in Novosti Sist. Vyssh. Rast.*, 23: 229 (1986).
subsect. Turanicae *N.S. Filatova in Novosti Sist. Vyssh. Rast.*, 23: 224 (1986).
caerulescens
subsp. densiflora *(Viv.) Gamisans ex M. Kerguélen & J. Lambinon in Lejeunia,* 120: 51 (1987): A. densiflora.
subsp. gargantae *J. Vallès-Xirau & J.A. Seoane-Camba in Candollea,* 42(1): 370 (1987)—Spain.
czukavinae *N.S. Filatova in Bot. Zhurn.*, 71(11): 1555 (1986)—U.S.S.R. (Middle Asia).

ARTEMISIA :–
dubia
var. subdigitata *(Mattf.) Y.R. Ling in Kew Bull.*, 42(2): 445 (1987): A. subdigitata.
grubovii *N.S. Filatova in Bot. Zhurn.*, 71(11): 1550 (1986)—China.
junatovii *N.S. Filatova in Bot. Zhurn.*, 71(11): 1552 (1986)—Xizang.
myriantha
var. pleiocephala *(Pamp.) Y.R. Ling in Kew Bull.*, 42(2): 446 (1987): A. pleiocephala.
thomsoniana *(C.B. Clarke) N.S. Filatova in Novosti Sist. Vyssh. Rast.*, 23: 239 (1986): A. maritima var. thomsoniana.
tridentata
var. wyomingensis *(Beetle & Young) S.L. Welsh in Great Basin Nat.*, 43(2): 215 (1983): A. tridentata subsp. wyomingensis.
xylorhiza *Krasch. ex N.S. Filatova in Bot. Zhurn.*, 71(11): 1553 (1986)—Mongolia.
zhaodongensis *G.Y. Chang & M.Y. Liou in Bull. Bot. Res. North-East. Forest. Inst.*, 7(3): 79 (1987)—China.

ARTHROPODIUM (Anthericac.).
dyeri *(Domin) N.H. Brittan in Fl. Australia,* 45: 467 (1987): Chamaescilla dyeri.

ARTHROSTYLIDIUM (Gramin.).
banaoense *L.J. Catasús Guerra in Acta Bot. Cubana,* 37: 6 (1987)—Cuba.

ARUM (Arac.).
italicum
f. majoricense *(L. Chodat) M. Mus, J.J. Pericás & J.A. Rosselló in Candollea,* 42(1): 397 (1987): A. majoricense.
purpureospathum *P.C. Boyce in Aroideana,* 10(1): 8 (1987)—Crete.

ARUNDINARIA (Gramin.).
longifimbriata *(S.Y. Chen) T.H. Wen in J. Bamboo Res.,* 6(3): 34 (1987), without basionym page: Pleioblastus longifimbriatus.
nanningensis *Q.H. Dai in J. Bamboo Res.*, 6(3): 35 (1987)—China.

ARUNDO (Gramin.).
laxa *(Host.) G. Wahlenberg, Fl. Carpat. Princip.:* 35 (1814 publ. 1815): Calamagrostis laxa.

ASCLEPIAS (Asclepiadac.).
compressidens *(N.E. Br.) A. Nicholas in Bothalia,* 17(1): 19 (1987): A. navicularis var. compressidens.
gordon-grayae *A. Nicholas in Bothalia,* 17(1): 17 (1987)—South Africa (Natal).

ASCOCENTRUM (Orchidac.).
himalaicum *(Deb, Sengupta & Malick) E.A. Christenson in Notes Roy. Bot. Gard. Edinburgh,* 44(2): 256 (1987): Saccolabium himalaicum.

ASCOCHILUS (Orchidac.).
mindanaensis *(Ames) E.A. Christenson in Indian Orchid J.*, 1(4): 151 (1985); *et in Selbyana,* 9(1): 167 (1986): Thrixspermum mindanaense.

ASPERULA (Rubiac.).
capillacea *(Lange) R. Vilm. ex M. Kerguélen in Lejeunia,* 120: 53 (1987): A. cynanchica var. capillacea.
syrticola *(Miq.) H.R. Toelken in Fl. S. Australia,* 2: 1063 (1986): Rubia syrticola.

ASPHODELUS (Asphodelac.).
fistulosus
subsp. nilotica *P. Ravenna in Herbertia,* 43(1): 40 (1987)—Egypt, Algeria.

ASPICARPA (Malpighiac.).
brevipes *(DC.) W.R. Anderson in Contrib. Univ. Mich. Herb.*, 16: 55 (1987): Banisteria brevipes.
harleyi *W.R. Anderson in Contrib. Univ. Mich. Herb.*, 16: 55 (1987)—Brazil.
schininii *W.R. Anderson in Contrib. Univ. Mich. Herb.*, 16: 59 (1987), nom. nov.: Camarea lanata *Chodat*.

ASPIDISTRA (Convallariac.).
acetabuliformis *Y. Wan & C.C. Huang in Acta Phytotax. Sin.*, 25(5): 396 (1987)—China.
austrosinensis *Y. Wan & C.C. Huang in Guihaia,* 7(3): 221 (1987)—China.
cruciformis *Y. Wan & X.H. Lu in Guihaia,* 7(3): 217 (1987)—China.
leyeensis *Y. Wan & C.C. Huang in Guihaia,* 7(3): 219 (1987)—China.
linearifolia *Y. Wan & C.C. Huang in Guihaia,* 7(3): 220 (1987)—China.
longipetala *S.Z. Huang in Guihaia,* 6(4): 273 (1986)—China.
subrotata *Y. Wan & C.C. Huang in Guihaia,* 7(3): 223 (1987)—China.
xilinensis *Y. Wan & X.H. Lu in Acta Phytotax. Sin.*, 25(5): 397 (1987)—China.

ASPIDOSTEMON J.G. Rohwer & H.G. Richter in Bot. Jahrb., 109(1): 71 (1987). *LAURACEAE.*
caudatum *J.G. Rohwer in Bot. Jahrb.*, 109(1): 74 (1987)—Madagascar.
dolichocarpum *(Kosterm.) J.G. Rohwer in Bot. Jahrb.,* 109(1): 74 (1987): Cryptocarya dolichocarpa.
glandulosum *J.G. Rohwer in Bot. Jahrb.*, 109(1): 75 (1987)—Madagascar.
humbertianum *(Kosterm.) J.G. Rohwer in Bot. Jahrb.,* 109(1): 78 (1987): Cryptocarya humbertiana.
inconspicuum *J.G. Rohwer in Bot. Jahrb.*, 109(1): 75 (1987)—Madagascar.
lacrimans *(Kosterm.) J.G. Rohwer in Bot. Jahrb.,* 109(1): 78 (1987): Cryptocarya lacrimans.

ASPIDOSTEMON :-
percoriaceum *(Kosterm.) J.G. Rohwer in Bot. Jahrb.,* 109(1): 75 (1987): Cryptocarya percoriacea.
perrieri *(Danguy) J.G. Rohwer in Bot. Jahrb.,* 109(1): 74 (1987): Cryptocarya perrieri.
scintillans *(Kosterm.) J.G. Rohwer in Bot. Jahrb.,* 109(1): 78 (1987): Cryptocarya scintillans.
synandra *J.G. Rohwer in Bot. Jahrb.,* 109(1): 78 (1987)—Madagascar.
trianthera *(Kosterm.) J.G. Rohwer in Bot. Jahrb.,* 109(1): 75 (1987): Cryptocarya trianthera.

ASPILIA (Compos.).
andrade-limae *J.U.M. dos Santos in Bol. Mus. Paraense Emílio Goeldi, Bot.,* 2(1): 5 (1985)—Brazil.
verticillata *J.U.M. dos Santos in Bol. Mus. Paraense Emílio Goeldi, Bot.,* 2(1): 9 (1985)—Brazil.

ASTER (Compos.).
cordifolius
var. sagittifolius *(Wedem. ex Willd.) A.G. Jones in Phytologia,* 63(2): 131 (1987): A. sagittifolius.
drummondii
var. parviceps *(Shinners) A.G. Jones in Phytologia,* 63(2): 131 (1987): A. texanus var. parviceps.
var. texanus *(Burgess) A.G. Jones in Phytologia,* 63(2): 131 (1987): A. texanus.
falcatus
var. commutatus *(Torrey & Gray) A.G. Jones in Phytologia,* 63(2): 131 (1987): A. multiflorus var. commutatus.
kingii
var. barnebyana *(Welsh & Goodrich) S.L. Welsh in Great Basin Nat.,* 43(2): 221 (1983): Machaeranthera kingii var. barnebyana.
lanceolatus
subsp. hesperius *(Gray) J.C. Semple & J.G. Chmielewski in Canad. J. Bot.,* 65(5): 1060 (1987): A. hesperius.
var. hirsuticaulis *J.C. Semple & J.G. Chmielewski in Canad. J. Bot.,* 65(5): 1058 (1987)—U.S.A. (Michigan).
var. interior *(Wiegand) J.C. Semple & J.G. Chmielewski in Canad. J. Bot.,* 65(5): 1058 (4 June 1987), with incorrect basionym vol. no.: A. interior.
var. interior *(Wieg.) A.G. Jones in Phytologia,* 63(2): 132 (1987): A. interior.
var. latifolius *J.C. Semple & J.G. Chmielewski in Canad. J. Bot.,* 65(5): 1060 (1987)—U.S.A. (Pennsylvania).
var. simplex *(Willd.) A.G. Jones in Phytologia,* 63(2): 132 (1987): A. simplex.
linosyris
subsp. armoricanus *(Rouy) M. Kerguélen in Lejeunia,* 120: 54 (1987): A. linosyris proles armoricanus.

ASTER :-
novi-belgii
var. johannensis *(Fern.) A.G. Jones in Phytologia,* 63(2): 132 (1987): A. johannensis.
var. tardiflorus *(L.) A.G. Jones in Phytologia,* 63(2): 132 (1987): A. tardiflorus.
occidentalis
var. fremontii *(Torrey & Gray) A.G. Jones in Phytologia,* 63(2): 132 (1987): A. ascendens var. fremontii.
puniceus
var. elliottii *(Torrey & Gray) A.G. Jones in Phytologia,* 63(2): 132 (1987): A. elliottii.
yakushimensis *(Kitamura) Soejima & T. Yahara in J. Fac. Sci. Univ. Tokyo, Bot.,* 14(2): 112 (1987): A. ageratoides subsp. yakushimensis.

ASTERIDIA (Compos.).
athrixioides
f. horripes *(J. Black) D.A. Cooke in Fl. S. Australia,* 3: 1500 (1986): Athrixia tenella var. horripes.

ASTEROGYNE (Palmae).
yaracuyense *A. Henderson & J.A. Steyermark in Brittonia,* 38(4): 309 (1986)—Venezuela.

ASTILBE (Saxifragac.).
glaberrima
var. saxatilis *(Nakai) H. Ohba in J. Fac. Sci. Univ. Tokyo, Bot.,* 14(2): 86 (1987): A. glaberrima lus. saxatilis.

ASTRACANTHA (Legumin.).
arnacantha
subsp. aitosensis *(Ivanisch.) U. Réer & D. Podlech in Mitt. Bot. Staatssamml. München,* 22: 529 (1986): Astragalus aitosensis.
subsp. arnacanthoides *(Boriss.) U. Réer & D. Podlech in Mitt. Bot. Staatssamml. München,* 22: 530 (1986): Tragacantha arnacanthoides.
echinus
subsp. arabica *I.C. Hedge & D. Podlech in Bot. Jahrb.,* 108(2–3): 260 (1987)—Saudi Arabia.
rumelica *(Bunge) U. Réer & D. Podlech in Mitt. Bot. Staatssamml. München,* 22: 542 (1986): Astragalus rumelicus.
subsp. taygetica *(Sirjaev) U. Réer & D. Podlech in Mitt. Bot. Staatssamml. München,* 22: 544 (1986): Astragalus rumelicus var. taygeticus.

ASTRAGALUS (Legumin.).
bibullatus *R.C. Barneby & E.L. Bridges in Brittonia,* 39(3): 359 (1987)—U.S.A. (Tennessee).
chaldiranicus *Kit Tan & F. Sorger in Aliso,* 11(4): 624 (1987)—Turkey.
chuskanus *R.C. Barneby & R. Spellenberg in Brittonia,* 39(2): 188 (1987)—U.S.A. (New Mexico, Arizona).

ASTRAGALUS :-
collenetteae *I.C. Hedge & D. Podlech in Bot. Jahrb.*, 108(2–3): 269 (1987), as 'collenettiae'—Saudi Arabia.
eliasianus *Kit Tan & F. Sorger in Aliso,* 11(4): 625 (1987)—Turkey.
ertterae *R.C. Barneby & J.R. Shevock in Aliso,* 11(4): 585 (1987)—U.S.A. (California).
hartvigii *Kit Tan in Aliso,* 11(4): 619 (1987)—Turkey.
isparticus *Kit Tan & F. Sorger in Aliso,* 11(4): 626 (1987)—Turkey.
kangalicus *Kit Tan & F. Sorger in Aliso,* 11(4): 626 (1987)—Turkey.
kitianus *F. Sorger in Aliso,* 11(4): 624 (1987)—Turkey.
lamalaensis *Z.C. Ni in Acta Phytotax. Sin.*, 25(3): 232 (1987)—Xizang.
maireanus *W. Greuter & Burdet in Willdenowia,* 16(2): 441 (1987)—Morocco.
nanfengensis *Z.C. Ni in Acta Phytotax. Sin.*, 25(3): 233 (1987)—Xizang.
neobarnebyanus *E. Gómez-Sosa in Brittonia,* 38(4): 427 (1986)—Peru.
ocakverdii *Kit Tan & F. Sorger in Pl. Syst. Evol.*, 154(1–2): 115 (1986)—Turkey.
renzianus *D. Podlech in Mitt. Bot. Staatssamml. München,* 22: 1 (1986)—Iran.
robertianus *Kit Tan & F. Sorger in Aliso,* 11(4): 622 (1987)—Turkey.
stridii *Kit Tan in Aliso,* 11(4): 621 (1987)—Turkey.
tragacanthoides
subsp. pinetorum *(Boiss.) D. Podlech in Bot. Jahrb.*, 108(2–3): 328 (1987): A. pinetorum.
tuus *Kit Tan in Aliso,* 11(4): 627 (1987)—Turkey.

ASTROLOBA (Aloeac.).
spiralis
subsp. foliolosa *(Haw.) L.E. Groen in Succulenta,* 66(7–8): 162 (1987): Aloe foliolosa.

ASYSTASIA (Acanthac.).
pinguifolia *T.J. Edwards in S. Afr. J. Bot.*, 53(3): 231 (1987)—Mozambique, Tanzania, South Africa (Natal).

ATHENAEA (Solanac.).
velutina *(Sendt.) W.G. D'Arcy in Solanaceae Newsl.*, 2(4): 16 (1986): Aureliana velutina.

ATRACTYLIS (Compos.).
subgen. Acarna *(Allioni) D. Petit in Bull. Soc. Bot. France, Lett. Bot.*, 134(2): 179 (1987): gen. Acarna.

ATRACTYLODES (Compos.).
comosa *(Sieb. ex Cass.) D. Petit in Bull. Soc. Bot. France, Lett. Bot.*, 134(2): 179 (1987): Atractylis comosa.
cuneata *(Boiss.) D. Petit in Bull. Soc. Bot. France, Lett. Bot.*, 134(2): 179 (1987): Atractylis cuneata.

ATRIPLEX (Chenopodiac.).
tornabenei
var. pedunculata *S. Castroviejo in An. Jard. Bot. Madrid,* 43(2): 475 (1986 publ. 1987)—Balearic Is.

ATROPA (Solanac.).
belladonna
subsp. caucasica *(Kreyer) V.E. Avetisyan in Biol. Zhurn. Arm.*, 34(9): 908 (1981): A. caucasica.

×**ATTABIGNYA** M.J. Balick, A.B. Anderson & J.T. de Medeiros-Costa in Brittonia, 39(1): 27 (1987). *PALMAE.* (Attalea × Orbignya).
minarum *M.J. Balick, A.B. Anderson & J.T. de Medeiros-Costa in Brittonia,* 39(1): 27 (1987)—Brazil.

ATTALEA (Palmae).
colenda *(O.F. Cook) H. Balslev & A. Henderson in Brittonia,* 39(1): 1 (1987), without exact basionym page: Ynesa colenda.

ATUNA (Chrysobalanac.).
cordata *Cockburn ex G.T. Prance in Brittonia,* 39(3): 364 (1987)—Borneo (Sabah).

AURICULARDISIA (Myrsinac.).
moraviana *C.L. Lundell in Phytologia,* 63(2): 74 (1987)—Costa Rica.
sarapiquiensis *C.L. Lundell in Phytologia,* 63(2): 74 (1987)—Costa Rica.
spathulata *C.L. Lundell in Phytologia,* 63(2): 75 (1987)—Costa Rica.
tarariae *(Lundell) C.L. Lundell in Phytologia,* 63(2): 75 (1987): Ardisia tarariae.
trichomata *C.L. Lundell in Phytologia,* 63(2): 75 (1987)—Costa Rica.
wilburiana *C.L. Lundell in Phytologia,* 63(2): 76 (1987)—Costa Rica.

AUSTRALINA (Urticac.).
pusilla
subsp. muelleri *(Wedd.) I. Friis & C.M. Wilmot-Dear in Nordic J. Bot.*, 7(2): 126 (1987): A. muelleri.

AVENA (Gramin.).
barbata
subsp. lusitanica *(Tab. Mor.) C. Romero Zarco in Lagascalia,* 14(1): 166 (1986): A. barbata subvar. lusitanica.
discolor *(Thuill.) A.-L.-S. Lejeune & R. Courtois, Comp. Fl. Belg.*, 1: 69 (1828): Aira discolor.
gaudiniana *(Boiss.) L. Bouvier, Fl. Alpes Suisse Savoie, ed. 1:* 741 (1878): Trisetum gaudinianum.
sulcata *Gay ex Boissier, Elench. Pl. Nov.:* 88 (1838)—Locality not stated.

AVENULA (Gramin.).
vasconica *(Sennen ex St.-Yves) C. Gervais in Bull. Soc. Neuchâtel. Sci. Nat.*, 104: 159 (1981): Avena pratensis var. vasconica.

AXINIPHYLLUM (Compos.).
durangense *B.L. Turner in Madroño,* 34(2): 165 (1987)—Mexico.

AZILIA I.C. Hedge & J.M. Lamond in Fl. Iran., 162: 386 (1987). *UMBELLIFERAE.*
eryngioides *(Pau) I.C. Hedge & J.M. Lamond in Fl. Iran.,* 162: 386 (1987): Prangos eryngioides.

B

BACCAUREA (Euphorbiac.).
henii *N.N. Thìn in J. Biol. (Vietnam),* 9(2): 37 (1987)—Vietnam.

BACCHARIS (Compos.).
polygama *L.A. Espinar in Kurtziana,* 9: 83 (1976)—Argentina.

BALIOSPERMUM (Euphorbiac.).
raziana *K.R. Keshava Murthy & S.N. Yoganarasimhan in Current Sci.,* 56(10): 486 (1987)—India.

BALSAMITA (Compos.).
major
subsp. balsamitoides *(Schultz-Bip.) M. Kerguélen in Lejeunia,* 120: 57 (1987): Tanacetum balsamitoides.

BAMBUSA (Gramin.).
latiflora *(Balansa) Nguyen to Quyen in Bot. Zhurn.,* 72(6): 829 (1987): B. vulgaris var. latiflora.
vinhphuensis *Nguyen to Quyen in Bot. Zhurn.,* 72(6): 828 (1987)—Vietnam.

BAPTISIA (Legumin.).
alba
var. macrophylla *(Larisey) D. Isely in Sida,* 11(4): 435 (1986): B. pendula var. macrophylla.

BAPTORHACHIS W.D. Clayton & S.A. Renvoize, Genera Graminum (Kew Bull. Addit. Ser. 13): 377, 298 (1986). *GRAMINEAE.*
foliacea *(Clayton) W.D. Clayton in Kew Bull.,* 42(2): 401 (1987): Stereochlaena foliacea.

BARBAREA (Crucif.).
minor
var. anfractuosa *P. Hartvig & A. Strid in Bot. Jahrb.,* 108(2–3): 316 (1987)—Turkey.

BARBOSELLA (Orchidac.).
dussii *(Cogn.) D.D. Dod in Moscosoa,* 4: 192 (1986): Pleurothallis dussii.

BARJONIA (Asclepiadac.).
harleyi *J. Fontella Pereira & N. Marquete Ferreira da Silva in Kew Bull.,* 42(3): 663 (1987)—Brazil.

BASTARDIASTRUM (Malvac.).
batesii *P.A. Fryxell & S.D. Koch in Aliso,* 11(4): 544 (1987)—Mexico.

BAUHINIA (Legumin.).
argentinensis
var. megasiphon *(Burkart) R.H. Fortunato in Darwiniana,* 27(1–4): 548 (1986): B. megasiphon.
forficata
subsp. pruinosa *(Vog.) R.H. Fortunato & Wunderlin in Darwiniana,* 27(1–4): 550 (1986): B. pruinosa.
pinheiroi *R.P. Wunderlin in Phytologia,* 62(2): 89 (1987)—Brazil.
scandens
var. horsfieldii *(Miq.) H. Ohashi in Enum. Fl. Pl. Nepal,* 2: 108 (1979): Lasiobema horsfieldii.

BEAUFORTIA (Myrtac.).
bicolor *A. Strid in Pl. Syst. Evol.,* 155(1–4): 339 (1987)—Australia (W. Australia).

BEGONIA (Begoniac.).
brandbygeana *L.B. Smith & D.C. Wasshausen in Fl. Ecuador,* 25(133): 54 (1986)—Ecuador.
breedlovei *K. Burt-Utley in Brittonia,* 38(4): 336 (1986)—Mexico.
chiapensis *K. Burt-Utley in Brittonia,* 38(4): 333 (1986)—Mexico.
holmnielseniana *L.B. Smith & D.C. Wasshausen in Fl. Ecuador,* 25(133): 62 (1986)—Ecuador.
jaliscana *K. Burt-Utley in Brittonia,* 38(4): 335 (1986)—Mexico.
lugonis *L.B. Smith & D.C. Wasshausen in Fl. Ecuador,* 25(133): 55 (1986)—Ecuador.
lyman-smithii *K. Burt-Utley & J.F. Utley in Brittonia,* 39(1): 59 (1987)—Mexico.
napoensis *L.B. Smith & D.C. Wasshausen in Fl. Ecuador,* 25(133): 63 (1986)—Ecuador.
oellgaardii *L.B. Smith & D.C. Wasshausen in Fl. Ecuador,* 25(133): 59 (1986)—Ecuador.
pectennervia *L.B. Smith & D.C. Wasshausen in Fl. Ecuador,* 25(133): 59 (1986)—Ecuador.
umbraculifolia *Y. Wan & B.N. Chang in Acta Phytotax. Sin.,* 25(4): 322 (1987)—China.

BEILSCHMIEDIA (Laurac.).
deomalica *S.S.R. Bennet & S. Chandra in Indian J. Forest.,* 9(3): 273 (1986 publ. 1987)—India.
supraglandulosa *Y.K. Li in Guihaia,* 5(4): 344 (1985)—China.

BEISELIA L.L. Forman in Kew Bull., 42(1): 262 (1987). *BURSERACEAE.*
mexicana *L.L. Forman in Kew Bull.,* 42(1): 262 (1987)—Mexico.

BELOGLOTTIS (Orchidac.).
laxispica *P.M. Catling in Rhodora,* 89(857): 13 (1987)—Mexico.

BERBERIS (Berberidac.).
hajirensis *R.A. Qureshi & M.N. Chaudhri in Pakistan Syst.,* 3(1): 17 (1987)—Pakistan.
pakistanica *R.A. Qureshi & M.N. Chaudhri in Pakistan Syst.,* 3(1): 19 (1987)—Pakistan.
pseudotibetica *C.Y. Wu ex S.Y. Bao in Acta Phytotax. Sin.,* 25(2): 159 (1987), as 'pseudo-tibetica'—China.
tianbaoshanensis *S.Y. Bao in Acta Phytotax. Sin.,* 25(2): 158 (1987)—China.

BESLERIA (Gesneriac.).
arbusta *L.E. Skog in Op. Bot.,* 92: 231 (1987)—Peru.
steyermarkiorum *Wiehler ex L.E. Skog in Op. Bot.,* 92: 233 (1987)—Venezuela.

BETA (Chenopodiac.).
vulgaris
var. asiatica *V.I. Burenin in Byull. Vses. Ord. Lenina Inst. Rast. N.I. Vavilova,* 161: 69 (1986)—Cultivation.
var. rubra *V.I. Burenin in Byull. Vses. Ord. Lenina Inst. Rast. N.I. Vavilova,* 161: 70 (1986)—Cultivation.
var. rubrifolia *Krassochk. ex V.I. Burenin in Byull. Vses. Ord. Lenina Inst. Rast. N.I. Vavilova,* 161: 69 (1986)—Cultivation.
var. subrosea *Krassochk. ex V.I. Burenin in Byull. Vses. Ord. Lenina Inst. Rast. N.I. Vavilova,* 161: 70, line 12 (1986), but see ICBN(1983) Art.64.3—Cultivation.
var. subrosea *Krassochk. ex V.I. Burenin in Byull. Vses. Ord. Lenina Inst. Rast. N.I. Vavilova,* 161: 70, line 22 (1986), but see ICBN(1983) Art. 64.3—Cultivation.

BETULA (Betulac.).
murrayana *B.V. Barnes & B.P. Dancik in Canad. J. Bot.,* 63(2): 226 (1985)—U.S.A. (Michigan).
pubescens
subsp. czerepanovii *(Orlova) L. Hämet-Ahti in Phytocoenologia,* 15(4): 450 (1987): B. czerepanovii.
szaferi *Jentys-Szaferowa ex J. Staszkiewicz in Acta Soc. Bot. Pol.,* 55(3): 364 (1986)—Poland.

BHIDEA (Gramin.).
fischeri *P.V. Sreekumar & B.V. Shetty in Kew Bull.,* 42(3): 683 (1987)—India.

BILDERDYKIA (Polygonac.).
cilinodis
f. erecta *(Peck) C.F. Reed in Phytologia,* 63(5): 410 (1987): Polygonum cilinode var. erectum.

BILLBERGIA (Bromeliac.).
robert-readii *E. Gross & W. Rauh in J. Bromeliad Soc.,* 37(2): 57 (1987)—Peru.

BIOPHYTUM (Oxalidac.).
cowanii *T. Wendt in Brittonia,* 39(1): 133 (1987)—Mexico.

BISCUTELLA (Crucif.).
austriaca
subsp. hungarica *(Soó ex Soó) M. Peniašteková in Biológia, A (Czechoslovakia),* 41(5): 454 (1986): B. laevigata subsp. hungarica.
coronopifolia
var. pinnatifida *(Jord.) J.D. Olowokudejo in Feddes Repert.,* 97(9–10): 569 (1986): B. pinnatifida.
var. polyclada *(Jord.) J.D. Olowokudejo in Feddes Repert.,* 97(9–10): 569 (1986): B. polyclada.
glacialis
var. harana *J.D. Olowokudejo in Folia Geobot. Phytotax.,* 21(4): 402 (1986); *et in Feddes Repert.,* 97(9–10): 567 (Oct. 1986)— Spain.
lusitanica
var. macrocarpa *(Samp. ex Guinea) J.D. Olowokudejo in Feddes Repert.,* 97(9–10): 568 (1986): B. macrocarpa.
megacarpea
subsp. atlantica *(Maire) J.E. Hernández-Bermejo & M. Clemente-Muñoz in Lagascalia,* 14(2): 200 (1986), without basionym page: B. laevigata var. atlantica.
subsp. variegata *(Boiss. & Reuter) J.E. Hernández-Bermejo & M. Clemente-Muñoz in Lagascalia,* 14(2): 200 (1986): B. variegata.
sempervirens
var. elvira *J.D. Olowokudejo in Feddes Repert.,* 97(9–10): 566 (Oct. 1986)—Spain.
subsp. glacialis *(Boiss. & Reuter) J.E. Hernández-Bermejo & M. Clemente-Muñoz in Lagascalia,* 14(2): 201 (1986): B. laxa var. glacialis.
valentina
var. leptophylla *(Pau) J.D. Olowokudejo in Feddes Repert.,* 97(9–10): 567 (1986): B. leptophylla.
var. pinnata *J.D. Olowokudejo in Feddes Repert.,* 97(9–10): 568 (1986)—Spain.
var. tenuicaulis *(Jord.) J.D. Olowokudejo in Feddes Repert.,* 97(9–10): 567 (1986): B. tenuicaulis.
varia
subsp. gracilis *(Mach.-Laur.) M. Peniašteková in Biológia, A (Czechoslovakia),* 41(5): 455 (1986): B. laevigata subsp. gracilis.
subsp. guestphalica *(Mach.-Laur.) M. Peniašteková in Biológia, A (Czechoslovakia),* 41(5): 455 (1986): B. laevigata subsp. guestphalica.
subsp. kerneri *(Mach.-Laur. ex Heywood) M. Peniašteková in Biológia, A (Czechoslovakia),* 41(5): 455 (1986): B. laevigata subsp. kerneri.

BISCUTELLA :–
subsp. subaphylla *(Mach.-Laur.) M. Peniašteková in Biológia, A (Czechoslovakia),* 41(5): 455 (1986): B. laevigata subsp. subaphylla.
subsp. tenuifolia *(Bluff & Fingerh.) M. Peniašteková in Biológia, A (Czechoslovakia),* 41(5): 455 (1986): B. ambigua β. tenuifolia.
variegata
var. foliosa *(Mach.-Laur.) J.D. Olowokudejo in Feddes Repert.,* 97(9–10): 566 (1986): B. foliosa.
var. megacarpaea *(Boiss. & Reuter) J.D. Olowokudejo in Feddes Repert.,* 97(9–10): 566 (1986): B. megacarpaea.

BISSEA V. Fuentes in Rev. Jard. Bot. Nacion. Univ. Habana, 6(3): 12 (1985 publ. 1986), nom. nov.: Henoonia Grisebach (Goetzeac.).
myrtifolia *(Griseb.) V. Fuentes in Rev. Jard. Bot. Nacion. Univ. Habana,* 6(3): 12 (1985 publ. 1986): Henoonia myrtifolia.

BISTORTA (Polygonac.).
amplexicaulis
var. alba *A.H. Munshi & G.N. Javeid, Syst. Stud. Polygonac. Kashmir Himalaya (J. Econ. Taxon. Bot. Addit. Ser. 2):* 61 (1986)—India.
var. speciosa *(Meisn.) A.H. Munshi & G.N. Javeid, Syst. Stud. Polygonac. Kashmir Himalaya (J. Econ. Taxon. Bot. Addit. Ser. 2):* 60 (1986), without basionym ref.: Basionym not stated.

BLABEROPUS (Apocynac.).
mairei *(Levl.) Trân Đình Lý in Feddes Repert.,* 97(9–10): 616 (1986): Alstonia mairei.

BLEPHAROCALYX (Myrtac.).
eggersii *(Kiaerskou) L.R. Landrum in Fl. Neotrop.,* 45: 121 (1986): Marlieriopsis eggersii.

BOBEA (Rubiac.).
platypes *F.R. Fosberg in Phytologia,* 62(3): 179 (1987)—Hawaiian Is.

BOEBEROIDES (DC.) J.L. Strother in Sida, 11(4): 373 (1986): Dyssodia sect. Boeberoides (Compos.).
grandiflora *(DC.) J.L. Strother in Sida,* 11(4): 377 (1986): Dyssodia grandiflora.

BOERHAVIA (Nyctaginac.).
sinuata *(Meikle) W. Greuter & Burdet in Willdenowia,* 16(2): 448 (1987): Commicarpus sinuatus.

BOESENBERGIA (Zingiberac.).
acuminata *P. Sirirugsa in Nordic J. Bot.,* 7(4): 421 (1987)—Thailand.
aurantiaca *R.M. Smith in Notes Roy. Bot. Gard. Edinburgh,* 44(2): 226 (1987)—Borneo (Sabah).

BOESENBERGIA :–
basispicata *K. Larsen ex P. Sirirugsa in Nordic J. Bot.,* 7(4): 423 (1987)—Thailand.
flavoalba *R.M. Smith in Notes Roy. Bot. Gard. Edinburgh,* 44(2): 218 (1987)—Borneo (Sarawak, Sabah).
petiolata *P. Sirirugsa in Nordic J. Bot.,* 7(4): 423 (1987)—Thailand.
pulchella
var. attenuata *R.M. Smith in Notes Roy. Bot. Gard. Edinburgh,* 44(2): 225 (1987)—Borneo (Sabah).
siamensis *(Gagnep.) P. Sirirugsa in Nordic J. Bot.,* 7(4): 425 (1987): Gastrochilus siamense.
stenophylla *R.M. Smith in Notes Roy. Bot. Gard. Edinburgh,* 44(2): 230 (1987)—Borneo (Sabah, Sarawak).

BOGNERA (Arac.).
recondita *(Madison) S. Mayo & D.H. Nicolson in Taxon,* 33(4): 690 (1984): Ulearum reconditum.

BOLANTHUS (Caryophyllac.).
stenopetalus *P. Hartvig & A. Strid in Bot. Jahrb.,* 108(2–3): 318 (1987)—Turkey.

BOLTONIA (Compos.).
apalachicolensis *L.C. Anderson in Syst. Bot.,* 12(1): 133 (1987)—U.S.A. (Florida).

BONAMIA (Convolvulac.).
alatisemina *R.W. Johnson in Austrobaileya,* 2(4): 405 (1987)—Australia (W. Australia, N. Terr.).
deserticola *R.W. Johnson in Austrobaileya,* 2(4): 406 (1987)—Australia (N. Terr.).
erecta *R.W. Johnson in Austrobaileya,* 2(4): 406 (1987)—Australia (N. Terr., S. Australia, W. Australia).

BONGARDIA (Berberidac.).
margalla *R.R. Stewart ex R.A. Qureshi & M.N. Chaudhri in Pakistan Syst.,* 3(1): 21 (1987)—Pakistan.

BONNETIA (Guttif.).
bolivarensis *J.A. Steyermark in Ann. Missouri Bot. Gard.,* 74(1): 102 (1987)—Venezuela.
guaiquinimae *J.A. Steyermark in Ann. Missouri Bot. Gard.,* 74(1): 102 (1987)—Venezuela.
ptariensis *J.A. Steyermark in Ann. Missouri Bot. Gard.,* 74(1): 102 (1987)—Venezuela.
tristyla
subsp. nervosa *J.A. Steyermark in Ann. Missouri Bot. Gard.,* 74(1): 103 (1987)—Venezuela.

BORYA (Anthericac.).
jabirabela *D.M. Churchill in Fl. Australia,* 45: 468, 270 (1987)—Australia (N. Terr., W. Australia).
longiscapa *D.M. Churchill in Fl. Australia,* 45: 468, 272 (1987)—Australia (W. Australia).

BORZICACTUS (Cactac.).
celsianus
var. trollii *(Kupp.) G.D. Rowley in Excelsa,* 12: 35 (1986): Cereus trollii.

BOURRERIA (Boraginac.).
lyciacea *M. Thulin in Nordic J. Bot.,* 7(4): 415 (1987)—Somali Republic, Kenya.
nemoralis *(Gürke) M. Thulin in Nordic J. Bot.,* 7(4): 415 (1987): Ehretia nemoralis.
orbicularis *(Hutch. & E.A. Bruce) M. Thulin in Nordic J. Bot.,* 7(4): 415 (1987): Ehretia orbicularis.
petiolaris *(Lam.) M. Thulin in Nordic J. Bot.,* 7(4): 414 (1987): Ehretia petiolaris.
teitensis *(Gürke) M. Thulin in Nordic J. Bot.,* 7(4): 415 (1987): Ehretia teitensis.

BRACHIARIA (Gramin.).
lata
subsp. caboverdiana *H.J. Conert & C. Köhler in Senckenberg. Biol.,* 67(4–6): 432 (1987)—Cape Verde Is.
paucispicata *(Morong) W.D. Clayton in Kew Bull.,* 42(2): 401 (1987): Panicum paucispicatum.

BRACHYCHITON (Sterculiac.).
diversifolium *R. Brown, Pterocymbium:* 234 (4 June 1844)—Australia. †
incanum *R. Brown, Pterocymbium:* 234 (4 June 1844)—Australia. †
platanoides *R. Brown, Pterocymbium:* 234 (4 June 1844)—Australia. †
populneum *R. Brown, Pterocymbium:* 234 (4 June 1844)—Australia. †
ramiflorum *R. Brown, Pterocymbium:* 234 (4 June 1844)—Australia. †

BRACHYPODIUM (Gramin.).
gaditanum *S. Talavera in Lagascalia,* 14(1): 120 (1986)—Spain.
glaucovirens *(Murbeck) T. Durand & B.D. Jackson in Index Kewensis,* Suppl. 1(1): 62 (1886–1895 publ. 1902): B. sylvaticum subsp. glaucovirens.
pubescens *(Peterm.) S.G. Musaev in Bot. Zhurn.,* 72(1): 94 (1987): B. gracile β. pubescens.
ramosum *(L.) Beauvois, Ess. Agrostogr.:* 163 (1812): Bromus ramosus.

BRACHYTOME (Rubiac.).
pitardii *D.D. Tirvengadum in Bull. Mus. Nation. Hist. Nat., B, Adansonia,* Ser. 4, 8(3): 294 (1986)—Vietnam.
russellii *D.B. Deb & M. Gangopadhyay in Candollea,* 42(1): 353 (1987)—Burma.
wardii *D.B. Deb & M. Gangopadhyay in Candollea,* 42(1): 356 (1987)—Burma.

BRASSICA (Crucif.).
oleracea
var. kashmiriana *A.R. Naqshi & G.N. Javeid in J. Econ. Taxon. Bot.,* 9(1): 91 (1987)—Cultivation.

BRAYA (Crucif.).
forrestii
var. puberula *W.T. Wang in Acta Bot. Yunnanica,* 9(1): 19 (1987)—China.

BREWCARIA (Bromeliac.).
marahuacae *L.B. Smith, J.A. Steyermark & H. Robinson in Ann. Missouri Bot. Gard.,* 73(4): 714 (1986 publ. 1987)—Venezuela.

BRICKELLIA (Compos.).
microphylla
var. watsonii *(Robins.) S.L. Welsh in Great Basin Nat.,* 43(2): 230 (1983): B. watsonii.

BRIDELIA (Euphorbiac.).
chineensis *N.N. Thìn in J. Biol. (Vietnam),* 9(2): 37 (1987)—Vietnam.

BROCCHINIA (Bromeliac.).
oliva-estevae *J.A. Steyermark & L.B. Smith in Ann. Missouri Bot. Gard.,* 74(1): 85 (1987)—Venezuela.

BROMOPSIS (Gramin.).
alpina *(Malysch.) G.A. Peshkova in Novosti Sist. Vÿssh. Rast.,* 23: 26 (1986): Bromus sibiricus f. alpinus.
altaica *G.A. Peshkova in Novosti Sist. Vÿssh. Rast.,* 23: 26 (1986)—U.S.S.R. (Siberia, Middle Asia), Mongolia.
austrosibirica *G.A. Peshkova in Novosti Sist. Vÿssh. Rast.,* 23: 28 (1986)—U.S.S.R. (Siberia).
sibirica *(Drob.) G.A. Peshkova in Novosti Sist. Vÿssh. Rast.,* 23: 31 (1986), without basionym ref.: Basionym not stated.
taimyrensis *(Roshev.) G.A. Peshkova in Novosti Sist. Vÿssh. Rast.,* 23: 31 (1986): Bromus sibiricus var. taimyrensis.

BROMUS (Gramin.).
affinis *(Dumort.) B.D. Jackson in Index Kewensis,* 1: 341 (1753–1885 publ. 1893): B. mollis * (subsp.) affinis.
barobalianus *G. Singh in G. Singh & P. Kachroo, Forest Fl. Srinagar:* 129 (1987)—India.
coytaei *W. Curtis, Pract. Obs. Brit. Grasses, ed. 3:* 50 (1798)—Great Britain.
gunckelii *O. Matthei in Gayana, Bot.,* 43(1–4): 62 (1986)—Chile.

BRONGNIARTIA (Legumin.).
sousae *O. Dorado in Aliso,* 11(4): 631 (1987)—Mexico.

BRUGMANSIA (Solanac.).
amesianum *(Schultes) W.G. D'Arcy in Solanaceae Newsl.,* 2(4): 17 (1986): Methysticodendron amesianum.

BRYONOPSIS G.A.W. Arnott in Madras J. Lit. Sci., 12: 49 (1840). *CUCURBITACEAE.*

BUCHNERA (Scrophulariac.).
chimanimaniensis *D. Philcox in Kew Bull.*, 42(1): 208 (1987)—Zimbabwe.
chisumpae *D. Philcox in Kew Bull.*, 42(2): 384 (1987)—Zambia.
ebracteolata *D. Philcox in Kew Bull.*, 42(2): 384 (1987)—Zambia.
geminiflora *D. Philcox in Kew Bull.*, 42(2): 384 (1987)—Malawi.
laxiflora *D. Philcox in Kew Bull.*, 42(1): 208 (1987)—Zambia.
nervosa *D. Philcox in Kew Bull.*, 42(2): 384 (1987)—Zambia.
subglabra *D. Philcox in Kew Bull.*, 42(1): 208 (1987)—Zimbabwe.
wildii *D. Philcox in Kew Bull.*, 42(1): 208 (1987)—Zimbabwe.

BUDDLEJA (Buddlejac.).
crispa
var. glandulifera *(Marq.) S.Y. Bao in Fl. Xizangica,* 3: 899 (1986): B. tibetica var. glandulifera.
var. grandiflora *(Marq.) S.Y. Bao in Fl. Xizangica,* 3: 899 (1986): B. tibetica var. grandiflora.
minima *S.Y. Bao in Fl. Xizangica,* 3: 895 (1986)—Xizang.

BUFONIA (Caryophyllac.).
macropetala
subsp. willkommiana *(Boiss.) F. Amich García in An. Jard. Bot. Madrid,* 43(2): 467 (1986 publ. 1987): B. willkommiana.

BUGLOSSOIDES (Boraginac.).
arvensis
subsp. incrassatus *(Guss.) M. Kerguélen in Lejeunia,* 120: 59 (1987): Lithospermum incrassatum.

BULBINE (Asphodelac.).
glauca *(Raf.) E.M. Watson in Fl. Australia,* 45: 469 (1987): Nemopogon glaucum.
vagans *E.M. Watson in Fl. Australia,* 45: 470, 238 (1987)—Australia (Queensland).

BULBOPHYLLUM (Orchidac.).
acutebracteatum
var. rubrobrunneopapillosum *(De Wild.) J.J. Vermeulen in Bull. Jard. Bot. Nation. Belg.,* 56(3–4): 238 (1986): B. rubrobrunneopapillosum.
bidenticulatum
subsp. joyceae *J.J. Vermeulen in Orchid Monogr.,* 2: 90 (1987)—Kenya.
calyptratum
var. graminifolium *(Summerh.) J.J. Vermeulen in Bull. Jard. Bot. Nation. Belg.,* 56(3–4): 235 (1986): B. graminifolium.
var. lucifugum *(Summerh.) J.J. Vermeulen in Orchid Monogr.,* 2: 131 (1987): B. lucifugum.

BULBOPHYLLUM :–
carnosisepalum *J.J. Vermeulen in Bull. Jard. Bot. Nation. Belg.,* 56(3–4): 236 (1986)—Zaïre, Ivory Coast, Cameroun, Uganda.
cochleatum
var. bequaertii *(De Wild.) J.J. Vermeulen in Bull. Jard. Bot. Nation. Belg.,* 56(3–4): 230 (1986): B. bequaertii.
var. brachyanthum *(Summerh.) J.J. Vermeulen in Bull. Jard. Bot. Nation. Belg.,* 56(3–4): 231 (1986): B. bequaertii var. brachyanthum.
var. gravidum *(Lindley) J.J. Vermeulen in Bull. Jard. Bot. Nation. Belg.,* 56(3–4): 230 (1986): B. gravidum.
var. tenuicaule *(Lindley) J.J. Vermeulen in Bull. Jard. Bot. Nation. Belg.,* 56(3–4): 230 (1986): B. tenuicaule.
comatum
var. inflatum *(Rolfe) J.J. Vermeulen in Bull. Jard. Bot. Nation. Belg.,* 56(3–4): 232 (1986): B. inflatum.
dolabriforme *J.J. Vermeulen in Kew Bull.,* 42(1): 263 (1987)—Nigeria.
falcatum
var. bufo *(Lindley) J.J. Vermeulen in Bull. Jard. Bot. Nation. Belg.,* 56(3–4): 235 (1986), without basionym ref.: Basionym not stated.
var. velutinum *(Lindley) J.J. Vermeulen in Bull. Jard. Bot. Nation. Belg.,* 56(3–4): 235 (1986), without basionym ref.: Basionym not stated.
fuscum
var. melinostachyum *(Schlechter) J.J. Vermeulen in Bull. Jard. Bot. Nation. Belg.,* 56(3–4): 240 (1986): B. melinostachyum.
horridulum *J.J. Vermeulen in Bull. Jard. Bot. Nation. Belg.,* 56(3–4): 238 (1986)—Zaïre.
injoloense
subsp. pseudoxypterum *(J.J. Vermeulen) J.J. Vermeulen in Bull. Jard. Bot. Nation. Belg.,* 56(3–4): 232 (1986): B. injoloense var. pseudoxypterum.
inornatum *J.J. Vermeulen in Kew Bull.,* 42(1): 266 (1987)—Tanzania.
josephi
var. mahonii *(Rolfe) J.J. Vermeulen in Bull. Jard. Bot. Nation. Belg.,* 56(3–4): 232 (1986): B. mahonii.
kivuense *J.J. Vermeulen in Bull. Jard. Bot. Nation. Belg.,* 56(3–4): 231 (1986)—Rwanda.
resupinatum
var. filiforme *(Kraenzlin) J.J. Vermeulen in Bull. Jard. Bot. Nation. Belg.,* 56(3–4): 235 (1986): B. filiforme.
saltatorium
var. albociliatum *(Finet) J.J. Vermeulen in Bull. Jard. Bot. Nation. Belg.,* 56(3–4): 230 (1986): B. calamarium var. albociliatum.
var. calamarium *(Lindley) J.J. Vermeulen in Bull. Jard. Bot. Nation. Belg.,* 56(3–4): 230 (1986): B. calamarium.

BULBOPHYLLUM :–
sandersonii
subsp. stenopetalum *(Kraenzlin) J.J. Vermeulen in Bull. Jard. Bot. Nation. Belg.,* 56(3–4): 234 (1986): B. stenopetalum.
scaberulum
var. crotalicaudatum *J.J. Vermeulen in Orchid Monogr.,* 2: 118 (1987)—Tanzania.
var. fuerstenbergianum *(De Wild.) J.J. Vermeulen in Bull. Jard. Bot. Nation. Belg.,* 56(3–4): 234 (1986): Megaclinium fuerstenbergianum.
schinzianum
var. irigaleae *(Cribb & Perez-Vera) J.J. Vermeulen in Orchid Monogr.,* 2: 29 (1987): B. irigaleae.
var. phaeopogon *(Schlechter) J.J. Vermeulen in Bull. Jard. Bot. Nation. Belg.,* 56(3–4): 229 (1986): B. phaeopogon.
subligaculiferum *J.J. Vermeulen in Orchid Monogr.,* 2: 30 (1987)—Gabon.
unifoliatum
subsp. flectens *(Cribb & Taylor) J.J. Vermeulen in Orchid Monogr.,* 2: 156 (1987): B. flectens.
subsp. infracarinatum *(Williamson) J.J. Vermeulen in Orchid Monogr.,* 2: 155 (1987): B. infracarinatum.

BUMELIA (Sapotac.).
dominicana *R.D. Whetstone & T.A. Atkinson in Sida,* 11(4): 396 (1986), nom. nov.: Dipholis ferruginea *Ekman & O.C. Schmidt.*

BUNCHOSIA (Malpighiac.).
berlinii *W.R. Anderson in Contrib. Univ. Mich. Herb.,* 16: 61 (1987)—Peru.
plowmanii *W.R. Anderson in Contrib. Univ. Mich. Herb.,* 16: 66 (1987)—Peru, Ecuador.

BUNIUM (Umbellif.).
kandaharicum *K.H. Rechinger in Fl. Iran.,* 162: 255 (1987)—Afghanistan.
luristanicum *K.H. Rechinger in Fl. Iran.,* 162: 248 (1987)—Iran.

BUPLEURUM (Umbellif.).
freitagii *K.H. Rechinger in Fl. Iran.,* 162: 290 (1987)—Afghanistan, Pakistan.
kabulicum *K.H. Rechinger in Fl. Iran.,* 162: 293 (1987)—Afghanistan.

BURCHARDIA (Colchicac.).
bairdiae *G.J. Keighery in Fl. Australia,* 45: 470, 408 (1987)—Australia (W. Australia).
rosea *G.J. Keighery in Fl. Australia,* 45: 471, 410 (1987)—Australia (W. Australia).

×**BURKHARDTARA** hort. in Orchid Rev., 95(1126): centre page pull-out p. 8 (1987). *ORCHIDACEAE.* (Leochilus × Odontoglossum × Oncidium × Rodriguezia).

C

CABRERIELLA (Compos.).
oppositicordia *(Cuatr.) J. Cuatrecasas in Proc. Biol. Soc. Wash.,* 98(3): 626 (1985): Senecio oppositicordius.

CACTUS (Cactac.).
langsdorfii *J.G.C. Lehmann in Sem. Hort. Bot. Hamb.:* 17 (1826)—Locality not stated. †

CAESIA (Anthericac.).
callianthа *R.J.F. Henderson in Fl. Australia,* 45: 471, 286 (1987)—Australia (S. Australia, New South Wales, Victoria), Tasmania.
parviflora
var. minor *R.J.F. Henderson in Fl. Australia,* 45: 471, 285 (1987)—Australia (S. Australia, New South Wales, Victoria), Tasmania.
var. vittata *(R. Br.) R.J.F. Henderson in Fl. Australia,* 45: 472 (1987): C. vittata.

CALAMAGROSTIS (Gramin.).
nagensis *(Bor) B.P. Uniyal in J. Econ. Taxon. Bot.,* 8(1): 235 (1986): Agrostis nagensis.

CALAMINTHA (Labiat.).
caroli-henricana *Kit Tan & F. Sorger in Pl. Syst. Evol.,* 155(1–4): 100 (1987)—Turkey.
rouyana *(Briq.) J.J. Pericás & J.A. Rosselló in Candollea,* 42(1): 397 (1987): Satureja rouyana.

×**CALAMMOPHILA** A. Brand in Koch, Syn. Deutsch. Schweiz. Fl., ed. 3, 3: 2715 (1907). *GRAMINEAE.* (Ammophila × Calamagrostis).
baltica *(Flügge ex Schrader) A. Brand in Koch, Syn. Deutsch. Schweiz. Fl., ed. 3,* 3: 2715 (1907): Arundo baltica.

CALAMUS (Palmae).
dransfieldii *C. Renuka in Kew Bull.,* 42(2): 433 (1987)—India.
vattayila *C. Renuka in Current Sci.,* 56(19): 1012 (1987)—India.

CALANTHE (Orchidac.).
plantaginea
var. lushuiensis *K.Y. Lang & Z.H. Tsi in Acta Phytotax. Sin.,* 25(5): 404 (1987)—China.

CALATHIANA (Gentianac.).
terglouensis
subsp. schleicheri *(Vaccari) Á. Löve & D. Löve in Taxon,* 35(4): 898 (1986): Gentiana verna f. schleicheri.
verna
subsp. pontica *(Soltok.) Á. Löve & D. Löve in Taxon,* 35(4): 898 (1986): Gentiana pontica.
subsp. tergestina *(G. Beck) Á. Löve & D. Löve in Taxon,* 35(4): 898 (1986): Gentiana angulosa f. tergestina.

CALCEOLARIA (Scrophulariac.).
fabrisii *S.M. Botta & A.L. Cabrera in Darwiniana,* 27(1–4): 1 (1986)—Argentina, Bolivia.

CALEA (Compos.).
dalyi *J.F. Pruski & L.E. Urbatsch in Brittonia,* 39(2): 201 (1987)—Bolivia.
fruticosa *(Gardner) L.E. Urbatsch, A. Zlotsky & J.F. Pruski in Syst. Bot.,* 11(4): 506 (1986): Amphicalea fruticosa.
sickii *(Barroso) L.E. Urbatsch, A. Zlotsky & J.F. Pruski in Syst. Bot.,* 11(4): 504 (1986): Brasilia sickii.

CALLIANDRA (Legumin.).
kingii *Prain ex S.R. Paul & T. Husain in Feddes Repert.,* 97(11–12): 759 (1986)—Burma.

CALLITRICHE (Callitrichac.).
truncata
subsp. occidentalis *(Rouy) H.D. Schotsman in Lagascalia,* 14(1): 153 (1986): C. truncata race occidentalis.

CALOCHORTUS (Liliac.).
raichei *S. Farwig & V. Girard in Herbertia,* 43(1): 2 (1987)—U.S.A. (California).

CALTHA (Ranunculac.).
palustris
var. orientalisinensis *X.H. Guo in Acta Phytotax. Sin.,* 25(3): 241 (1987), as 'orientali-sinensis'—China.

CALYPTRANTHES (Myrtac.).
acevedoi *Alain H. Liogier in Phytologia,* 61(6): 353 (1986)—Puerto Rico.
banilejoana *Alain H. Liogier in Phytologia,* 61(6): 354 (1986)—Hispaniola (Dominican Republic).
guayabillo *Alain H. Liogier in Phytologia,* 61(6): 354 (1986)—Hispaniola (Dominican Republic).
jimenoana *Alain H. Liogier in Phytologia,* 61(6): 355 (1986)—Hispaniola (Dominican Republic).
limoncillo *Alain H. Liogier in Phytologia,* 61(6): 355 (1986)—Hispaniola (Dominican Republic).

CALYPTROGENIA (Myrtac.).
cuspidata *Alain H. Liogier in Phytologia,* 61(6): 356 (1986)—Hispaniola (Dominican Republic).

CALYSTEGIA (Convolvulac.).
spithamaea
var. pubescens *(A. Gray) C.F. Reed in Phytologia,* 63(5): 411 (1987): C. sepium var. pubescens.

CAMELLIA (Theac.).
apolyodonta *H.T. Chang & Q.M. Chen in Acta Sci. Nat. Univ. Sunyatseni,* 1987(3): 84 (1987)—China.

CAMELLIA :–
brevistyla
f. rubida *P.L. Chiu in J. Nanjing Forest. Univ.,* 1987(2): 22 (1987)—China.
changii *C.X. Ye in Acta Sci. Nat. Univ. Sunyatseni,* 1987(1): 17 (1987)—China.
chekiang-oleosa
f. tanglii *P.L. Chiu in J. Nanjing Forest. Univ.,* 1987(2): 24 (1987)—China.
chrysantha
f. longistyla *S.L. Mo & Y.C. Zhong in Guihaia,* 5(4): 355 (1985)—China.
limonia
f. obovata *S.L. Mo & Y.C. Zhong in Guihaia,* 5(4): 355 (1985)—China.
longgangensis
var. patens *S.L. Mo & Y.C. Zhong in Guihaia,* 5(4): 354 (1985)—China.
odorata *L.S. Xie & Z.Y. Zhang in Acta Bot. Yunnanica,* 9(1): 57 (1987)—China.
parvipetala *J.Y. Liang & Z.M. Su in Guihaia,* 5(4): 357 (1985)—China.
pubescens *Chang & C.X. Ye in Acta Sci. Nat. Univ. Sunyatseni,* 1987(1): 18 (1987)—China.
pyxidiacea *Z.R. Xu, F.P. Chen & C.Y. Deng in Guihaia,* 7(1): 19 (1987)—China.
quinquebracteata *Chang & C.X. Ye in Acta Sci. Nat. Univ. Sunyatseni,* 1987(1): 20 (1987)—China.
quinqueloculosa *S.L. Mo & Y.C. Zhong in Guihaia,* 5(4): 353 (1985)—China.
rhytidophylla *Y.K. Li & M.Z. Yang in Guihaia,* 7(1): 13 (1987)—China.
rubimuricata *Chang & Z.R. Xu in Guihaia,* 5(4): 347 (1985)—China.
siangensis *T.K. Paul & M.P. Nayar in Bull. Bot. Surv. India,* 27(1–4): 92 (1985 publ. 1987)—India.
sinensis
subsp. buisanensis *(Sasaki) S.Y. Lu & Y.P. Yang in Quart. J. Chinese Forest.,* 20(1): 106 (1987): C. buisanensis.
trichoclada
f. leucantha *P.L. Chiu in J. Nanjing Forest. Univ.,* 1987(2): 26 (1987)—China.

CAMPANULA (Campanulac.).
lycica *Kit Tan & F. Sorger in Pl. Syst. Evol.,* 155(1–4): 96 (1987)—Turkey.
yildirimlii *Kit Tan & F. Sorger in Pl. Syst. Evol.,* 154(1–2): 124 (1986)—Turkey.

CAMPOMANESIA (Myrtac.).
espiritosantensis *L.R. Landrum in Brittonia,* 39(2): 245 (1987)—Brazil.

CAMPYLOTROPIS (Legumin.).
macrocarpa
var. giraldii *(Schindl.) K.T. Fu ex P.Y. Fu in Bull. Bot. Res. North-East. Forest. Inst.,* 7(4): 42 (1987): Lespedeza giraldii.
f. giraldii *(Schindl.) P.Y. Fu in Bull. Bot. Res. North-East. Forest. Inst.,* 7(4): 42 (1987): Lespedeza giraldii.

CAMPYLOTROPIS :-
f. hupehensis *(Pamp.) P.Y. Fu in Bull. Bot. Res. North-East. Forest. Inst.*, 7(4): 43 (1987): C. macrocarpa var. hupehensis.
f. lanceolata *P.Y. Fu in Bull. Bot. Res. North-East. Forest. Inst.*, 7(4): 41 (1987)—China.
f. longepedunculata *(Rick.) P.Y. Fu in Bull. Bot. Res. North-East. Forest. Inst.*, 7(4): 43 (1987): C. longepedunculata.
f. microphylla *K.T. Fu ex P.Y. Fu in Bull. Bot. Res. North-East. Forest. Inst.*, 7(4): 43 (1987)—China.
polyantha
f. macrophylla *P.Y. Fu in Bull. Bot. Res. North-East. Forest. Inst.*, 7(4): 46 (1987)—China.
f. souliei *(Schindl.) P.Y. Fu in Bull. Bot. Res. North-East. Forest. Inst.*, 7(4): 46 (1987): C. souliei.
var. tomentosa *P.Y. Fu in Bull. Bot. Res. North-East. Forest. Inst.*, 7(4): 46 (1987)—China.
tenuiramea *P.Y. Fu in Bull. Bot. Res. North-East. Forest. Inst.*, 7(4): 37 (1987)—China.
tomentosipetiolata *P.Y. Fu in Bull. Bot. Res. North-East. Forest. Inst.*, 7(4): 47 (1987)—China.
wenshaaica *P.Y. Fu in Bull. Bot. Res. North-East. Forest. Inst.*, 7(4): 35 (1987)—China.
yajiangensis *P.Y. Fu in Bull. Bot. Res. North-East. Forest. Inst.*, 7(4): 32 (1987)—China.
var. deronica *P.Y. Fu in Bull. Bot. Res. North-East. Forest. Inst.*, 7(4): 34 (1987)—China, Xizang.
yunnanensis
var. filipes *(Rick.) P.Y. Fu in Bull. Bot. Res. North-East. Forest. Inst.*, 7(4): 52 (1987): C. filipes.
var. zhongdenensis *P.Y. Fu in Bull. Bot. Res. North-East. Forest. Inst.*, 7(4): 52 (1987)—China.

CANISTRUM (Bromeliac.).
camacaensis *G. Martinelli & E.M.C. Leme in Bradea*, 4(33): 256 (1986)—Brazil.

CANTHIUM (Rubiac.).
subgen. Bullockia *D.M. Bridson in Kew Bull.*, 42(3): 630 (1987).
fadenii *D.M. Bridson in Kew Bull.*, 42(3): 632 (1987)—Kenya.
impressinervium *D.M. Bridson in Kew Bull.*, 42(3): 633 (1987)—Tanzania.
pseudosetiflorum *D.M. Bridson in Kew Bull.*, 42(3): 635 (1987)—Kenya, Ethiopia, Uganda, Tanzania.
setiflorum
subsp. telidosma *(K. Schum.) D.M. Bridson in Kew Bull.*, 42(3): 634 (1987): Plectronia telidosma.
vanwykii *P.M. Tilney & P.D.F. Kok in S. Afr. J. Bot.*, 53(1): 98 (1987)—South Africa (Natal, Cape Province).

CARALLIA (Rhizophorac.).
eugenioidea
var. sumatrensis *M. Ito & M. Hotta in Acta Phytotax. Geobot.*, 37(4–6): 147 (1986)—Sumatra.

CARDAMINE (Crucif.).
basisagittata *W.T. Wang in Acta Bot. Yunnanica*, 9(1): 10 (1987)—China.
bijiangensis *W.T. Wang in Acta Bot. Yunnanica*, 9(1): 12 (1987)—China.
caroides *C.Y. Wu in Acta Bot. Yunnanica*, 9(1): 16 (1987)—China.
dentata
var. palustris *(Wimm. & Grab.) K.S. Khatri in Feddes Repert.*, 97(5–6): 281 (1986): C. pratensis var. palustris.
flexuosoides *W.T. Wang in Acta Bot. Yunnanica*, 9(1): 12 (1987)—China.
var. glabricaulis *W.T. Wang in Acta Bot. Yunnanica*, 9(1): 13 (1987)—China.
griffithii
var. pentaloba *W.T. Wang in Acta Bot. Yunnanica*, 9(1): 16 (1987)—China.
heishuiensis *W.T. Wang in Acta Bot. Yunnanica*, 9(1): 15 (1987)—China.
hydrocotyloides *W.T. Wang in Acta Bot. Yunnanica*, 9(1): 8 (1987)—China.
levicaulis *W.T. Wang in Acta Bot. Yunnanica*, 9(1): 9 (1987)—China.
longistyla *W.T. Wang in Acta Bot. Yunnanica* 9(1): 14 (1987)—Xizang.
muliensis *W.T. Wang in Acta Bot. Yunnanica*, 9(1): 11 (1987)—China.
pratensis
subsp. major *P. Tomšovic in Folia Geobot. Phytotax.*, 21(4): 429 (1986)—Czechoslovakia.
×smejkalii *P. Tomšovic in Folia Geobot. Phytotax.*, 21(4): 430 (1986)—Czechoslovakia. (C. matthiolii × pratensis subsp. pratensis).
truncatolobata *W.T. Wang in Acta Bot. Yunnanica*, 9(1): 14 (1987)—China.
weixiensis *W.T. Wang in Acta Bot. Yunnanica*, 9(1): 9 (1987)—China.

CARDUUS (Compos.).
pulchellus *Greene ex P.A. Rydberg, Fl. Colorado:* 400, 401 (1906)—U.S.A. (Colorado).

CAREX (Cyperac.).
subsect. Intumescentes *F.J. Menapace, D.E. Wujek & A.A. Reznicek in Canad. J. Bot.*, 64(11): 2787 (1986).
austroamericana *G.A. Wheeler in Brittonia*, 38(4): 319 (1986)—Argentina, Chile, Tierra del Fuego.
×cariei *P. Aubin in Bull. Mens. Soc. Linn. Lyon*, 56(1): 18 (1987)—France. (C. pallescens × panicea).
depressa
subsp. basilaris *(Jordan) M. Kerguélen in Lejeunia*, 120: 62 (1987): C. basilaris.

CAREX :–
echinata
subsp. townsendii *(Mackenzie) A.A. Reznicek in Contrib. Univ. Mich. Herb.,* 16: 207 (1987): C. townsendii.
impressinervia *C.T. Bryson, R. Kral & J.R. Manhart in Rhodora,* 89(860): 357 (1987)—U.S.A. (Alabama, Mississippi, N. Carolina).
longii
var. meridionalis *(Kük.) G.A. Wheeler in Aliso,* 11(4): 536 (1987): C. albolutescens var. meridionalis.
nairii *N. Ghildyal & U.C. Bhattacharyya in J. Econ. Taxon. Bot.,* 8(2): 479 (1986)—India.
paniculata
subsp. calderae *(Hansen) K. Lewejohann & W. Lobin in Senckenberg. Biol.,* 67(4–6): 438 (1987): C. calderae.
subsp. hansenii *K. Lewejohann & W. Lobin in Senckenberg. Biol.,* 67(4–6): 440 (1987)—Cape Verde Is.
pleioneura *G.A. Wheeler in Aliso,* 11(4): 533 (1987)—Chile, Argentina.
pubicarpa *(J. Rohlena) A.W. Hill in Index Kewensis,* Supp. 6: 37 (1916–1920 publ. 1926): C. glauca var. pubicarpa.
transandina *G.A. Wheeler in Brittonia,* 38(4): 317 (1986)—Chile.

×**CARPENTERARA** hort. in Orchid Rev., 95(1123): centre page pull-out p. 9 (1987). *ORCHIDACEAE.* (Baptistonia × Odontoglossum × Oncidium).

CARPINUS (Corylac.).
caroliniana
subsp. virginiana *(Marshall) J.J. Furlow in Syst. Bot.,* 12(3): 429 (1987): C. betulus virginiana.
henryana
var. oblongifolia *(Hu) K. Rushforth in Plantsman,* 8(4): 249 (1987): C. turczaninowii var. oblongifolia.
var. simplicidentata *(Hu) K. Rushforth in Plantsman,* 8(4): 249 (1987): C. simplicidentata.
luochengensis *J.Y. Liang in Guihaia,* 6(4): 275 (1986)—China.
tropicalis
subsp. mexicana *J.J. Furlow in Syst. Bot.,* 12(3): 431 (1987)—Mexico.

CARTERELLA E.E. Terrell in Brittonia, 39(2): 250 (1987). *RUBIACEAE.*
alexanderae *(A. Carter) E.E. Terrell in Brittonia,* 39(2): 250 (1987): Bouvardia alexanderae.

CARUM (Umbellif.).
rupicola *P. Hartvig & A. Strid in Bot. Jahrb.,* 108(2–3): 302 (1987)—Turkey.

CARYA (Juglandac.).
dabieshanensis *M.C. Liu & Z.J. Li in J. Zhejiang Forest. Coll.,* 1(1): 41 (1984), with two types cited—China.

CARYOCAR (Caryocarac.).
harlingii *G.T. Prance & F. Encarnación in Op. Bot.,* 92: 182 (1987)—Peru, Colombia.

CASSIA (Legumin.).
acuta *Meyen ex T. Vogel, Syn. Gen. Cassiae:* 42 (1837)—Chile. †

CASSINE (Celastrac.).
albens *(Retz.) A.J.G.H. Kostermans in Gard. Bull. Singapore,* 39(2): 178 (1986 publ. 1987): Schrebera albens.
balae *A.J.G.H. Kostermans in Gard. Bull. Singapore,* 39(2): 185 (1986 publ. 1987)—Sri Lanka.
congylos *A.J.G.H. Kostermans in Gard. Bull. Singapore,* 39(2): 187 (1986 publ. 1987)—Sri Lanka.
grossa *(Wall. ex Roxb.) A.J.G.H. Kostermans in Gard. Bull. Singapore,* 39(2): 178 (1986 publ. 1987): Euonymus grossa.
koordersii *A.J.G.H. Kostermans in Gard. Bull. Singapore,* 39(2): 188 (1986 publ. 1987)—Java.
obiensis *A.J.G.H. Kostermans in Gard. Bull. Singapore,* 39(2): 190 (1986 publ. 1987)—Moluccas.

CASSIPOUREA (Rhizophorac.).
evrardii *J.J. Floret in Bull. Jard. Bot. Nation. Belg.,* 56(3–4): 447 (1986)—Zaïre.

CASTANOPSIS (Fagac.).
carlesii
var. taiwanensis *C.F. Shen in C.F. Shen & T.S. Liu, Taxon. & Pollen Morph. Fagaceae Taiwan:* 9 (1984)—Taiwan.
eyrei
var. brachyacantha *(Hay.) C.F. Shen in C.F. Shen & T.S. Liu, Taxon. & Pollen Morph. Fagaceae Taiwan:* 11 (1984): C. brachyacantha.
sessilis *(Nakai) C.F. Shen in C.F. Shen & T.S. Liu, Taxon. & Pollen Morph. Fagaceae Taiwan:* 15 (1984): C. carlesii var. sessilis.

×**CATASANDRA** hort. in Orchid Rev., 95(1129): centre page pull-out p. 10 (1987). *ORCHIDACEAE.* (Catasetum × Galeandra).

CATHA (Celastrac.).
abbottii *A.E. van Wyk & M. Prins in S. Afr. J. Bot.,* 53(3): 202 (1987)—South Africa (Cape Province, Natal).
edulis *(Vahl) S. Endlicher, Enchiridion Bot.:* 575 (1841): Celastrus edulis. †

CATTLEYA (Orchidac.).
labiata
var. albo-oculata *(Cogn.) J.A. Fowlie in Orchid Dig.,* 51(3): 111 (1987): C. labiata f. albo-oculata.

CATTLEYA :–
var. albo-oculata *(Cogn.) L.C. Menezes in Orchid Dig.,* 51(3): 138 (1987): C. labiata f. albo-oculata.
var. purpureostriata *(Cogn.) L.C. Menezes in Orchid Dig.,* 51(3): 138 (1987): C. labiata f. purpureo-striata.
mendelii
var. morganae *(Warner & Williams) G.J. Braem, Cattleya II, Unifoliate Cattleyas:* 58 (1986): C. morganae.
mossiae
var. wageneri *(Rchb. f.) G.J. Braem, Cattleya II, Unifoliate Cattleyas:* 65 (1986), without basionym page: C. wageneri.
trichopiliochila
var. virginalis *(Linden & André) G.J. Braem, Cattleya II, Unifoliate Cattleyas:* 84 (1986): C. virginalis.

CAULINIA (Najadac.).
orientalis *(Triest & Uotila) N.N. Tsvelev in Sosud. Rast. Sovet. Dal'nego Vostoka,* 2: 344 (1987): Najas orientalis.

CAVENDISHIA (Ericac.).
chocoensis
var. hirsuta *J.L. Luteyn in Op. Bot.,* 92: 113 (1987)—Colombia.

CAYRATIA (Vitac.).
cardiophylla *B.R. Jackes in Austrobaileya,* 2(4): 371 (1987)—Australia (Queensland).
kiujiangensis *C.Y. Wu in Fl. Xizangica,* 3: 228 (1986)—China, Xizang.
maritima *B.R. Jackes in Austrobaileya,* 2(4): 366 (1987)—Australia (Queensland, N. Terr.).

CELTIS (Ulmac.).
pallida
var. discolor *A.T. Hunziker & N.M. Dottori in Kurtziana,* 9: 130 (1976)—Argentina.

CENCHRUS (Gramin.).
robustus *R.D. Webster, Australian Paniceae (Poac.):* 27 (1987)—Australia (Queensland).

CENTAUREA (Compos.).
boissieri
subsp. transmalvana *(Emb. & Maire) I. Breitwieser & D. Podlech in Mitt. Bot. Staatssamml. München,* 22: 52 (1986): C. boissieri var. transmalvana.
debdouensis *I. Breitwieser & D. Podlech in Mitt. Bot. Staatssamml. München,* 22: 54 (1986)—Morocco.
delicatula *I. Breitwieser & D. Podlech in Mitt. Bot. Staatssamml. München,* 22: 59 (1986)—Tunisia, Algeria.
leucophaea
subsp. controversa *(Briq. & Cavill.) M. Kerguélen in Lejeunia,* 120: 65 (1987): C. paniculata subsp. controversa.

CENTAUREA :–
pomeliana
subsp. rouxiana *(Maire) I. Breitwieser & D. Podlech in Mitt. Bot. Staatssamml. München,* 22: 74 (1986): C. pomeliana var. rouxiana.
resupinata
subsp. lagascae *(Nyman) I. Breitwieser & D. Podlech in Mitt. Bot. Staatssamml. München,* 22: 90 (1986): C. lagascae.
subsp. rifana *(Emb. & Maire) I. Breitwieser & D. Podlech in Mitt. Bot. Staatssamml. München,* 22: 87 (1986): C. boissieri var. rifana.
subsp. simulans *(Emb. & Maire) I. Breitwieser & D. Podlech in Mitt. Bot. Staatssamml. München,* 22: 85 (1986): C. spachii var. simulans.
subsp. spachii *(Schultz Bip. ex Willk.) I. Breitwieser & D. Podlech in Mitt. Bot. Staatssamml. München,* 22: 82 (1986): C. spachii.
subsp. vulnerariifolia *(Pomel) I. Breitwieser & D. Podlech in Mitt. Bot. Staatssamml. München,* 22: 79 (1986): C. vulnerariifolia.

CENTAURIUM (Gentianac.).
quadrifolium
subsp. parviflorum *(Willk.) J. Pedrol in An. Jard. Bot. Madrid,* 43(1): 31 (1986): Erythraea gypsicola var. parviflora.

CENTRANTHUS (Valerianac.).
macrosiphon
var. andalusii *R. Fanlo in Lagascalia,* 14(1): 6 (1986), without type—Spain.
nevadensis
subsp. battandieri *(Maire) J. Fernández Casas & J. Molero in Treb. Inst. Bot. Barcelona,* 11: 30 (1987): C. battandieri.

CENTROLEPIS (Centrolepidac.).
eremica *D.A. Cooke in Fl. S. Australia,* 4: 1826 (1986)—Australia (S. Australia, W. Australia, N. Terr., New South Wales).

CEPHALOCROTON (Euphorbiac.).
incanus *M.G. Gilbert in Kew Bull.,* 42(2): 365 (1987)—Nigeria, Ethiopia.

CERASTIUM (Caryophyllac.).
durmitoreum *(J. Rohlena) A.W. Hill in Index Kewensis,* Supp. 6: 41 (1916–1920 publ. 1926): Cerastium lanigerum var. durmitoreum.
nigrescens
subsp. arcticum *(Lange) P.S. Lusby in Watsonia,* 16(3): 296 (1987): C. arcticum.
ramosissimum
f. carpetanum *(Lomax) E. Rico Hernández in An. Jard. Bot. Madrid,* 43(2): 470 (1986 publ. 1987): C. carpetanum.

CERASUS (Rosac.).
maackii *(Rupr.) G.V. Eremin & V.S. Simagin in Byull. Vses. Ord. Lenina Inst. Rast. N.I. Vavilova,* 166: 47 (1986): Prunus maackii.

CERCOCARPUS (Rosac.).
mexicanus *J. Henrickson in Syst. Bot.,* 12(2): 293 (1987)—Mexico.
rzedowskii *J. Henrickson in Syst. Bot.,* 12(2): 296 (1987)—Mexico.

CEREUS (Cactac.).
calcirupicola *(Ritt.) C.T. Rizzini in Rev. Brasil. Biol.,* 46(4): 782 (1986 publ. 1987): Piptanthocereus calcirupicola.
var. albicans *C.T. Rizzini in Rev. Brasil. Biol.,* 46(4): 784 (1986 publ. 1987)—Brazil.
var. pluricostatus *(Ritt.) C.T. Rizzini in Rev. Brasil. Biol.,* 46(4): 784 (1986 publ. 1987), without basionym page: Piptanthocereus calcirupicola var. pluricostatus.

CERINTHE (Boraginac.).
glabra
subsp. longiflora *(Viv.) M. Kerguélen in Lejeunia,* 120: 68 (1987): C. longiflora.

CEROPEGIA (Asclepiadac.).
schaijesiorum *F. Malaisse in Bull. Jard. Bot. Nation. Belg.,* 56(3–4): 491 (1986), as 'schaijesorum'—Zaïre.

CERVARIA (Umbellif.).
caucasica *(M.B.) M.G. Pimenov in Fl. Iran.,* 162: 453 (1987): Selinum caucasicum.
cervariifolia *(C.A. Mey.) M.G. Pimenov in Fl. Iran.,* 162: 452 (1987): Peucedanum cervariifolium.

CHAENORHINUM (Scrophulariac.).
origanifolium
subsp. cotiellae *P. Montserrat Recoder & G. Montserrat Martí in An. Jard. Bot. Madrid,* 43(1): 44 (1986)—Spain.
villosum
subsp. granatensis *(Willk.) B. Valdés in Lagascalia,* 14(1): 94 (1986): Linaria granatensis.

CHAMAEANTHUS (Orchidac.).
brigittae *N. Hallé in Bull. Mus. Nation. Hist. Nat., B, Adansonia,* Ser. 4, 8(3): 228 (1986)—New Caledonia.
florenciae *N. Hallé in Bull. Mus. Nation. Hist. Nat., B, Adansonia,* Ser. 4, 8(3): 227 (1986)—New Caledonia.

CHAMAECRISTA (Legumin.).
axilliflora *H.S. Irwin & R.C. Barneby in Brittonia,* 39(1): 7 (1987)—Brazil.
desvauxii
var. latistipula *(Benth.) G.P. Lewis, Legumes of Bahia:* 77 (1987): Cassia latistipula.

CHAMAECRISTA :-
fasciculata
var. depressa *(Pollard) C.F. Reed in Phytologia,* 63(5): 410 (1987): Cassia depressa.
var. macrosperma *(Fern.) C.F. Reed in Phytologia,* 63(5): 410 (1987): Cassia fasciculata var. macrosperma.
gumminans *H.S. Irwin & R.C. Barneby in Brittonia,* 39(1): 9 (1987)—Brazil.
nictitans
var. hebecarpa *(Fern.) C.F. Reed in Phytologia,* 63(5): 411 (1987): Cassia nictitans var. hebecarpa.
var. leiocarpa *(Fern.) C.F. Reed in Phytologia,* 63(5): 411 (1987): Cassia nictitans var. leiocarpa.
ramosa
var. curvifolia *(Vog.) G.P. Lewis, Legumes of Bahia:* 84 (1987): Cassia curvifolia.

CHAMAECYTISUS (Legumin.).
triflorus
subsp. pumilus *(De Not.) W. Greuter & Burdet in Willdenowia,* 16(2): 441 (1987): Cytisus pumilus.

CHAMAEDOREA (Palmae).
murriensis *G. Galeano-Garces in Principes,* 31(3): 143 (1987), nom. nov.: C. macroloba *Galeano-Garces.*

CHAMAELAUCIUM (Myrtac.).
marchantii *A. Strid in Pl. Syst. Evol.,* 155(1–4): 341 (1987)—Australia (W. Australia).

CHAMAELEON (Compos.).
macrophyllus *(Desf.) D. Petit in Bull. Soc. Bot. France, Lett. Bot.,* 134(2): 179 (1987): Atractylis macrophylla.

CHAMAESCILLA (Anthericac.).
corymbosa
var. latifolia *(F. Muell.) R.J.F. Henderson in Fl. Australia,* 45: 473 (1987): Caesia spiralis var. latifolia.
var. paradoxa *(Endl.) R.J.F. Henderson in Fl. Australia,* 45: 473 (1987): Caesia paradoxa.

CHEIROSTYLIS (Orchidac.).
seidenfadeniana *C. Sathish Kumar & F.N. Rasmussen in Nordic J. Bot.,* 7(4): 409 (1987)—India.

CHENOPODIUM (Chenopodiac.).
amurense *M.S. Ignatov in Byull. Mosk. Obshch. Ispȳt. Prir., Biol.,* 91(3): 111 (1986)—U.S.S.R. (Soviet Far East).
×dadakovae *F. Dvořák in Feddes Repert.,* 97(11–12): 793 (1986)—Czechoslovakia. (C. strictum × suecicum).
×fallax *(Aellen) F. Dvořák in Feddes Repert.,* 97(11–12): 793 (1986): C. album subsp. fallax. (C. album × strictum × suecicum).

CHENOPODIUM :-
f. mirabile *F. Dvořák in Feddes Repert.,* 97(11–12): 798 (1986), as 'mirabilis'—Czechoslovakia.
f. procerum *F. Dvořák in Feddes Repert.,* 97(11–12): 797 (1986), as 'procera'—Czechoslovakia.
×humiliforme *(J. Murr) F. Dvořák in Feddes Repert.,* 97(11–12): 783 (1986): C. ficifolium var. humiliforme. (C. album × ficifolium × strictum).
×jedlickae *F. Dvořák in Feddes Repert.,* 97(11–12): 780 (1986)—Czechoslovakia. (C. album × ficifolium).
×jehlikii *F. Dvořák in Feddes Repert.,* 97(11–12): 787 (1986)—Czechoslovakia. (C. ficifolium × opulifolium).
leptophyllum *(Moq.) B.D. Jackson in Index Kewensis,* 1: 513 (1753–1885 publ. 1893): C. album var. leptophyllum.
×podperae *F. Dvořák in Feddes Repert.,* 97(11–12): 790 (1986)—Czechoslovakia. (C. diversifolium × striatiforme).
pseudoborbasii
f. aellenii *F. Dvořák in Feddes Repert.,* 97(11–12): 815 (1986)—Czechoslovakia.
f. albiforme *F. Dvořák in Feddes Repert.,* 97(11–12): 815 (1986)—Czechoslovakia.
f. borbasiiforme *F. Dvořák in Feddes Repert.,* 97(11–12): 816 (1986), as 'borbasiiformis'—Czechoslovakia.
f. longipedicellatum *F. Dvořák in Feddes Repert.,* 97(11–12): 815 (1986), as 'longipedicellata'—Czechoslovakia.
f. ramosum *F. Dvořák in Feddes Repert.,* 97(11–12): 815 (1986), as 'ramosa'—Czechoslovakia.
×pseudostriatum
nothovar. erosum *(J. Murr) F. Dvořák in Biológia, A (Czechoslovakia),* 39(1): 68 (1984): C. striatum f. erosum.
×rhombicum *(J. Murr) F. Dvořák in Feddes Repert.,* 97(11–12): 821 (1986): C. vulvaria var. rhombicum.
×smardae *F. Dvořák in Feddes Repert.,* 97(11–12): 787 (1986)—Czechoslovakia. (C. album × ficifolium × suecicum).
×unarii *F. Dvořák in Feddes Repert.,* 97(11–12): 824 (1986), as 'unari'—Czechoslovakia. (C. album × opulifolium × pedunculare × suecicum).

CHILOGLOTTIS (Orchidac.).
sylvestris *D.L. Jones & M.A. Clements in Proc. Roy. Soc. Queensl.,* 98: 123 (1987)—Australia (Queensland).
truncata *D.L. Jones & M.A. Clements in Proc. Roy. Soc. Queensl.,* 98: 124 (1987)—Australia (Queensland).

CHILOSCHISTA (Orchidac.).
trudelii *G. Seidenfaden in Orchidee,* 38(6): 310 (1987)—Thailand.

CHIMAPHILA (Ericac.).
umbellata
subsp. domingensis *(S.F. Blake) L.J. Dorr in Sida,* 11(4): 370 (1986): C. domingensis.

CHIMONOBAMBUSA (Gramin.).
ningnanica *J.R. Xue & L.Z. Gao in J. Bamboo Res.,* 6(2): 13 (1987)—China.
sichuanensis *(Yi) T.H. Wen in J. Bamboo Res.,* 6(3): 33 (1987): Sinobambusa sichuanensis.
tuberculata *J.R. Xue & L.Z. Gao in J. Bamboo Res.,* 6(2): 11 (1987)—China.

CHIONANTHUS (Oleac.).
bumelioides
var. lanceolata *(Knobl.) Alain H. Liogier in Phytologia,* 63(1): 66 (1987): Linociera lanceolata.
hainanensis *(Merr. & Chun) B.M. Miao in Invest. Stud. Nat., Mus. Hist. Nat. Shanghaiense,* 7: 18 (1987): Linociera hainanensis.
leucocladus *(Merr. & Chun) B.M. Miao in Invest. Stud. Nat., Mus. Hist. Nat. Shanghaiense,* 7: 18 (1987): Linociera leucoclada.
longiflorus *(Li) B.M. Miao in Invest. Stud. Nat., Mus. Hist. Nat. Shanghaiense,* 7: 18 (1987): Linociera longiflora.
ramiflorus
var. grandiflorus *B.M. Miao in Invest. Stud. Nat., Mus. Hist. Nat. Shanghaiense,* 7: 17 (1987)—China.

CHIONOHEBE (Scrophulariac.).
glabra *(Cheesem.) M.J. Heads in Bot. Soc. Otago Newsl.,* 5: 4 (1987): Veronica thomsonii var. glabra.

CHIRITA (Gesneriac.).
yungfuensis *W.T. Wang in Guihaia,* 7(2): 125 (1987)—China.

CHLOROLEUCON (Legumin.).
dumosum *(Benth.) G.P. Lewis, Legumes of Bahia:* 165 (1987): Pithecellobium dumosum.
foliolosum *(Benth.) G.P. Lewis, Legumes of Bahia:* 166 (1987): Pithecellobium foliolosum.
glazioui *(Benth.) G.P. Lewis, Legumes of Bahia:* 166 (1987): Pithecellobium glazioui.

CHOMELIA (Rubiac.).
stergiosii *J.A. Steyermark in Ann. Missouri Bot. Gard.,* 74(1): 105 (1987)—Venezuela.

CHONDROPYXIS D.A. Cooke in Fl. S. Australia, 3: 1612 (1986). *COMPOSITAE.*
halophila *D.A. Cooke in Fl. S. Australia,* 3: 1613 (1986)—Australia (S. Australia, W. Australia).

CHROMOLAENA (Compos.).
impetiolaris *(Griseb.) D.H. Nicolson in Phytologia,* 62(3): 164 (1987): Eupatorium impetiolare.

CHROMOLAENA :–
macrodon *(DC.) D.H. Nicolson in Phytologia,* 62(3): 164 (1987): Eupatorium macrodon.

CHRYSOTHAMNUS (Compos.).
nauseosus
var. abbreviatus *(Jones) S.L. Welsh in Great Basin Nat.,* 43(2): 241 (1983): Bigelovia leiosperma var. abbreviata.
var. arenarius *(L.C. Anderson) S.L. Welsh in Great Basin Nat.,* 43(2): 241 (1983): C. nauseosus subsp. arenarius.
var. glareosus *(Jones) S.L. Welsh in Great Basin Nat.,* 43(2): 241 (1983): Bigelovia glareosa.
var. iridis *(L.C. Anderson) S.L. Welsh in Great Basin Nat.,* 43(2): 242 (1983): C. nauseosus subsp. iridis.
var. nitidus *(L.C. Anderson) S.L. Welsh in Great Basin Nat.,* 43(2): 242 (1983): C. nauseosus subsp. nitidus.

CHUSQUEA (Gramin.).
effusa *S.A. Renvoize in Kew Bull.,* 42(4): 924 (1987)—Brazil.
elegans *S.A. Renvoize in Kew Bull.,* 42(4): 924 (1987)—Brazil.

CINNA (Gramin.).
agrostoidea *Beauvois, Ess. Agrostogr.:* 147 (1812), nom. nov.: Agrostis arundinacea *L.*

CIRSIUM (Compos.).
arizonicum
var. nidulum *(Jones) S.L. Welsh in Great Basin Nat.,* 43(2): 248 (1983): Cnicus nidulus.
calcareum
var. bipinnatum *(Eastw.) S.L. Welsh in Great Basin Nat.,* 43(2): 249 (1983): Cnicus drummondii var. bipinnatum.
var. pulchellum *(Greene ex Rydb.) S.L. Welsh in Great Basin Nat.,* 43(2): 249 (1983): Carduus pulchellus.
chikushiense
var. yakusimense *(Masamune) T. Yahara in J. Fac. Sci. Univ. Tokyo, Bot.,* 14(2): 113 (1987): C. yakusimense.
davisianum *Kit Tan & F. Sorger in Pl. Syst. Evol.,* 154(1–2): 122 (1986)—Turkey.
eliasianum *Kit Tan & F. Sorger in Pl. Syst. Evol.,* 154(1–2): 123 (1986)—Turkey.
japonicum
var. okiense *H. Koyama & Murata in Acta Phytotax. Geobot.,* 37(4–6): 188 (1986)—Japan.
neomexicanum
var. utahense *(Petrak) S.L. Welsh in Great Basin Nat.,* 43(2): 251 (1983): C. utahense.
undulatum
var. tracyi *(Rydb.) S.L. Welsh in Great Basin Nat.,* 43(2): 254 (1983): Carduus tracyi.

CISTUS (Cistac.).
×cyprius
nothovar. ellipticus *J.-P. Demoly in Bull. Soc. Bot. Centre-Ouest, n.s.,* 16: 93 (1985), contrary to Art. H5.2 ICBN (1983)—Cultivation. (C. ladanifer f. sulcatus × laurifolius var. laurifolius).
ladanifer
f. bicolor *J.-P. Demoly in Bull. Soc. Bot. Centre-Ouest, n.s.,* 16: 93 (1985)—Portugal.
var. sulcatus *J.-P. Demoly in Bull. Soc. Bot. Centre-Ouest, n.s.,* 16: 93 (1985)—Portugal.
f. sulcatus *J.-P. Demoly in Bull. Soc. Bot. Centre-Ouest, n.s.,* 16: 93 (1985), without latin descr.—Locality not stated.
×purpureus
var. argenteus *J.-P. Demoly in Bull. Soc. Bot. Centre-Ouest, n.s.,* 16: 93 (1985)—Cultivation. (C. creticus f. creticus × ladanifer f. albiflorus).
nothof. argenteus *J.-P. Demoly in Bull. Soc. Bot. Centre-Ouest, n.s.,* 16: 93 (1985), without latin descr.—Locality not stated.
nothof. stictus *J.-P. Demoly in Bull. Soc. Bot. Centre-Ouest, n.s.,* 16: 93 (1985)—Cultivation. (C. creticus f. creticus × ladanifer f. ladanifer).
×ralletii *J.-P. Demoly in Bull. Soc. Centre-Ouest, n.s.,* 16: 93 (1985)—Cultivation. (C. creticus var. creticus × symphytifolius var. symphytifolius).

CLADOCARPA (Cucurbitac.).
capitata *H. St. John in Phytologia,* 63(3): 185 (1987)—Hawaiian Is.
discoidea *H. St. John in Phytologia,* 63(3): 185 (1987)—Hawaiian Is.
harrisoniae *H. St. John in Phytologia,* 63(3): 185 (1987), as 'harrisonae' Hawaiian Is.
herbstii *H. St. John in Phytologia,* 63(3): 185 (1987)—Hawaiian Is.
julesii *H. St. John in Phytologia,* 63(3): 185 (1987)—Hawaiian Is.
kawelaensis *H. St. John in Phytologia,* 63(3): 185 (1987), without type—Hawaiian Is.
lanaiensis *H. St. John in Phytologia,* 63(3): 186 (1987)—Hawaiian Is.
lehuaensis *H. St. John in Phytologia,* 63(3): 186 (1987)—Hawaiian Is.
munroi *H. St. John in Phytologia,* 63(3): 186 (1987)—Hawaiian Is.
pauciramosa *H. St. John in Phytologia,* 63(3): 186 (1987)—Hawaiian Is.
umbellata *H. St. John in Phytologia,* 63(3): 186 (1987)—Hawaiian Is.
waimeaensis *H. St. John in Phytologia,* 63(3): 186 (1987)—Hawaiian Is.

CLARKIA (Onagrac.).
concinna
subsp. automixa *R.N. Bowman in Madroño,* 34(1): 41 (1987)—U.S.A. (California).

CLAVIJA (Theophrastac.).
minor *B. Ståhl in Nordic J. Bot.*, 6(6): 771 (1986)—Colombia.
repanda *B. Ståhl in Nordic J. Bot.*, 6(6): 769 (1986)—Ecuador.

CLAYTONIA (Portulacac.).
palustris *J.R. Swanson & W.A. Kelley in Madroño*, 34(2): 155 (1987)—U.S.A. (California).

×**CLEISOFINETIA** hort. in Orchid Rev., 95(1127): centre page pull-out p. 9 (1987). *ORCHIDACEAE*. (Cleisocentron × Neofinetia).

CLEISTOCACTUS (Cactac.).
acanthurus *(Vaupel) D. Hunt in Bradleya*, 5: 92 (1987): Cereus acanthurus.
aureispinus *(Ritter) D. Hunt in Bradleya*, 5: 92 (1987): Winteria aureispina.
fieldianus *(Britton & Rose) D. Hunt in Bradleya*, 5: 92 (1987): Borzicactus fieldianus.
fossulatus *R. Mottram in Chileans*, 13(43): 30 (1985)—Bolivia.
samaipatanus *(Cardenas) D. Hunt in Bradleya*, 5: 92 (1987): Bolivicereus samaipatanus.

CLEISTOCHLOA (Gramin.).
rigida *(S.T. Blake) R.D. Webster, Australian Paniceae (Poac.)*: 30 (1987): Dimorphochloa rigida.
rigida *(S.T. Blake) W.D. Clayton in Kew Bull.*, 42(2): 401 (5 June 1987): Dimorphochloa rigida.

CLEMATIS (Ranunculac.).
subgen. Campanella *M. Tamura in Acta Phytotax. Geobot.*, 38: 38 (1987).
sect. Campanella *M. Tamura in Acta Phytotax. Geobot.*, 38: 38 (1987).
subgen. Viorna *(Reichb.) M. Tamura in Acta Phytotax. Geobot.*, 38: 40 (1987): gen. Viorna.
brachyura
var. hexasepala *Y.N. Lee in Korean J. Bot.*, 25(4): 177 (1982)—Korea.
henryi
var. mollis *W.T. Wang in Bull. Bot. Res. North-East. Forest. Inst.*, 7(2): 99 (1987)—China.
orientalis
var. uniflora *M. Tamura in Acta Phytotax. Geobot.*, 37(4–6): 156 (1986)—Nepal.
taeguensis *Y.N. Lee in Korean J. Bot.*, 25(4): 175 (1982)—Korea.

CLERETUM (Aizoac.).
herrei *(Schwantes) H.-D. Ihlenfeldt & M. Struck in Beitr. Biol. Pfl.*, 61(3): 437 (1986 publ. 1987), without exact basionym page: Micropterum herrei.
lyratifolium *H.-D. Ihlenfeldt & M. Struck in Beitr. Biol. Pfl.*, 61(3): 437 (1986 publ. 1987)—South Africa (Cape Province).

CLERMONTIA (Campanulac.).
albimontis *H. St. John in Phytologia*, 63(5): 350 (1987)—Hawaiian Is.
bicolorata *H. St. John in Phytologia*, 63(5): 350 (1987)—Hawaiian Is.
epilosa *H. St. John in Phytologia*, 63(5): 350 (1987)—Hawaiian Is.
glabra *H. St. John in Phytologia*, 63(5): 350 (1987)—Hawaiian Is.
gracilis *H. St. John in Phytologia*, 63(5): 350 (1987)—Hawaiian Is.
hualalaiensis *H. St. John in Phytologia*, 63(5): 351 (1987)—Hawaiian Is.
kahuaensis *H. St. John in Phytologia*, 63(5): 351 (1987)—Hawaiian Is.
kakeana
var. rosea *(Hillebr.) H. St. John in Phytologia*, 63(5): 351 (1987): C. macrocarpa var. rosea.
kipahuluensis *H. St. John in Phytologia*, 63(5): 351 (1987)—Hawaiian Is.
mauiensis *H. St. John in Phytologia*, 63(5): 351 (1987)—Hawaiian Is.
pluriflora *H. St. John in Phytologia*, 63(5): 351 (1987)—Hawaiian Is.
spatulata *H. St. John in Phytologia*, 63(5): 352 (1987)—Hawaiian Is.
subteralbulus *H. St. John in Phytologia*, 63(5): 352 (1987)—Hawaiian Is.
viridis *H. St. John in Phytologia*, 63(5): 352 (1987)—Hawaiian Is.
waikoluensis *H. St. John in Phytologia*, 63(5): 352 (1987)—Hawaiian Is.

CLERODENDRUM (Verbenac.).
japonicum
f. album *(P'ei) H.N. Moldenke in Phytologia*, 61(5): 332 (1986): C. japonicum var. album.
kaempferi
f. album *(P'ei) H.N. Moldenke in Phytologia*, 58(3): 196 (1985), without basionym ref.; *et in Phytologia*, 61(6): 398 (1986), with basionym ref.: C. japonicum var. album.
laevifolium
var. pubiflorum *Bakh. ex H.N. Moldenke in Phytologia*, 61(7): 461 (1987)—Borneo (Sarawak), Java, Sumatra.

CLIDEMIA (Melastomatac.).
ayangannensis *J.J. Wurdack in Brittonia*, 39(2): 161 (1987)—Guyana.
granvillei *J.J. Wurdack in Brittonia*, 39(2): 163 (1987)—French Guiana.
venosa *(Gleason) J.J. Wurdack in Brittonia*, 39(2): 164 (1987): Henriettella venosa.

CLINOPODIUM (Labiat.).
multicaule
var. yakusimense *(Masamune) T. Yahara in J. Fac. Sci. Univ. Tokyo, Bot.*, 14(2): 107 (1987): Satureja yakusimensis.

CLINOPODIUM :-
vulgare
var. neogaeum *(Fern.) C.F. Reed in Phytologia,* 63(5): 412 (1987): Satureja vulgaris var. neogaea.

CLUSIA (Guttif.).
cylindrica *B.E. Hammel in Selbyana,* 9(1): 112 (1986)—Costa Rica, Nicaragua, Panama.
duartei *B. Maguire in Moscosoa,* 4: 215 (1986)—Venezuela.
troncosii *B. Maguire in Moscosoa,* 4: 215 (1986)—Venezuela.

COCCOCYPSELUM (Rubiac.).
croatii *J.A. Steyermark in Ann. Missouri Bot. Gard.,* 74(1): 106 (1987)—Venezuela.

COCCOLOBA (Polygonac.).
liportizii *J. Gómez-Laurito & N. Zamora in Phytologia,* 62(1): 7 (1987)—Costa Rica.

COCCULUS (Menispermac.).
prainianus *(Diels) A. Pramanik & K. Thothathri in Kew Bull.,* 42(3): 706 (1987): Pericampylus prainianus.

CODONOPSIS (Campanulac.).
convolvulacea
subsp. vinciflora *(Kom.) D.Y. Hong in Fl. Xizangica,* 4: 582 (1985), comb. already publ. see I.K. Supp. 18: C. vinciflora.

COELACHYRUM (Gramin.).
annuum *T.A. Cope & L. Boulos in Kew Bull.,* 42(4): 919 (1987)—Egypt.

COELIDIUM (Legumin.).
flavum *R. Granby in Nordic J. Bot.,* 7(1): 51 (1987)—South Africa (Cape Province).

COELOSPERMUM (Rubiac.).
crassifolium *J.T. Johansson in Blumea,* 32(2): 321 (1987)—New Caledonia.

COILOSTIGMA (Ericac.).
subgen. Anomalosepala *E.G.H. Oliver in Bothalia,* 17(2): 168 (1987).
zeyherianum
var. tenuifolium *(Klotzsch) E.G.H. Oliver in Bothalia,* 17(2): 167 (1987), without basionym ref.: Basionym not stated.

COINCYA (Crucif.).
transtagana *(Coutinho) M. Clemente & J.E. Hernández-Bermejo in Lagascalia,* 14(1): 138 (1986): Sinapis longirostra subsp. transtagana.

COLA (Sterculiac.).
cordifolia *R. Brown, Pterocymbium:* 237 (4 June 1844)—W. Tropical Africa. †

COMACLINIUM (Compos.).
montanum *(Benth.) J.L. Strother in Sida,* 11(4): 377 (1986): Clomenocoma montana.

COMAROSTAPHYLIS (Ericac.).
arbutoides
subsp. costaricensis *(Small) G.M. Diggs in Brittonia,* 38(4): 344 (1986): C. costaricensis.
discolor
subsp. rupestris *(Rob. & Seaton) G.M. Diggs in Brittonia,* 38(4): 345 (1986): Arctostaphylos rupestris.
polifolia
subsp. minor *(Small) G.M. Diggs in Brittonia,* 38(4): 346 (1986): C. minor.
spinulosa *(Mart. & Gal.) G.M. Diggs in Brittonia,* 38(4): 347 (1986): Arctostaphylos spinulosa.
subsp. glandulifera *G.M. Diggs in Brittonia,* 38(4): 348 (1986)—Mexico.

COMMELINA (Commelinac.).
mathewsii *(C.B. Clarke) R.B. Faden & D.R. Hunt in Ann. Missouri Bot. Gard.,* 74(1): 122 (1987): Phaeosphaerion mathewsii.
rufipes
var. glabrata *(D.R. Hunt) R.B. Faden & D.R. Hunt in Ann. Missouri Bot. Gard.,* 74(1): 122 (1987): Commelinopsis glabrata.

COMPARETTIA (Orchidac.).
×maloi *I. Bock in Orchidee,* 37(6): 262 (1986)—Ecuador. (C. falcata × speciosa).

CONOSTYLIS (Haemodorac.).
subgen. Androstemma *(Lindley) S.D. Hopper in Fl. Australia,* 45: 456 (1987): gen. Androstemma.
subgen. Bicolorata *S.D. Hopper in Fl. Australia,* 45: 456, 67 (1987).
subgen. Brachycaulon *(Benth.) S.D. Hopper in Fl. Australia,* 45: 457 (1987): C. sect. Brachycaulon.
sect. Divaricata *S.D. Hopper in Fl. Australia,* 45: 457, 102 (1987).
subgen. Greenia *(Geer.) S.D. Hopper in Fl. Australia,* 45: 457 (1987): C. sect. Greenia.
subgen. Pendula *S.D. Hopper in Fl. Australia,* 45: 457, 86 (1987).
aculeata
subsp. cygnorum *S.D. Hopper in Fl. Australia,* 45: 457, 71 (1987)—Australia (W. Australia).
subsp. echinissima *S.D. Hopper in Fl. Australia,* 45: 458, 74 (1987)—Australia (W. Australia).
subsp. gracilis *S.D. Hopper in Fl. Australia,* 45: 458, 71 (1987)—Australia (W. Australia).
subsp. septentrionora *S.D. Hopper in Fl. Australia,* 45: 458, 74 (1987)—Australia (W. Australia).
subsp. spinuligera *(F. Muell. ex Benth.) S.D. Hopper in Fl. Australia,* 45: 458 (1987): C. spinuligera.

CONOSTYLIS :–
albescens *S.D. Hopper in Fl. Australia,* 45: 458, 66 (1987)—Australia (W. Australia).
angustifolia *S.D. Hopper in Fl. Australia,* 45: 458, 108 (1987)—Australia (W. Australia).
argentea *(J. Green) S.D. Hopper in Fl. Australia,* 45: 459 (1987): C. androstemma subsp. argentea.
candicans
subsp. calcicola *S.D. Hopper in Fl. Australia,* 45: 459, 76 (1987)—Australia (W. Australia).
subsp. flavifolia *S.D. Hopper in Fl. Australia,* 45: 459, 77 (1987)—Australia (W. Australia).
subsp. procumbens *S.D. Hopper in Fl. Australia,* 45: 459, 77 (1987)—Australia (W. Australia).
canteriata *S.D. Hopper in Fl. Australia,* 45: 459, 95 (1987)—Australia (W. Australia).
caricina
subsp. elachys *S.D. Hopper in Fl. Australia,* 45: 460, 98 (1987)—Australia (W. Australia).
crassinerva
subsp. absens *S.D. Hopper in Fl. Australia,* 45: 460, 94 (1987)—Australia (W. Australia).
dielsii
subsp. teres *S.D. Hopper in Fl. Australia,* 45: 460, 100 (1987)—Australia (W. Australia).
festucacea
subsp. filifolia *(F. Muell.) S.D. Hopper in Fl. Australia,* 45: 460 (1987): C. filifolia.
hiemalis *S.D. Hopper in Fl. Australia,* 45: 460, 106 (1987)—Australia (W. Australia).
latens *S.D. Hopper in Fl. Australia,* 45: 461, 95 (1987)—Australia (W. Australia).
lepidospermoides *S.D. Hopper in Fl. Australia,* 45: 461, 80 (1987)—Australia (W. Australia).
micrantha *S.D. Hopper in Fl. Australia,* 45: 461, 92 (1987)—Australia (W. Australia).
pauciflora
subsp. euryrhipis *S.D. Hopper in Fl. Australia,* 45: 461, 80 (1987)—Australia (W. Australia).
resinosa *S.D. Hopper in Fl. Australia,* 45: 461, 106 (1987)—Australia (W. Australia).
rogeri *S.D. Hopper in Fl. Australia,* 45: 462, 101 (1987)—Australia (W. Australia).
seminuda *S.D. Hopper in Fl. Australia,* 45: 462, 106 (1987)—Australia (W. Australia).
seorsiflora
subsp. longissima *S.D. Hopper in Fl. Australia,* 45: 462, 86 (1987)—Australia (W. Australia).
subsp. trichophylla *S.D. Hopper in Fl. Australia,* 45: 462, 86 (1987)—Australia (W. Australia).
setigera
subsp. dasys *S.D. Hopper in Fl. Australia,* 45: 463, 89 (1987)—Australia (W. Australia).

CONOSTYLIS :–
teretifolia
subsp. planescens *S.D. Hopper in Fl. Australia,* 45: 463, 92 (1987)—Australia (W. Australia).
tomentosa *S.D. Hopper in Fl. Australia,* 45: 463, 108 (1987)—Australia (W. Australia).

CONSOLIDA (Ranunculac.).
kandaharica *M. Iranshahr in Pl. Syst. Evol.,* 155(1–4): 56 (1987)—Afghanistan.
lorestanica *M. Iranshahr in Pl. Syst. Evol.,* 155(1–4): 55 (1987)—Iran.

CONVOLVULUS (Convolvulac.).
eyreanus *R.W. Johnson in Austrobaileya,* 2(4): 408 (1987)—Australia (S. Australia).
microsepalus *R.W. Johnson in Austrobaileya,* 2(4): 410 (1987)—Australia (S. Australia).

COPIAPOA (Cactac.).
cinerea
var. gigantea *(Backeberg) N.P. Taylor in Bradleya,* 5: 92 (1987): C. gigantea.

COPROSMA (Rubiac.).
niphophila *A.E. Orchard in Brunonia,* 9(2): 134 (1987)—Australia (New South Wales), New Zealand.
perpusilla
subsp. subantarctica *A.E. Orchard in Brunonia,* 9(2): 133 (1987)—Macquarie Is., New Zealand.

CORDIA (Boraginac.).
subgen. Myxa *N. Taroda in Rev. Brasil. Bot.,* 9(1): 39 (1986 publ. 1987).
sect. Superbiflorae *N. Taroda in Rev. Brasil. Bot.,* 9(1): 40 (1986 publ. 1987).
dentata *J.L.M. Poiret in Lamarck, Encycl. Meth., Bot.,* 7: 48 (1806)—Netherlands Antilles.
macvaughii *J.S. Miller in Syst. Bot.,* 11(4): 579 (1986)—Mexico.

CORISPERMUM (Chenopodiac.).
ladakhianum *C. Grey-Wilson & B.M. Wadhwa in Kew Bull.,* 42(2): 471 (1987)—India.
maynense *M.S. Ignatov in Byull. Mosk. Obshch. Ispȳt. Prir., Biol.,* 91(3): 112 (1986)—U.S.S.R. (Soviet Far East).
ochotense *M.S. Ignatov in Byull. Mosk. Obshch. Ispȳt. Prir., Biol.,* 91(3): 113 (1986)—U.S.S.R. (Soviet Far East).

CORNUS (Cornac.).
sect. Japonicae *(Nakai) Q.Y. Xiang in Bull. Bot. Res. North-East. Forest. Inst.,* 7(2): 44 (1987): Cynoxylon subsect. Japonicae.
subgen. Sinocornus *Q.Y. Xiang in Acta Phytotax. Sin.,* 25(2): 128 (1987).
subgen. Syncarpea *(Nakai) Q.Y. Xiang in Acta Phytotax. Sin.,* 25(2): 128 (1987): Benthamia subgen. Syncarpea.
canadensis
f. semivirescens *J. Cayouette in Nat. Canad.,* 113(4): 333 (1986)—Canada (Quebec).

CORNUS :-
capitata
subsp. angustata *(Chun) Q.Y. Xiang in Bull. Bot. Res. North-East. Forest. Inst.,* 7(2): 40 (1987): C. kousa var. angustata.
subsp. brevipedunculata *(Fang & Hsieh) Q.Y. Xiang in Bull. Bot. Res. North-East. Forest. Inst.,* 7(2): 39 (1987): Dendrobenthamia brevipedunculata.
subsp. emeiensis *(Fang & Hsieh) Q.Y. Xiang in Bull. Bot. Res. North-East. Forest. Inst.,* 7(2): 39 (1987): Dendrobenthamia emeiensis.
hongkongensis
subsp. elegans *(Fang & Hsieh) Q.Y. Xiang in Bull. Bot. Res. North-East. Forest. Inst.,* 7(2): 43 (1987): Dendrobenthamia elegans.
subsp. ferruginea *(Wu) Q.Y. Xiang in Bull. Bot. Res. North-East. Forest. Inst.,* 7(2): 44 (1987): C. ferruginea.
subsp. gigantea *(Hand.-Mazz.) Q.Y. Xiang in Bull. Bot. Res. North-East. Forest. Inst.,* 7(2): 43 (1987): C. hongkongensis var. gigantea.
var. jinyunensis *(Fang & W.K. Hu) Q.Y. Xiang in Bull. Bot. Res. North-East. Forest. Inst.,* 7(2): 43 (1987): Dendrobenthamia jinyunensis.
subsp. melanotricha *(Pojark.) Q.Y. Xiang in Bull. Bot. Res. North-East. Forest. Inst.,* 7(2): 42 (1987): Cynoxylon melanotricha.
subsp. tonkinensis *(Fang) Q.Y. Xiang in Bull. Bot. Res. North-East. Forest. Inst.,* 7(2): 42 (1987): Dendrobenthamia tonkinensis.
kousa
subsp. chinensis *(Osborn) Q.Y. Xiang in Bull. Bot. Res. North-East. Forest. Inst.,* 7(2): 46 (1987): C. kousa var. chinensis.
var. leucotricha *(Fang & Hsieh) Q.Y. Xiang in Bull. Bot. Res. North-East. Forest. Inst.,* 7(2): 47 (1987): Dendrobenthamia japonica var. leucotricha.
multinervosa *(Pojark.) Q.Y. Xiang in Bull. Bot. Res. North-East. Forest. Inst.,* 7(2): 47 (1987): Cynoxylon multinervosa.
ulotricha
var. leptophylla *W.K. Hu ex P.C. Li in Fl. Xizangica,* 3: 528 (1986)—Xizang.

CORONILLA (Legumin.).
aurantiaca *(J. Rohlena) A.W. Hill in Index Kewensis,* Supp. 6: 53 (1916–1920 publ. 1926): C. vaginalis var. aurantiaca.
valentina
subsp. speciosa *(Uhrová) Quézel & Santa ex W. Greuter & Burdet in Willdenowia,* 16(2): 442 (1987): C. speciosa.

CORRIGIOLA (Molluginac.).
litoralis
subsp. foliosa *(Pérez Lara) J.A. Devesa in Lagascalia,* 14(1): 131 (1986): C. telephiifolia subvar. foliosa.

CORYANTHES (Orchidac.).
pegiae *G.A. Romero in Selbyana,* 9(1): 147 (1986)—Venezuela.
seegeri *G. Gerlach in Orchidee,* 38(2): 87 (1987)—Peru.

CORYDALIS (Papaverac.).
sect. Hamatae *C.Y. Wu & Z.Y. Su in Acta Bot. Yunnanica,* 8(4): 407 (1986).
binderae
subsp. pseudohamata *(Fedde) Z.Y. Su in Acta Bot. Yunnanica,* 8(4): 410 (1986): C. pseudohamata.
capnoides
subsp. tenella *(Kar. & Kir.) R.V. Kamelin in Byull. Mosk. Obshch. Ispȳt. Prir., Biol.,* 91(6): 91 (1986): C. tenella.
foetida *C.Y. Wu & Z.Y. Su in Acta Bot. Yunnanica,* 9(1): 39 (1987)—China.
var. brevipedunculata *Z.Y. Su in Acta Bot. Yunnanica,* 9(1): 41 (1987)—China.
hamata
var. ramosa *Z.Y. Su in Acta Bot. Yunnanica,* 8(4): 409 (1986)—China.
muliensis *C.Y. Wu & Z.Y. Su in Acta Bot. Yunnanica,* 8(4): 411 (1986)—China.
racemosa
var. ecalcarata *Z.Y. Su in Acta Bot. Yunnanica,* 9(1): 37 (1987)—China.

CORYLOPSIS (Hamamelidac.).
glabrescens
var. gotoana *(Makino) T. Yamanaka in Acta Phytotax. Geobot.,* 37(4–6): 104 (1986): C. gotoana.
f. gotoana *(Makino) T. Yamanaka in Acta Phytotax. Geobot.,* 37(4–6): 104 (1986): C. gotoana.
f. pubescens *(Nakai) T. Yamanaka in Acta Phytotax. Geobot.,* 37(4–6): 104 (1986): C. coreana var. pubescens.

CORYNOTHECA (Anthericac.).
asperata *R.J.F. Henderson in Fl. Australia,* 45: 473, 304 (1987)—Australia (N. Terr., W. Australia).
flexuosissima *R.J.F. Henderson in Fl. Australia,* 45: 474, 304 (1987)—Australia (W. Australia).
licrota *R.J.F. Henderson in Fl. Australia,* 45: 474 (1987), nom. nov.: C. lateriflora var. laevisperma *Gauba.*
micrantha
var. acanthoclada *(F. Muell.) R.J.F. Henderson in Fl. Australia,* 45: 474 (1987): Caesia acanthoclada.
var. divaricata *R.J.F. Henderson in Fl. Australia,* 45: 474, 303 (1987)—Australia (S. Australia, W. Australia, N. Terr., Queensland).
var. elongata *R.J.F. Henderson in Fl. Australia,* 45: 475, 303 (1987)—Australia (W. Australia).

CORYNOTHECA :–
var. gracilis *R.J.F. Henderson in Fl. Australia,* 45: 475, 302 (1987)—Australia (W. Australia).
var. panda *R.J.F. Henderson in Fl. Australia,* 45: 475, 303 (1987)—Australia (W. Australia).
pungens *R.J.F. Henderson in Fl. Australia,* 45: 475, 304 (1987)—Australia (W. Australia).

CORYTHOPHORA (Lecythidac.).
rimosa
subsp. rubra *S.A. Mori in Mem. New York Bot. Gard.,* 44: 31 (1987)—French Guiana, Brazil.

COURSETIA (Legumin.).
sect. Cracca *(Benth.) M. Lavin in C.H. Stirton (ed.), Advances in Legume System.,* 3: 59 (1987): gen. Cracca.
sect. Neocracca *(O. Kuntze) M. Lavin in C.H. Stirton (ed.), Advances in Legume System.,* 3: 59 (1987): gen. Neocracca.
sect. Poissonia *(Baillon) M. Lavin in C.H. Stirton (ed.), Advances in Legume System.,* 3: 58 (1987): gen. Poissonia.
apantensis *M. Sousa S. in Ann. Missouri Bot. Gard.,* 73(4): 732 (1986 publ. 1987)—Nicaragua.
caribaea *(Jacq.) M. Lavin in C.H. Stirton (ed.), Advances in Legume System.,* 3: 63 (1987): Galega caribaea.
elliptica *M. Sousa S. & V.E. Rudd in Ann. Missouri Bot. Gard.,* 73(4): 732 (1986 publ. 1987)—Costa Rica, Nicaragua.
ferruginea *(Kunth) M. Lavin in C.H. Stirton (ed.), Advances in Legume System.,* 3: 63 (1987): Robinia ferruginea.
glabella *(A. Gray) M. Lavin in C.H. Stirton (ed.), Advances in Legume System.,* 3: 63 (1987): Cracca edwardsii var. glabella.
heterantha *(Griseb.) M. Lavin in C.H. Stirton (ed.), Advances in Legume System.,* 3: 64 (1987): Tephrosia heterantha.
hypoleuca *(Speg.) M. Lavin in C.H. Stirton (ed.), Advances in Legume System.,* 3: 63 (1987): Chiovendaea hypoleuca.
oaxacensis *M. Sousa S. & V. Rudd in Syst. Bot.,* 12(1): 108 (1987)—Mexico.
paniculata *M. Sousa S. & M. Lavin in Syst. Bot.,* 12(1): 113 (1987)—Mexico.
paucifoliolata *M. Sousa S. in Ann. Missouri Bot. Gard.,* 73(4): 734 (1986 publ. 1987)—Nicaragua.
polyphylla
var. acutifolia *M. Sousa S. & M. Lavin in Syst. Bot.,* 12(1): 111 (1987)—Nicaragua.
var. breviloba *M. Sousa S. & M. Lavin in Syst. Bot.,* 12(1): 112 (1987)—Honduras, Guatemala, Nicaragua.
pumila *(Rose) M. Lavin in C.H. Stirton (ed.), Advances in Legume System.,* 3: 63 (1987): Benthamantha pumila.
robinioides *M. Sousa S. & M. Lavin in Syst. Bot.,* 12(1): 110 (1987)—Mexico.

COURTENIA R Brown, Pterocymbium: 236 (4 June 1844). *STERCULIACEAE.* †
afzelii *R. Brown, Pterocymbium:* 236 (4 June 1844)—Tropical Africa. †
heterophylla *R. Brown, Pterocymbium:* 236 (4 June 1844)—Nigeria. †
triloba *R. Brown, Pterocymbium:* 236 (4 June 1844)—W. Tropical Africa. †

COUTAPORTLA (Rubiac.).
pailensis *J.A. Villarreal Q. in Sida,* 12(1): 223 (1987)—Mexico.

CRANICHIS (Orchidac.).
amplectens *D.D. Dod in Moscosoa,* 4: 179 (1986)—Hispaniola (Dominican Republic).
galatea *D.D. Dod in Moscosoa,* 4: 181 (1986)—Hispaniola (Haiti).

CRATAEGUS (Rosac.).
curvisepala
nothovar. angustiloba *T. Petauer in Biol. Vestn.,* 28(2): 79 (1980 publ. 1981)—Jugoslavia.
nothovar. latiloba *T. Petauer in Biol. Vestn.,* 28(2): 79 (1980 publ. 1981)—Jugoslavia.
nothovar. naviculiformis *T. Petauer in Biol. Vestn.,* 28(2): 79 (1980 publ. 1981)—Jugoslavia.
var. quadriloba *T. Petauer in Biol. Vestn.,* 28(2): 77 (1980 publ. 1981)—Jugoslavia.
subsp. zlatnensis *T. Petauer in Biol. Vestn.,* 28(2): 77 (1980 publ. 1981)—Jugoslavia.
monogyna
subsp. acutiloba *(Kerner) T. Baranec in Biológia, A (Czechoslovakia),* 38(9): 857 (1983), without basionym page: C. monogyna var. acutiloba.
var. plesivecensis *(Hrab.-Uhr.) T. Baranec in Biológia, A (Czechoslovakia),* 38(9): 861 (1983): C. monogyna subsp. plesivecensis.
rosiformis
nothosubsp. zlatnensis *(Petauer) M. Kerguélen & J. Lambinon in Lejeunia,* 120: 75 (1987): C. curvisepala subsp. zlatnensis.

CRATEROSTIGMA (Scrophulariac.).
alatum *F.N. Hepper in Kew Bull.,* 42(4): 945 (1987)—Kenya.
longicarpum *F.N. Hepper in Kew Bull.,* 42(4): 945 (1987)—Kenya, Ethiopia, Tanzania.

CREMANTHODIUM (Compos.).
botrycephalum *S.W. Liu in Fl. Xizangica,* 4: 850 (1985)—Xizang.
brunneopilosum *S.W. Liu in Fl. Xizangica,* 4: 851 (1985), as 'brunneo-pilosum'—Xizang.
conaense *S.W. Liu in Fl. Xizangica,* 4: 845 (1985)—Xizang.
ellisii *(Hook. f.) S.W. Liu in Fl. Xizangica,* 4: 854 (1985), comb. already publ. see I.K. Supp. 18: Werneria ellisii.
glandulipilosum *Y.L. Chen in Fl. Xizangica,* 4: 852 (1985)—Xizang.

CREMANTHODIUM :-
hirtiflorum *S.W. Liu in Fl. Xizangica,* 4: 850 (1985)—Xizang.
laciniatum *Ling & Y.L. Chen in Fl. Xizangica,* 4: 855 (1985)—Xizang.
lingulatum *S.W. Liu in Fl. Xizangica,* 4: 849 (1985)—Xizang.
microphyllum *S.W. Liu in Fl. Xizangica,* 4: 856 (1985)—Xizang.
nervosum *S.W. Liu in Fl. Xizangica,* 4: 847 (1985)—Xizang.
obovatum *Ling & S.W. Liu in Fl. Xizangica,* 4: 852 (1985)—Xizang.
phyllodineum *S.W. Liu in Fl. Xizangica,* 4: 844 (1985)—Xizang.
pinnatisectum *(Ludlow) Y.L. Chen & S.W. Liu in Fl. Xizangica,* 4: 841 (1985): C. campanulatum var. pinnatisectum.
plantagineum
var. ramosum *(Ling) Ling & S.W. Liu in Fl. Xizangica,* 4: 853 (1985): C. discoideum subsp. ramosum.
puberulum *S.W. Liu in Fl. Xizangica,* 4: 848 (1985)—Xizang.
trilobum *S.W. Liu in Fl. Xizangica,* 4: 847 (1985)—Xizang.
yadongense *S.W. Liu in Fl. Xizangica,* 4: 847 (1985)—Xizang.

CREPIDOSPERMUM (Burserac.).
sect. Hemicrepidospermum *(Swart) D.C. Daly in Brittonia,* 39(1): 55 (1987): gen. Hemicrepidospermum.
cuneifolium *(Cuatrec.) D.C. Daly in Brittonia,* 39(1): 55 (1987): Hemicrepidospermum cuneifolium.
prancei *D.C. Daly in Brittonia,* 39(1): 56 (1987)—Brazil, Peru.

CREPIS (Compos.).
dachhigamensis *G. Singh in G. Singh & P. Kachroo, Forest Fl. Srinagar:* 127 (1987)—India.
runcinata
var. glauca *(Nutt.) S.L. Welsh in Great Basin Nat.,* 43(2): 258 (1983): Crepidium glaucum.

CRICIUMA T.R. Soderstrom & X. Londoño in Amer. J. Bot., 74(1): 35 (1987). *GRAMINEAE.*
asymmetrica *T.R. Soderstrom & X. Londoño in Amer. J. Bot.,* 74(1): 35 (1987)—Brazil.

CRINUM (Amaryllidac.).
calamistratum *J. Bogner & H. Heine in Aqua Planta,* 1987(4): 127 (1987)—Cameroun.
woolliamsii *L.S. Hannibal in Herbertia,* 43(1): 14 (1987)—New Guinea (Papua New Guinea).

CROIZATIA (Euphorbiac.).
panamensis *G.L. Webster in Syst. Bot.,* 12(1): 7 (1987)—Panama, Colombia.

CROTON (Euphorbiac.).
chamelensis *E.J. Lott in Brittonia,* 39(2): 302 (1987)—Mexico.
ituzaingensis *L.Z. Ahumada in Veröff. Geobot. Inst. Rübel,* 91: 256 (1986)—Argentina.
landoltii *L.Z. Ahumada in Veröff. Geobot. Inst. Rübel,* 91: 252 (1986)—Argentina.
robustus
var. serratus *T. Chakrabarty, N.P. Balakrishnan & D. Gupta in J. Econ. Taxon. Bot.,* 9(2): 251 (1987)—Burma.

CTENANTHE (Marantac.).
amabilis *(Morren) H. Kennedy & D.H. Nicolson in Brittonia,* 39(1): 139 (1987): Stromanthe amabilis.

CUATRESIA (Solanac.).
colombiana *A.T. Hunziker in Op. Bot.,* 92: 79 (1987)—Colombia.
exiguiflora *(D'Arcy) A.T. Hunziker in Op. Bot.,* 92: 75 (1987): Witheringia exiguiflora.
fosteriana *A.T. Hunziker in Op. Bot.,* 92: 80 (1987)—Peru.
garciae *A.T. Hunziker in Op. Bot.,* 92: 76 (1987)—Colombia.
harlingiana *A.T. Hunziker in Op. Bot.,* 92: 75 (1987)—Ecuador.
riparia
var. cuspidata *(Dunal) A.T. Hunziker in Op. Bot.,* 92: 78 (1987): Sicklera cuspidata.

CURCUMA (Zingiberac.).
exigua *N. Liu in Guihaia,* 7(1): 15 (1987)—Cultivated in China.
oligantha
var. lutea *(R. Ansari, V.J. Nair & N.C. Nair) K.G. Bhat in Indian J. Forest.,* 10(1): 68 (1987): C. cannanorensis var. lutea.
yunnanensis *N. Liu & S.J. Chen in Guihaia,* 7(1): 16 (1987)—Cultivated in China.

CURTIA (Gentianac.).
confusa *E.H.M. Grothe & P.J.M. Maas in Proc. Kon. Nederl. Akad. Wetensch., C,* 87(1): 42 (1984)—Brazil.
tenuifolia
subsp. tenella *(Martius) E.H.M. Grothe & P.J.M. Maas in Proc. Kon. Nederl. Akad. Wetensch., C,* 87(1): 40 (1984): Schuebleria tenella.

CUSCUTA (Convolvulac.).
macrolepis *R.C. Fang & S.H. Huang in Fl. Xizangica,* 4: 33 (1985)—Xizang.
planiflora
subsp. godronii *(Desmoulins) M. Kerguélen in Lejeunia,* 120: 75 (1987): C. godronii.
scandens
var. bidentis *(Berthiot) M. Kerguélen in Lejeunia,* 120: 76 (1987): C. bidentis.

CUSSONIA (Araliac.).
paniculata
var. sinuata *W.F. Reyneke & P.D.F. Kok in S. Afr. J. Bot.,* 53(5): 321 (1987)—South Africa (Transvaal, Orange Free State, Natal, Cape Province), Botswana, Swaziland, Lesotho.

CUTANDIA (Gramin.).
incrassata *(Salzm. ex Loisel.) Battandier & Trabut, Fl. Algérie, Monocotylédones:* 238 (1895): Festuca incrassata.

CUVIERA (Rubiac.).
subgen. Globulostylis *(Wernham) B. Verdcourt in Kew Bull.,* 42(1): 189 (1987): gen. Globulostylis.
minor *(Wernham) B. Verdcourt in Kew Bull.,* 42(1): 189 (1987): Globulostylis minor.
talbotii *(Wernham) B. Verdcourt in Kew Bull.,* 42(1): 189 (1987): Globulostylis talbotii.

CYANEA (Campanulac.).
hanaensis *H. St. John in Phytologia,* 62(5): 433 (1987)—Hawaiian Is.

CYBIANTHUS (Myrsinac.).
subgen. Grammadenia *(Bentham) J.J. Pipoly in Mem. New York Bot. Gard.,* 43: 47 (1987): gen. Grammadenia.
costaricanus
subsp. acuminatus *(Lundell) J.J. Pipoly in Mem. New York Bot. Gard.,* 43: 52 (1987): Grammadenia acuminata.
subsp. morii *(Lundell) J.J. Pipoly in Mem. New York Bot. Gard.,* 43: 53 (1987): Grammadenia morii.
subsp. panamensis *(Lundell) J.J. Pipoly in Mem. New York Bot. Gard.,* 43: 53 (1987): Grammadenia panamensis.
lineatus *(Bentham) J.J. Pipoly in Mem. New York Bot. Gard.,* 43: 64 (1987): Grammadenia lineata.
magnus *(Mez) J.J. Pipoly in Mem. New York Bot. Gard.,* 43: 56 (1987): Grammadenia magna.
subsp. asymmetricus *(Mez) J.J. Pipoly in Mem. New York Bot. Gard.,* 43: 57 (1987): Grammadenia asymmetrica.
marginatus *(Bentham) J.J. Pipoly in Mem. New York Bot. Gard.,* 43: 60 (1987): Grammadenia marginata.
parasiticus *(Swartz) J.J. Pipoly in Mem. New York Bot. Gard.,* 43: 59 (1987): Ardisia parasitica.
piresii *J.J. Pipoly in Mem. New York Bot. Gard.,* 43: 65 (1987)—Brazil, Venezuela.
ptariensis *(Steyermark) J.J. Pipoly in Mem. New York Bot. Gard.,* 43: 66 (1987): Grammadenia ptariensis.

CYLACTIS (Rosac.).
saxatilis *(L.) A. Löve in Taxon,* 36(3): 660 (1987): Rubus saxatilis.

CYLINDROCLINE (Compos.).
lorencei *A.J. Scott in Kew Bull.,* 42(2): 476 (1987)—Mauritius.

CYMBIDIUM (Orchidac.).
formosanum
f. albiflorum *S.S. Ying in Quart. J. Chinese Forest.,* 20(2): 54 (1987), without type—Taiwan.
sinense
var. pallidiflorum *S.S. Ying in Quart. J. Chinese Forest.,* 20(2): 54 (1987)—Taiwan.
var. taiwanianum *S.S. Ying in Quart. J. Chinese Forest.,* 20(2): 54 (1987)—Taiwan.
tortisepalum
f. albiflorum *S.S. Ying in Quart. J. Chinese Forest.,* 20(2): 54 (1987), without type—Taiwan.

CYMBOPAPPUS (Compos.).
piliferus *(Thell.) B. Nordenstam in Op. Bot.,* 92: 147 (1987): Matricaria pilifera.

CYMOPTERUS (Umbellif.).
evertii *R.L. Hartman & R.S. Kirkpatrick in Brittonia,* 38(4): 420 (1986)—U.S.A. (Wyoming).

CYNANCHUM (Asclepiadac.).
acerifolium *(Lasser & Maguire) G. Morillo in Mem. Soc. Cienc. Nat. La Salle,* 37(107): 119 (1977): Ditassa acerifolia.
avilense *G. Morillo in Mem. Soc. Cienc. Nat. La Salle,* 37(107): 119 (1977), nom. nov.: C. mucronatum *HBK.*
boldinghii *(Schlechtd.) Van der Aa & A.L. Stoffers in Proc. Kon. Nederl. Akad. Wetensch., C,* 84(3): 309 (1981): Metastelma boldinghii.
maccartii
var. latifolium *Turner ex J. Henrickson in Sida,* 12(1): 98 (1987)—Mexico.
oxyphyllum *(Turcz.) G. Morillo in Mem. Soc. Cienc. Nat. La Salle,* 37(107): 119 (1977): Ditassa oxyphylla.
pringlei *(Gray) J. Henrickson in Sida,* 12(1): 91 (1987): Metastelma pringlei.
tepuiense *G. Morillo in Mem. Soc. Cienc. Nat. La Salle,* 37(107): 119 (1977), nom. nov.: Blepharodon ciliatum *Moldenke.*
xeroneuron *(Schltr.) G. Morillo in Mem. Soc. Cienc. Nat. La Salle,* 37(107): 119 (1977): Ditassa xeroneura.

CYNODON (Gramin.).
dactylon
var. affinis *(Caro & Sánchez) C. Romero Zarco in Lagascalia,* 14(1): 171 (1986): C. affinis.

CYNOGLOSSUM (Boraginac.).
cheirifolium
var. antiatlanticum *J. Molero & J.M. Montserrat in Treb. Inst. Bot. Barcelona,* 11: 25 (1987)—Morocco.

CYNOMORIUM (Balanophorac.).
coccineum
subsp. songaricum *(Rupr.) J. Léonard in Bull. Jard. Bot. Nation. Belg.,* 56(3–4): 304 (1986): C. songaricum.

CYPEROCHLOA M. Lazarides & L. Watson in Brunonia, 9(2): 216 (1987). *GRAMINEAE.*
hirsuta *M. Lazarides & L. Watson in Brunonia,* 9(2): 216 (1987)—Australia (W. Australia).

CYPERUS (Cyperac.).
subcastaneus *D.A. Simpson in Kew Bull.,* 42(4): 897 (1987)—Brazil.

CYPHIA (Campanulac.).
couroublei *P. Bamps & F. Malaisse in Bull. Jard. Bot. Nation. Belg.,* 56(3–4): 488 (1986)—Zaïre.

CYRTOSIA (Orchidac.).
integra *(Rolfe ex Downie) L.A. Garay in Bot. Mus. Leafl. Harvard Univ.,* 30(4): 232 (1986): Galeola integra.
minahassae *(Schltr.) L.A. Garay in Bot. Mus. Leafl. Harvard Univ.,* 30(4): 233 (1986): Galeola minahassae.
nana *(Rolfe ex Downie) L.A. Garay in Bot. Mus. Leafl. Harvard Univ.,* 30(4): 233 (1986): Galeola nana.
septentrionalis *(Rchb. f.) L.A. Garay in Bot. Mus. Leafl. Harvard Univ.,* 30(4): 233 (1986): Galeola septentrionalis.

CYRTOSTYLIS (Orchidac.).
robusta *D.L. Jones & M.A. Clements in Lindleyana,* 2(3): 157 (1987)—Australia (W. Australia, S. Australia, Victoria).
tenuissima *(Nicholls & Goadby) D.L. Jones & M.A. Clements in Lindleyana,* 2(3): 157 (1987): Acianthus tenuissimus.

D

DACTYLAEA Fedde ex H. Wolff in Fedde, Repert, 27: 304 (1930), descr. gen.-spec. *UMBELLIFERAE.* †

×DACTYLEUCORCHIS (Orchidac.).
bulgarica *D. Lindig in Orchidee,* 38(2): 101 (1987)—Bulgaria. (Dactylorhiza cordigera × Leucorchis albida).

DACTYLIS (Gramin.).
hyrcana *(Tzvel.) S.G. Musaev in Bot. Zhurn.,* 72(1): 94 (1987): D. glomerata subsp. hyrcana.
pendula *(Dumort.) B.D. Jackson in Index Kewensis,* 1: 709 (1753–1885 publ. 1893): D. glomerata * (subsp.) pendula.

DACTYLORHIZA (Orchidac.).
cordigera
var. vermionica *B. & E. Willing in Ber. Arbeitskr. Heim. Orch.,* 4(1): 36 (1987)—Greece.
×fourkensis *B. & E. Willing in Ber. Arbeitskr. Heim. Orch.,* 4(1): 61 (1987)—Greece. (D. baumanniana × sambucina).
×kerasovinensis *B. & E. Willing in Ber. Arbeitskr. Heim. Orch.,* 3(2): 217 (1986 publ. 1987)—Greece. (D. pindica × saccifera).
pindica *B. & E. Willing in Ber. Arbeitskr. Heim. Orch.,* 3(2): 215 (1986 publ. 1987)—Greece.
×souflikensis *B. & E. Willing in Ber. Arbeitskr. Heim. Orch.,* 4(1): 60 (1987)—Greece. (D. baumanniana × pindica).
×vorasica *B. & E. Willing in Ber. Arbeitskr. Heim. Orch.,* 4(1): 63 (1987)—Greece. (D. cordigera × sambucina).

DAHLIA (Compos.).
congestifolia *P.D. Sørensen in Rhodora,* 89(858): 198 (1987)—Mexico.

DALEA (Legumin.).
adenopoda *(Rydb.) D. Isely in Brittonia,* 38(4): 353 (1986): Kuhnistera adenopoda.

DAMNACANTHUS (Rubiac.).
indicus
var. okinawensis *S. Hatusima in J. Phytogeogr. Taxon.,* 34(1): 9 (1986)—Ryukyu Is.
okinawensis *S. Hatusima in J. Phytogeogr. Taxon.,* 34(1): 9 (1986)—Ryukyu Is.

DAMNXANTHODIUM J.L. Strother in Syst. Bot., 12(1): 41 (1987). *COMPOSITAE.*
calvum *(Greenman) J.L. Strother in Syst. Bot.,* 12(1): 41 (1987): Perymenium (?) calvum.

DAMPIERA (Goodeniac.).
dysantha *(Benth.) M.T.M. Rajput & R.C. Carolin in Fl. S. Australia,* 3: 1386 (1986): D. rosmarinifolia var. dysantha.
lanceolata
var. insularis *M.T.M. Rajput & R.C. Carolin in Fl. S. Australia,* 3: 1387 (1986)—Australia (S. Australia).
var. intermedia *M.T.M. Rajput & R.C. Carolin in Fl. S. Australia,* 3: 1387 (1986)—Australia (S. Australia).

DANTHONIA (Gramin.).
melanathera *(Hackel) L.M. Bernardello in Kurtziana,* 10: 249 (1977): D. cirrata var. melanathera.

DAPHNE (Thymelaeac.).
emeiensis *C.Y. Chang in Guihaia,* 6(4): 267 (1986)—China.
erosiloba *C.Y. Chang in Guihaia,* 6(4): 265 (1986)—China.
tripartita *H.F. Zhou ex C.Y. Chang in Guihaia,* 6(4): 268 (1986)—China.

DAPHNOPSIS (Thymelaeac.).
correae *K. Barringer & L.I. Nevling in Phytologia,* 61(6): 364 (1986)—Panama.
costaricensis *K. Barringer & Grayum in Phytologia,* 61(6): 363 (1986)—Costa Rica.
megacarpa *L.I. Nevling & K. Barringer in Phytologia,* 61(6): 361 (1986)—Mexico.
witsbergeri *L.I. Nevling, Matekaitis & K. Barringer in Phytologia,* 61(6): 362 (1986)—Salvador.

DARWINIOTHAMNUS (Compos.).
alternifolius *J.E. Lawesson & H. Adsersen in Op. Bot.,* 92: 10 (1987)—Galapagos Is.
lancifolius
subsp. glabriusculus *(Stewart) J.E. Lawesson & H. Adsersen in Op. Bot.,* 92: 13 (1987), without basionym page: Erigeron lancifolius var. glabriusculus.

DATURA (Solanac.).
lanosa *Barclay ex R.A. Bye in Phytologia,* 61(3): 204 (1986)—Mexico.

DELISSEA (Campanulac.).
aculeatiflora *(Rock) H. St. John in Phytologia,* 63(2): 79 (1987): Cyanea aculeatiflora.
acuminata
var. calycina *(Hosaka) H. St. John in Phytologia,* 63(2): 80 (1987): Cyanea acuminata var. calycina.
f. latifolia *(Wawra) H. St. John in Phytologia,* 63(2): 79 (1987): D. acuminata var. latifolia.
alba *H. St. John in Phytologia,* 63(5): 339 (1987)—Hawaiian Is.
albilineata *H. St. John in Phytologia,* 63(5): 339 (1987)—Hawaiian Is.
angustifolia
f. elliptica *(Rock) H. St. John in Phytologia,* 63(2): 80 (1987): Cyanea angustifolia f. elliptica.
var. hillebrandii *(Rock) H. St. John in Phytologia,* 63(2): 80 (1987): Cyanea angustifolia var. hillebrandii.
var. isabella *(E. Wimm.) H. St. John in Phytologia,* 63(2): 80 (1987): Cyanea angustifolia var. isabella.
var. lanaiensis *(Rock) H. St. John in Phytologia,* 63(2): 80 (1987): Cyanea angustifolia var. lanaiensis.
f. lanaiensis *(Rock) H. St. John in Phytologia,* 63(2): 80 (1987): Cyanea angustifolia var. lanaiensis.
var. racemosa *(Hillebr.) H. St. John in Phytologia,* 63(2): 80 (1987): Cyanea angustifolia var. racemosa.

DELISSEA :–
f. subpubescens *(E. Wimm.) H. St. John in Phytologia,* 63(2): 80 (1987): Cyanea angustifolia f. subpubescens.
var. tomentella *(Hillebr.) H. St. John in Phytologia,* 63(2): 80 (1987): Cyanea angustifolia var. tomentella.
asarifolia *(St. John) H. St. John in Phytologia,* 63(2): 81 (1987): Cyanea asarifolia.
atra *(Hillebr.) H. St. John in Phytologia,* 63(2): 81 (1987): Cyanea atra.
var. lobata *(Rock) H. St. John in Phytologia,* 63(2): 81 (1987): Cyanea atra var. lobata.
baldwinii *(Forbes & Munro) H. St. John in Phytologia,* 63(2): 81 (1987): Cyanea baldwinii.
bicolor *(St. John) H. St. John in Phytologia,* 63(2): 81 (1987): Cyanea bicolor.
bishopii *(Rock) H. St. John in Phytologia,* 63(2): 81 (1987): Cyanea bishopii.
bondiana *(Rock) H. St. John in Phytologia,* 63(2): 81 (1987): Cyanea pilosa var. bondiana.
brevipedicellata *H. St. John in Phytologia,* 63(5): 339 (1987)—Hawaiian Is.
bryanii *(Rock) H. St. John in Phytologia,* 63(2): 81 (1987): Cyanea bryanii.
carlsonii *(Rock) H. St. John in Phytologia,* 63(2): 81 (1987): Cyanea carlsonii.
cataracta *H. St. John in Phytologia,* 63(5): 339 (1987)—Hawaiian Is.
chartacea *H. St. John in Phytologia,* 63(5): 340 (1987)—Hawaiian Is.
chockii *(Rock) H. St. John in Phytologia,* 63(2): 81 (1987): Cyanea chockii.
christensenii *H. St. John in Phytologia,* 63(5): 340 (1987)—Hawaiian Is.
comata *(Hillebr.) H. St. John in Phytologia,* 63(2): 81 (1987): Cyanea comata.
copelandii *(Rock) H. St. John in Phytologia,* 63(2): 81 (1987): Cyanea copelandii.
coriacea
var. degeneriana *(E. Wimm.) H. St. John in Phytologia,* 63(2): 82 (1987): Cyanea coriacea var. degeneriana.
var. deltoidea *H. St. John in Phytologia,* 63(5): 340 (1987)—Hawaiian Is.
var. gratiosa *(E. Wimm.) H. St. John in Phytologia,* 63(2): 82 (1987): Cyanea coriacea f. gratiosa.
var. hardyi *(Rock) H. St. John in Phytologia,* 63(2): 82 (1987): Cyanea hardyi.
var. haupuensis *H. St. John in Phytologia,* 63(5): 340 (1987)—Hawaiian Is.
var. lumahaiensis *H. St. John in Phytologia,* 63(5): 340 (1987)—Hawaiian Is.
var. serratifolia *(Rock) H. St. John in Phytologia,* 63(2): 82 (1987): Cyanea coriacea var. serratifolia.
coronata *(E. Wimm.) H. St. John in Phytologia,* 63(2): 82 (1987): Cyanea coronata.
crispohirta *(E. Wimm.) H. St. John in Phytologia,* 63(2): 82 (1987): Cyanea crispohirta.
decumbens *H. St. John in Phytologia,* 63(5): 340 (1987)—Hawaiian Is.

DELISSEA :-
degeneriana *(E. Wimm.) H. St. John in Phytologia,* 63(2): 82 (1987): Cyanea degeneriana.
deltoidea *H. St. John in Phytologia,* 63(5): 340 (1987)—Hawaiian Is.
densiflora *(Rock) H. St. John in Phytologia,* 63(2): 82 (1987): Cyanea pilosa var. densiflora.
dentata *(E. Wimm.) H. St. John in Phytologia,* 63(2): 82 (1987): Cyanea dentata.
denticulata *H. St. John in Phytologia,* 63(5): 341 (1987)—Hawaiian Is.
divergens *H. St. John in Phytologia,* 63(5): 341 (1987)—Hawaiian Is.
dunbariae *(Rock) H. St. John in Phytologia,* 63(2): 82 (1987): Cyanea dunbarii.
duploserrata *H. St. John in Phytologia,* 63(5): 341 (1987)—Hawaiian Is.
eleeleensis *H. St. John in Phytologia,* 63(5): 341 (1987)—Hawaiian Is.
excurrens *H. St. John in Phytologia,* 63(5): 342 (1987)—Hawaiian Is.
fauriei *(H. Lévl.) H. St. John in Phytologia,* 63(5): 342 (1987): Cyanea fauriei.
fernaldii *(Rock) H. St. John in Phytologia,* 63(2): 82 (1987): Cyanea fernaldii.
ferox *(Hillebr.) H. St. John in Phytologia,* 63(2): 82 (1987): Cyanea ferox.
var. laevicalyx *(Skottsb.) H. St. John in Phytologia,* 63(2): 82 (1987): Cyanea ferox var. laevicalyx.
floribunda *(E. Wimm.) H. St. John in Phytologia,* 63(2): 83 (1987): Cyanea floribunda.
forbesii *H. St. John in Phytologia,* 63(2): 83 (1987), nom. nov.: Cyanea undulata *Forbes.*
fruticosa *H. St. John in Phytologia,* 63(5): 342 (1987)—Hawaiian Is.
gayana *(Rock) H. St. John in Phytologia,* 63(2): 83 (1987): Cyanea gayana.
var. duvelii *(Rock) H. St. John in Phytologia,* 63(2): 83 (1987): Cyanea gayana var. duvelii.
var. wainihaensis *(Rock) H. St. John in Phytologia,* 63(2): 83 (1987): Cyanea gayana var. wainihaensis.
gibsonii *(Hillebr.) H. St. John in Phytologia,* 63(2): 83 (1987): Cyanea gibsonii.
giffardii *(Rock) H. St. John in Phytologia,* 63(2): 83 (1987): Cyanea giffardii.
glabra *(E. Wimm.) H. St. John in Phytologia,* 63(2): 83 (1987): Cyanea knudsenii var. glabra.
glabrifolia *H. St. John in Phytologia,* 63(5): 342 (1987)—Hawaiian Is.
grimesiana *(Gaud.) H. St. John in Phytologia,* 63(2): 83 (1987): Cyanea grimesiana.
var. cylindrocalyx *(Rock) H. St. John in Phytologia,* 63(2): 83 (1987): Cyanea grimesiana var. cylindrocalyx.
var. hirsutifolia *(Rock) H. St. John in Phytologia,* 63(2): 83 (1987): Cyanea grimesiana var. hirsutifolia.

DELISSEA :-
var. lydgatei *(Rock) H. St. John in Phytologia,* 63(2): 83 (1987): Cyanea grimesiana var. lydgatei.
var. mauiensis *(Rock) H. St. John in Phytologia,* 63(2): 84 (1987): Cyanea grimesiana var. mauiensis.
var. munroi *(Hosaka) H. St. John in Phytologia,* 63(2): 84 (1987): Cyanea grimesiana var. munroi.
var. obatae *(St. John) H. St. John in Phytologia,* 63(2): 83 (1987): Cyanea grimesiana var. obatae.
habenata *H. St. John in Phytologia,* 63(5): 342 (1987)—Hawaiian Is.
haleakalaensis *(St. John) H. St. John in Phytologia,* 63(2): 84 (1987): Cyanea haleakalaensis.
hamatiflora *(Rock) H. St. John in Phytologia,* 63(2): 84 (1987): Cyanea hamatiflora.
hirtella
var. striata *(E. Wimm.) H. St. John in Phytologia,* 63(2): 84 (1987): Cyanea hirtella var. striata.
var. subglabra *(E. Wimm.) H. St. John in Phytologia,* 63(2): 84 (1987): Cyanea hirtella var. subglabra.
holophylla *(Hillebr.) H. St. John in Phytologia,* 63(2): 84 (1987): Cyanea holophylla.
var. obovata *(Rock) H. St. John in Phytologia,* 63(2): 84 (1987): Cyanea holophylla var. obovata.
horrida *(Rock) H. St. John in Phytologia,* 63(2): 84 (1987): Cyanea ferox var. horrida.
humilis *(H. Wawra) H. St. John in Phytologia,* 63(5): 342 (1987): Cyanea humilis.
hypoleuca *H. St. John in Phytologia,* 63(5): 342 (1987)—Hawaiian Is.
iliahiatilis *H. St. John in Phytologia,* 63(5): 343 (1987)—Hawaiian Is.
impedicellata *H. St. John in Phytologia,* 63(5): 343 (1987)—Hawaiian Is.
inermis *H. St. John in Phytologia,* 63(5): 343 (1987)—Hawaiian Is.
kahiliensis *H. St. John in Phytologia,* 63(5): 343 (1987)—Hawaiian Is.
keaensis *H. St. John in Phytologia,* 63(5): 343 (1987)—Hawaiian Is.
kealawelaensis *H. St. John in Phytologia,* 63(5): 343 (1987)—Hawaiian Is.
knudsenii *(Rock) H. St. John in Phytologia,* 63(2): 84 (1987): Cyanea knudsenii.
kolekoleensis *H. St. John in Phytologia,* 63(5): 344 (1987)—Hawaiian Is.
larrisonii *(Rock) H. St. John in Phytologia,* 63(2): 84 (1987): Cyanea larrisonii.
latibasilaris *H. St. John in Phytologia,* 63(5): 344 (1987)—Hawaiian Is.
leiophylla *H. St. John in Phytologia,* 63(5): 344 (1987)—Hawaiian Is.
leptostegia *(A. Gray) H. St. John in Phytologia,* 63(2): 84 (1987): Cyanea leptostegia.
var. velutina *(Skottsb.) H. St. John in Phytologia,* 63(2): 85 (1987): Cyanea leptostegia var. velutina.

DELISSEA :-

ligulata *H. St. John in Phytologia,* 63(5): 344 (1987)—Hawaiian Is.

limahuliensis *H. St. John in Phytologia,* 63(5): 344 (1987)—Hawaiian Is.

linearifolia *(Rock) H. St. John in Phytologia,* 63(2): 85 (1987): Cyanea linearifolia.

lobata *(H. Mann) H. St. John in Phytologia,* 63(2): 85 (1987): Cyanea lobata.

var. hamakuae *(Rock) H. St. John in Phytologia,* 63(2): 85 (1987): Cyanea lobata var. hamakuae.

longiantherae *H. St. John in Phytologia,* 63(5): 345 (1987)—Hawaiian Is.

longicalyx *H. St. John in Phytologia,* 63(5): 345 (1987)—Hawaiian Is.

longipedunculata *(Rock) H. St. John in Phytologia,* 63(2): 85 (1987): Cyanea longipedunculata.

longissima *(Rock) H. St. John in Phytologia,* 63(2): 85 (1987): Cyanea scabra var. longissima.

lumahaiensis *H. St. John in Phytologia,* 63(5): 345 (1987)—Hawaiian Is.

lydgatei *H. St. John in Phytologia,* 63(5): 345 (1987), nom. nov.: Cyanea undulata *Forbes.*

maceldowneyi *(Rock) H. St. John in Phytologia,* 63(2): 85 (1987): Cyanea maceldowneyi.

macrostegia *(Hillebr.) H. St. John in Phytologia,* 63(2): 85 (1987): Cyanea macrostegia.

var. parvibracteata *(Rock) H. St. John in Phytologia,* 63(2): 85 (1987): Cyanea macrostegia var. parvibracteata.

var. viscosa *(Rock) H. St. John in Phytologia,* 63(2): 85 (1987): Cyanea macrostegia var. viscosa.

magnifica *(E. Wimm.) H. St. John in Phytologia,* 63(2): 85 (1987): Cyanea magnifica.

mariana *(E. Wimm.) H. St. John in Phytologia,* 63(2): 85 (1987): Cyanea mariana.

marksii *(Rock) H. St. John in Phytologia,* 63(2): 85 (1987): Cyanea marksii.

megacarpa *(Rock) H. St. John in Phytologia,* 63(2): 86 (1987): Cyanea pilosa var. megacarpa.

membranacea *(Rock) H. St. John in Phytologia,* 63(2): 86 (1987): Cyanea membranacea.

multiramosa *H. St. John in Phytologia,* 63(5): 345 (1987)—Hawaiian Is.

multispicata *(H. Lévl.) H. St. John in Phytologia,* 63(2): 86 (1987): Cyanea multispicata.

napaliensis *H. St. John in Phytologia,* 63(5): 345 (1987)—Hawaiian Is.

nelsonii *(St. John) H. St. John in Phytologia,* 63(2): 86 (1987): Cyanea nelsonii.

nigra *H. St. John in Phytologia,* 63(5): 346 (1987)—Hawaiian Is.

noli-me-tangere *(Rock) H. St. John in Phytologia,* 63(2): 86 (1987): Cyanea noli-me-tangere.

occultans *(St. John) H. St. John in Phytologia,* 63(2): 86 (1987): Cyanea occultans.

DELISSEA :-

ovatisepala *(E. Wimm.) H. St. John in Phytologia,* 63(2): 86 (1987): Cyanea ovatisepala.

palakea *(C.N. Forbes) H. St. John in Phytologia,* 63(2): 86 (1987): Cyanea palakea.

paliensis *H. St. John in Phytologia,* 63(5): 346 (1987)—Hawaiian Is.

parva *H. St. John in Phytologia,* 63(5): 346 (1987)—Hawaiian Is.

perlmanii *H. St. John in Phytologia,* 63(5): 346 (1987), as 'perlmannii'—Hawaiian Is.

pluriflora *H. St. John in Phytologia,* 63(5): 346 (1987)—Hawaiian Is.

procera *(Hillebr.) H. St. John in Phytologia,* 63(2): 87 (1987): Cyanea procera.

profuga *(C.N. Forbes) H. St. John in Phytologia,* 63(2): 87 (1987): Cyanea profuga.

pulchra *(Rock) H. St. John in Phytologia,* 63(2): 87 (1987): Cyanea pulchra.

purpurea *H. St. John in Phytologia,* 63(5): 347 (1987)—Hawaiian Is.

pycnocarpa *(Hillebr.) H. St. John in Phytologia,* 63(2): 87 (1987): D. arborea var. pycnocarpa.

quercifolia *(Hillebr.) H. St. John in Phytologia,* 63(2): 87 (1987): Cyanea solanacea var. quercifolia.

var. atropurpurea *(E. Wimm.) H. St. John in Phytologia,* 63(2): 87 (1987): Cyanea quercifolia var. atropurpurea.

ramosa *H. St. John in Phytologia,* 63(5): 347 (1987)—Hawaiian Is.

remyi *(Rock) H. St. John in Phytologia,* 63(2): 87 (1987): Cyanea remyi.

robinsonii *H. St. John in Phytologia,* 63(5): 347 (1987)—Hawaiian Is.

rockii *(E. Wimm.) H. St. John in Phytologia,* 63(2): 87 (1987): Cyanea rockii.

rollandioides *(Rock) H. St. John in Phytologia,* 63(2): 87 (1987): Cyanea rollandioides.

scabra *(Hillebr.) H. St. John in Phytologia,* 63(2): 88 (1987): Cyanea scabra.

var. variabilis *(Rock) H. St. John in Phytologia,* 63(2): 88 (1987): Cyanea scabra var. variabilis.

scopuli *H. St. John in Phytologia,* 63(5): 347 (1987)—Hawaiian Is.

serratifolia *(Rock) H. St. John in Phytologia,* 63(5): 347 (1987): Cyanea coriacea var. serratifolia.

shipmanii *(Rock) H. St. John in Phytologia,* 63(2): 88 (1987): Cyanea shipmanii.

simplex *H. St. John in Phytologia,* 63(5): 348 (1987), as 'simples'—Hawaiian Is.

f. maculata *H. St. John in Phytologia,* 63(5): 348 (1987)—Hawaiian Is.

solanacea *(Hillebr.) H. St. John in Phytologia,* 63(2): 88 (1987): Cyanea solanacea.

solenocalyx *(Hillebr.) H. St. John in Phytologia,* 63(2): 88 (1987): Cyanea solenocalyx.

var. latifolia *(E. Wimm.) H. St. John in Phytologia,* 63(2): 88 (1987): Cyanea solenocalyx var. latifolia.

DELISSEA :–
spathulata *(Hillebr.) H. St. John in Phytologia,* 63(2): 88 (1987): Cyanea coriacea var. spathulata.
stictophylla *(Rock) H. St. John in Phytologia,* 63(2): 88 (1987): Cyanea stictophylla.
var. inermis *(Rock) H. St. John in Phytologia,* 63(2): 88 (1987): Cyanea stictophylla var. inermis.
subacuminata *H. St. John in Phytologia,* 63(5): 348 (1987)—Hawaiian Is.
subintegra *H. St. John in Phytologia,* 63(5): 348 (1987)—Hawaiian Is.
submuricata *(E. Wimm.) H. St. John in Phytologia,* 63(2): 89 (1987): Cyanea submuricata.
subsessilis *H. St. John in Phytologia,* 63(5): 348 (1987)—Hawaiian Is.
superba *(Cham.) H. St. John in Phytologia,* 63(2): 89 (1987): Lobelia superba.
var. velutina *(Rock) H. St. John in Phytologia,* 63(2): 89 (1987): Cyanea superba var. velutina.
sylvestris *(A. Heller) H. St. John in Phytologia,* 63(2): 89 (1987): Cyanea sylvestris.
var. eriantha *(Skottsb.) H. St. John in Phytologia,* 63(2): 89 (1987): Cyanea eriantha.
tritomantha *(A. Gray) H. St. John in Phytologia,* 63(2): 89 (1987): Cyanea tritomantha.
var. lydgatei *(Rock) H. St. John in Phytologia,* 63(2): 89 (1987): Cyanea tritomantha var. lydgatei.
truncata *(Rock) H. St. John in Phytologia,* 63(2): 89 (1987): Rollandia truncata.
var. juddii *(C.N. Forbes) H. St. John in Phytologia,* 63(2): 89 (1987): Cyanea juddii.
umbrosa *H. St. John in Phytologia,* 63(5): 348 (1987)—Hawaiian Is.
virgata *H. St. John in Phytologia,* 63(5): 349 (1987)—Hawaiian Is.
wailauensis *(Rock) H. St. John in Phytologia,* 63(2): 89 (1987): Cyanea wailauensis.
waioliensis *H. St. John in Phytologia,* 63(5): 349 (1987)—Hawaiian Is.
waipaensis *H. St. John in Phytologia,* 63(5): 349 (1987)—Hawaiian Is.

DELPHINIUM (Ranunculac.).
sect. Albocoerulea *N.I. Malyutin in Bot. Zhurn.,* 72(5): 689 (1987).
subsect. Albocoerulea *N.I. Malyutin in Bot. Zhurn.,* 72(5): 689 (1987).
subsect. Bicoloria *(Rydb.) N.I. Malyutin in Bot. Zhurn.,* 72(5): 687 (1987): D. sect. Bicoloria.
subsect. Caroliniana *N.I. Malyutin in Bot. Zhurn.,* 72(5): 691 (1987).
subsect. Echinata *(Ewan) N.I. Malyutin in Bot. Zhurn.,* 72(5): 689 (1987): D. ser. Echinatae.
subsect. Exaltata *N.I. Malyutin in Bot. Zhurn.,* 72(5): 688 (1987).

DELPHINIUM :–
sect. Grumosa *N.I. Malyutin in Bot. Zhurn.,* 72(5): 689 (1987).
subsect. Kolobopetala *N.I. Malyutin in Bot. Zhurn.,* 72(5): 687 (1987).
subsect. Macrocentra *N.I. Malyutin in Bot. Zhurn.,* 72(5): 690 (1987).
sect. Microcentra *N.I. Malyutin in Bot. Zhurn.,* 72(5): 689 (1987).
sect. Oligophyllon *N.I. Malyutin in Bot. Zhurn.,* 72(5): 691 (1987).
subsect. Pogonantha *N.I. Malyutin in Bot. Zhurn.,* 72(5): 690 (1987).
conaense *W.T. Wang in Acta Phytotax. Sin.,* 25(1): 34 (1987)—Xizang.
denudatum
var. latifolium *R.A. Qureshi & M.N. Chaudhri in Pakistan Syst.,* 3(1): 11 (1987)—Pakistan.
var. linearifolium *R.A. Qureshi & M.N. Chaudhri in Pakistan Syst.,* 3(1): 11 (1987)—Pakistan.
dolichocentroides
var. parvifolium *W.T. Wang in Bull. Bot. Res. North-East. Forest. Inst.,* 7(2): 96 (1987)—China.
incisolobulatum *W.T. Wang in Acta Phytotax. Sin.,* 25(1): 35 (1987)—Xizang.
leiostachyum *W.T. Wang in Bull. Bot. Res. North-East. Forest. Inst.,* 7(2): 96 (1987)—China.
tabatae *M. Tamura in Acta Phytotax. Geobot.,* 37(4–6): 156 (1986)—Nepal.
uncinatum
var. glabrum *R.A. Qureshi & M.N. Chaudhri in Pakistan Syst.,* 3(1): 9 (1987)—Pakistan.

DEMAVENDIA M.G. Pimenov in Fl. Iran., 162: 450 (1987). *UMBELLIFERAE.*
pastinacifolia *(Boiss. & Hausskn.) M.G. Pimenov in Fl. Iran.,* 162: 450 (1987): Peucedanum pastinacifolium.

DENDRANTHEMA (Compos.).
seticuspe *(Maxim.) S. Kitamura in Acta Phytotax. Sin.,* 38: 380 (1987): Pyrethrum seticuspe.
f. boreale *(Makino) S. Kitamura in Acta Phytotax. Geobot.,* 38: 381 (1987): Chrysanthemum indicum var. boreale.

DENDROBENTHAMIA (Cornac.).
longipedunculata *S.S. Chang & X. Chen in Guihaia,* 7(2): 128 (1987)—China.
qianxinanica *S.S. Chang & X. Chen in Guihaia,* 7(2): 127 (1987)—China.

DENDROBIUM (Orchidac.).
crystallinum
var. hainanense *S.J. Cheng & C.Z. Tang in Acta Bot. Austro Sinica,* 2: 26 (1986)—Hainan.

DENDROCALAMUS (Gramin.).
bambusoides *J.R. Xue & D.Z. Li in J. Bamboo Res.,* 6(2): 16 (1987)—China.

DENDROCHILUM (Orchidac.).
viride *G. Seidenfaden in Op. Bot.*, 89: 85 (1986)—Thailand.

DENDROPEMON (Loranthac.).
silvae *A.T. Leiva Sánchez in Rev. Jard. Bot. Nacion. Univ. Habana*, 6(3): 19 (1985 publ. 1986)—Cuba.

DENDROPHTHORA (Viscac.).
bulbifera *J. Kuijt in Fl. Ecuador*, 24(32B): 77 (1986)—Ecuador.
capillaris *J. Kuijt in Fl. Ecuador*, 24(32B): 85 (1986)—Ecuador.
dalstroemii *J. Kuijt in Fl. Ecuador*, 24(32B): 82 (1986)—Ecuador.
densifolia *J. Kuijt in Fl. Ecuador*, 24(32B): 73 (1986)—Ecuador, Peru.
fastigiata *J. Kuijt in Fl. Ecuador*, 24(32B): 82 (1986)—Ecuador.
lueri *J. Kuijt in Fl. Ecuador*, 24(32B): 69 (1986)—Ecuador, Colombia.
ovata *J. Kuijt in Fl. Ecuador*, 24(32B): 102 (1986)—Ecuador.
pearcei *(Rusby) J. Kuijt in Fl. Ecuador*, 24(32B): 79 (1986): Phoradendron pearcei.
polyantha *J. Kuijt in Fl. Ecuador*, 24(32B): 71 (1986)—Ecuador.
sumacoi *J. Kuijt in Fl. Ecuador*, 24(32B): 105 (1986)—Ecuador.
tenuifolia *J. Kuijt in Fl. Ecuador*, 24(32B): 109 (1986)—Ecuador.
thomasii *J. Kuijt in Fl. Ecuador*, 24(32B): 97 (1986)—Ecuador.
variabilis *J. Kuijt in Fl. Ecuador*, 24(32B): 109 (1986)—Ecuador.

DENDROSENECIO (Compos.).
brassiciformis *(R.E. & T.C.E. Fr.) D.J. Mabberley in F. Vuilleumier & M. Monasterio (eds.), High Alt. Trop. Biogeog.*: 100 (1986): S. brassiciformis.
johnstonii
subsp. adnivalis *(Stapf) D.J. Mabberley in F. Vuilleumier & M. Monasterio (eds.), High Alt. Trop. Biogeog.*: 100 (1986): Senecio adnivalis.
keniensis *(Bak.f.) D.J. Mabberley in F. Vuilleumier & M. Monasterio (eds.), High Alt. Trop. Biogeog.*: 100 (1986): S. keniensis.

DENDROTROPHE (Santalac.).
punctata *C.Y. Wu & D.D. Tao in Acta Phytotax. Sin.*, 25(5): 405 (1987)—China.

DEPPEA (Rubiac.).
splendens *D.E. Breedlove & D.H. Lorence in Phytologia*, 63(1): 43 (1987)—Mexico.

DESCHAMPSIA (Gramin.).
cespitosa
subsp. pyrenaica *A.M. Romo in Folia Bot. Misc.*, 5: 25 (1986)—Spain.

DESCHAMPSIA :–
subsp. austrobohemica *(Deyl) H.J. Conert in G. Hegi, Illustr. Fl. Mitteleur.*, ed. 3, 1(3: Lief. 4): 308 (1987): D. austrobohemica.

DESMODIUM (Legumin.).
heterocarpon
var. ovalifolium *(Wall. ex Prain) Rugayah in Berita Biol.*, 3(5): 205 (1986); *et in Reinwardtia*, 10(3): 382 (1987): D. polycarpum var. ovalifolium.
unibotryosum *C. Chen & X.J. Cui in Acta Bot. Yunnanica*, 9(3): 307 (1987), nom. nov.: Hedysarum diffusum *Roxb.*

DEVERRA (Umbellif.).
denudata *(Viv.) R. Pfisterer & D. Podlech in Mitt. Bot. Staatssamml. München*, 22: 587 (1986): Pituranthos denudatus.
subsp. aphylla *(Cham. & Schlechtend.) R. Pfisterer & D. Podlech in Mitt. Bot. Staatssamml. München*, 22: 589 (1986): Bubon aphyllus.
scoparia
subsp. tripolitana *(Andreansky) R. Pfisterer & D. Podlech in Mitt. Bot. Staatssamml. München*, 22: 594 (1986): Pituranthos tripolitanus.
triradiata
subsp. intermedia *(Chevall.) R. Pfisterer & D. Podlech in Mitt. Bot. Staatssamml. München*, 22: 599 (1986): D. intermedia.
subsp. musilii *(Chrtek, Osborn. & Sourkova) R. Pfisterer & D. Podlech in Mitt. Bot. Staatssamml. München*, 22: 598 (1986): D. musilii.

DIAMENA P. Ravenna in Op. Bot., 92: 185 (1987). *ANTHERICACEAE.*
stenantha *(Rav.) P. Ravenna in Op. Bot.*, 92: 188 (1987): Anthericum stenanthum.

DIANELLA (Phormiac.).
atraxis *R.J.F. Henderson in Fl. Australia*, 45: 476, 215 (1987)—Australia (Queensland).
brevipedunculata *R.J.F. Henderson in Fl. Australia*, 45: 476, 202 (1987)—Australia (Queensland).
caerulea
var. aquilonia *R.J.F. Henderson in Fl. Australia*, 45: 477, 211 (1987)—Australia (Queensland).
var. assera *R.J.F. Henderson in Fl. Australia*, 45: 477, 212 (1987)—Australia (Queensland, New South Wales).
var. cinerascens *R.J.F. Henderson in Fl. Australia*, 45: 477, 212 (1987)—Australia (New South Wales).
var. petasmatodes *R.J.F. Henderson in Fl. Australia*, 45: 478, 213 (1987)—Australia (Queensland, New South Wales).
var. producta *R.J.F. Henderson in Fl. Australia*, 45: 478, 211 (1987)—Australia (Queensland, New South Wales).

DIANELLA :–
var. protensa *R.J.F. Henderson in Fl. Australia,* 45: 478, 213 (1987)—Australia (Queensland, New South Wales).
var. vannata *R.J.F. Henderson in Fl. Australia,* 45: 478, 213 (1987)—Australia (Queensland, New South Wales), New Guinea (Papua New Guinea).
crinoides *R.J.F. Henderson in Fl. Australia,* 45: 479, 207 (1987)—Australia (Queensland, New South Wales).
incollata *R.J.F. Henderson in Fl. Australia,* 45: 480, 197 (1987)—Australia (Queensland).
longifolia
var. fragrans *R.J.F. Henderson in Fl. Australia,* 45: 480, 205 (1987)—Australia (Queensland).
var. grandis *R.J.F. Henderson in Fl. Australia,* 45: 481, 205 (1987)—Australia (Queensland, S. Australia, New South Wales, Victoria).
var. porracea *R.J.F. Henderson in Fl. Australia,* 45: 481, 206 (1987)—Australia (Queensland, New South Wales, S. Australia).
var. stupata *R.J.F. Henderson in Fl. Australia,* 45: 481, 207 (1987)—Australia (Queensland, New South Wales).
var. surculosa *R.J.F. Henderson in Fl. Australia,* 45: 481, 207 (1987)—Australia (Queensland, New South Wales).
nervosa *R.J.F. Henderson in Fl. Australia,* 45: 482, 208 (1987)—Australia (Queensland, New South Wales).
pavopennacea *R.J.F. Henderson in Fl. Australia,* 45: 482, 218 (1987)—Australia (Queensland).
var. major *R.J.F. Henderson in Fl. Australia,* 45: 482, 219 (1987)—Australia (Queensland).
var. robusta *R.J.F. Henderson in Fl. Australia,* 45: 483, 219 (1987)—Australia (Queensland).
prunina *R.J.F. Henderson in Fl. Australia,* 45: 483, 202 (1987)—Australia (New South Wales).
revoluta
var. divaricata *(R. Br.) R.J.F. Henderson in Fl. Australia,* 45: 484 (1987): D. divaricata.
var. minor *R.J.F. Henderson in Fl. Australia,* 45: 484, 201 (1987)—Australia (Queensland).
var. tenuis *R.J.F. Henderson in Fl. Australia,* 45: 484, 201 (1987)—Australia (Queensland).
var. vinosa *R.J.F. Henderson in Fl. Australia,* 45: 484, 201 (1987)—Australia (Queensland, New South Wales).

DIANTHUS (Caryophyllac.).
costae
subsp. algetanus *(Graells ex Williams) M. Laínz, Muñoz Garmendia & Soriano in An. Jard. Bot. Madrid,* 43(2): 473 (1986 publ. 1987): D. algetanus.

DIANTHUS :–
subsp. multiaffinis *(Pau) M. Laínz in An. Jard. Bot. Madrid,* 43(2): 473 (1986 publ. 1987): D. multiaffinis.
subsp. turolensis *(Pau) M. Laínz & Muñoz Garmendia in An. Jard. Bot. Madrid,* 43(2): 473 (1986 publ. 1987): D. turolensis.
furcatus
subsp. benearnensis *(H. Loret) M. Kergeúlen in Lejeunia,* 120: 79 (1987): D. benearnensis.
subsp. requienii *(Godron) M. Kerguélen in Lejeunia,* 120: 80 (1987): D. requienii.
goerkii *P. Hartvig & A. Strid in Bot. Jahrb.,* 108(2–3): 321 (1987)—Turkey.
inoxianus *M.J. Gallego in Lagascalia,* 14(1): 71 (1986)—Spain.
laricifolius
subsp. marizii *(Samp.) Franco ex M. Laínz in An. Jard. Bot. Madrid,* 43(1): 196 (1986): D. graniticus var. marizii. †
subsp. merinoi *(Laínz) M. Laínz in An. Jard. Bot. Madrid,* 43(1): 197 (1986): D. merinoi.
pungens
subsp. brachyanthus *(Boiss.) Bernal, Fernández Casas, G. López, M. Laínz & Muñoz Garmendia in An. Jard. Bot. Madrid,* 44(1): 180 (1987): D. brachyanthus.
subacaulis
subsp. ruscinonensis *(Boiss.) G. Bosc & M. Kerguélen in Lejeunia,* 120: 80 (1987): D. brachyanthus var. ruscinonensis.
sylvestris
var. godronianus *(Jordan) M. Kerguélen in Lejeunia,* 120: 81 (1987): D. godronianus.

DIAPHANANTHE (Orchidac.).
bilobata *(Summerh.) F.N. Rasmussen in Norw. J. Bot.,* 21(3): 229 (1974): Sarcorhynchus bilobatus.
orientalis *(Mansf.) F.N. Rasmussen in Norw. J. Bot.,* 21(3): 229 (1974): Sarcorhynchus orientalis.
polyantha *(Kränzl.) F.N. Rasmussen in Norw. J. Bot.,* 21(3): 229 (1974): Mystacidium polyanthum.

DICENTRA (Papaverac.).
macrocapnos *D. Prain in J. Asiat. Soc. Bengal,* 65: 11, 12 (1896)—India.

DICHANTHIUM (Gramin.).
tuberculatum *(Hack.) G. Singh in J. Econ. Taxon. Bot.,* 8(2): 497 (1986): Andropogon tuberculatus.

DICHOPOGON (Anthericac.).
capillipes *(Endl.) N.H. Brittan in Fl. Australia,* 45: 485 (1987): Arthropodium capillipes.
preissii *(Endl.) N.H. Brittan in Fl. Australia,* 45: 486 (1987): Arthropodium preissii.
tyleri *N.H. Brittan in Fl. Australia,* 45: 486, 345 (1987)—Australia (W. Australia).

DICHROA (Hydrangeac.).
ser. Mollissimae *S.M. Hwang in Acta Phytotax. Sin.*, 25(5): 386 (1987).
sect. Silvicola *S.M. Hwang in Acta Phytotax. Sin.*, 25(5): 388 (1987).
yunnanensis *S.M. Hwang in Acta Phytotax. Sin.*, 25(5): 387 (1987)—China.

DICTYOPHYLLARIA L.A. Garay in Bot. Mus. Leafl. Harvard Univ., 30(4): 231 (1986). *ORCHIDACEAE.*
dietschiana *(Edwall) L.A. Garay in Bot. Mus. Leafl. Harvard Univ.*, 30(4): 231 (1986): Vanilla dietschiana.

DIDYMODOXA (Urticac.).
capensis *(L.f.) I. Friis & C.M. Wilmot-Dear in Nordic J. Bot.*, 7(2): 126 (1987): Urtica capensis.
var. integrifolia *(Wedd.) I. Friis & C.M. Wilmot-Dear in Nordic J. Bot.*, 7(2): 126 (1987): Australina integrifolia.

DIEFFENBACHIA (Arac.).
liesneri *T.B. Croat in Aroideana,* 9(1–4): 62 (1986)—Venezuela.
longipistila *T.B. Croat in Aroideana,* 9(1–4): 65 (1986)—Venezuela.

DIETERLEA E.J. Lott in Brittonia, 38(4): 407 (1986). *CUCURBITACEAE.*
fusiformis *E.J. Lott in Brittonia,* 38(4): 407 (1986)—Mexico.

DIGITALIS (Scrophulariac.).
dubia
var. palaui *(Garcias & Marcos) J.J. Pericás & J.A. Rosselló in Candollea,* 42(1): 395 (1987): D. dubia subsp. palaui.
purpurea
subsp. bocquetii *B. Valdés in Lagascalia,* 14(1): 96 (1986)—Spain.

DIGITARIA (Gramin.).
dusenii *Chase ex S.A. Renvoize in Kew Bull.*, 42(4): 923 (1987)—Brazil.
paspaliformis *J. Woods, Tourist's Fl.*: 396 (1850)—France.
×umfolozi *D.W. Hall in Turrialba,* 33(3): 327 (1983)—Cultivation ex U.S.A. (Florida). (D. setivalva × valida).

DILOPHIA (Crucif.).
fontana
var. trichocarpum *W.T. Wang in Acta Bot. Yunnanica,* 9(1): 3 (1987)—Xizang, China, Xinjiang.

DIMERANDRA (Orchidac.).
buenaventurae *(Krzl.) E.S. Siegerist in Bot. Mus. Leafl. Harvard Univ.*, 30(4): 205 (1986): Telipogon buenaventurae.
carnosiflora *E.S. Siegerist in Bot. Mus. Leafl. Harvard Univ.*, 30(4): 206 (1986)—Peru, Brazil.

DIMERANDRA :-
elegans *(Focke) E.S. Siegerist in Bot. Mus. Leafl. Harvard Univ.*, 30(4): 207 (1986): Isochilus elegans.
latipetala *E.S. Siegerist in Bot. Mus. Leafl. Harvard Univ.*, 30(4): 211 (1986)—Nicaragua, Guatemala, Costa Rica, Panama, Colombia.
tenuicaulis *(Rchb. f.) E.S. Siegerist in Bot. Mus. Leafl. Harvard Univ.*, 30(4): 213 (1986): Epidendrum stenopetalum var. tenuicaule.

DIMORPHANDRA (Legumin.).
cuprea
subsp. ferruginea *(Ducke) M.F. da Silva in Fl. Neotrop.*, 44: 94 (1986): D. ferruginea.
subsp. velutina *(Ducke) M.F. da Silva in Fl. Neotrop.*, 44: 97 (1986): D. velutina.
macrostachya
subsp. congestiflora *(Sprague & Sandwith) M.F. da Silva in Fl. Neotrop.*, 44: 109 (1986): D. congestiflora.
subsp. glabrifolia *(Ducke) M.F. da Silva in Fl. Neotrop.*, 44: 106 (1986): D. glabrifolia.

DIORA P. Ravenna in Op. Bot., 92: 189 (1987). *ANTHERICACEAE.*
cajamarcaensis *(Poelln.) P. Ravenna in Op. Bot.*, 92: 191 (1987): Anthericum cajamarcaense.

DIOSCOREA (Dioscoreac.).
spongiosa *J.Q. Xi, M. Mizuno & W.L. Zhao in Acta Phytotax. Sin.*, 25(1): 52 (1987)—China.

DIOSPYROS (Ebenac.).
crassinervis
subsp. urbaniana *(Leonard) Alain H. Liogier in Phytologia,* 63(1): 65 (1987): Maba urbaniana.

DIPLOCLISIA (Menispermac.).
glaucescens
var. kunstleri *(King) A. Pramanik & K. Thothathri in Kew Bull.*, 42(3): 705 (1987): Cocculus kunstleri.

DIPLOGLOTTIS (Sapindac.).
berniana *S.T. Reynolds in Austrobaileya,* 2(4): 328 (1987)—Australia (Queensland).
obovata *S.T. Reynolds in Austrobaileya,* 2(4): 330 (1987)—Australia (Queensland).

DIPLOMERIS (Orchidac.).
josephii *A. Nageswara Rao & M.S. Swaminathan in Indian Orchid J.*, 2(1): 5 (1987)—India.

DIPLOTAENIA (Umbellif.).
damavandica *Mozaffarian, I.C. Hedge & J.M. Lamond in Fl. Iran.*, 162: 373 (1987)—Iran.

DIPODIUM (Orchidac.).
pulchellum *D.L. Jones & M.A. Clements in Proc. Roy. Soc. Queensl.*, 98: 128 (1987)—Australia (Queensland).

DIPODIUM :–
variegatum *M.A. Clements & D.L. Jones in Proc. Roy. Soc. Queensl.*, 98: 128 (1987)—Australia (Queensland, New South Wales, Victoria).

DIPTEROCARPUS (Dipterocarpac.).
andamanicus *(King) P.K. Tewary & A.K. Sarkar in Indian J. Forest.*, 10(1): 63 (1987): D. turbinatus var. andamanica.

DISCOCACTUS (Cactac.).
buenekeri *W.R. Abraham in Kakt. And. Sukk.*, 38(11): 284 (1987)—Brazil.
kimnachii *G.D. Rowley in Brit. Cact. Succ. J.*, 5(3): 84 (1987), nom. nov.: Nopalxochia horichii *Kimn.*
pseudolatispinus *L. Diers & E. Esteves Pereira in Kakt. And. Sukk.*, 38(10): 247 (1987)—Brazil.
ramulosus
var. angustissimus *(Web.) Kimnach in Cact. Succ. J. (U.S.A.)*, 59(2): 67 (1987): Rhipsalis angustissimus.

DISPHYMA (Aizoac.).
crassifolium
subsp. clavellatum *(Haw.) R.J. Chinnock in Fl. S. Australia*, 1: 194 (1986): Mesembryanthemum clavellatum.

DISPOROPSIS (Convallariac.).
jinfushanensis *Z.Y. Liu in Acta Phytotax. Sin.*, 25(1): 67 (1987)—China.

DISSOSPERMA J. Soják in Čas. Nár. Muz. (Prague), 154(1): 55 (1985 publ. 1986). *PAPAVERACEAE.*
claviculata *(L.) J. Soják in Čas. Nár. Muz. (Prague)*, 154(1): 55 (1985 publ. 1986): Fumaria claviculata.
subsp. picta *(Sampaio) J. Soják in Čas. Nár. Muz. (Prague)*, 154(1): 55 (1985 publ. 1986): Corydalis claviculata var. picta.

DISTEPHANUS (Compos.).
angolensis *(O. Hoffmann) H. Robinson & B. Kahn in Proc. Biol. Soc. Wash.*, 99(3): 498 (1986): Newtonia angolensis.
angulifolius *(DC.) H. Robinson & B. Kahn in Proc. Biol. Soc. Wash.*, 99(3): 499 (1986): Vernonia angulifolia.
anisochaetoides *(Sond) H. Robinson & B. Kahn in Proc. Biol. Soc. Wash.*, 99(3): 499 (1986): Vernonia anisochaetoides.
antandroy *(H. Humbert) H. Robinson & B. Kahn in Proc. Biol. Soc. Wash.*, 99(3): 499 (1986): Vernonia antandroy.
cloiselii *(Moore) H. Robinson & B. Kahn in Proc. Biol. Soc. Wash.*, 99(3): 499 (1986): Vernonia cloiselii.
divaricatus *(Steetz) H. Robinson & B. Kahn in Proc. Biol. Soc. Wash.*, 99(3): 499 (1986): Gongrothamnus divaricatus.
eriophyllus *(Drake) H. Robinson & B. Kahn in Proc. Biol. Soc. Wash.*, 99(3): 499 (1986): Vernonia eriophylla.

DISTEPHANUS :–
forrestii *(Anthony) H. Robinson & B. Kahn in Proc. Biol. Soc. Wash.*, 99(3): 499 (1986): Vernonia forrestii.
garnierianus *(Klatt) H. Robinson & B. Kahn in Proc. Biol. Soc. Wash.*, 99(3): 499 (1986): Vernonia garnieriana.
glandulicinctus *(H. Humb.) H. Robinson & B. Kahn in Proc. Biol. Soc. Wash.*, 99(3): 499 (1986): Vernonia glandulicincta.
glutinosus *(DC.) H. Robinson & B. Kahn in Proc. Biol. Soc. Wash.*, 99(3): 499 (1986): Vernonia glutinosa.
lastellei *(Drake) H. Robinson & B. Kahn in Proc. Biol. Soc. Wash.*, 99(3): 499 (1986): Vernonia lastellei.
mahafaly *(H. Humb.) H. Robinson & B. Kahn in Proc. Biol. Soc. Wash.*, 99(3): 499 (1986): Vernonia mahafaly.
majungensis *(H. Humb.) H. Robinson & B. Kahn in Proc. Biol. Soc. Wash.*, 99(3): 499 (1986): Vernonia majungensis.
malacophytus *(Baker) H. Robinson & B. Kahn in Proc. Biol. Soc. Wash.*, 99(3): 499 (1986): Vernonia malacophyta.
manambolensis *(H. Humb.) H. Robinson & B. Kahn in Proc. Biol. Soc. Wash.*, 99(3): 500 (1986): Vernonia manambolensis.
mangokensis *(H. Humb.) H. Robinson & B. Kahn in Proc. Biol. Soc. Wash.*, 99(3): 500 (1986): Vernonia mangokensis.
nummulariaefolius *(Klatt) H. Robinson & B. Kahn in Proc. Biol. Soc. Wash.*, 99(3): 500 (1986): Decaneurum (Gymnanthemum) nummulariaefolium.
ochroleucus *(Baker) H. Robinson & B. Kahn in Proc. Biol. Soc. Wash.*, 99(3): 500 (1986): Vernonia ochroleuca.
polygalaefolius *(Less.) H. Robinson & B. Kahn in Proc. Biol. Soc. Wash.*, 99(3): 500 (1986): Vernonia polygalaefolia.
rochonioides *(H. Humb.) H. Robinson & B. Kahn in Proc. Biol. Soc. Wash.*, 99(3): 500 (1986): Vernonia rochonioides.
streptocladus *(Baker) H. Robinson & B. Kahn in Proc. Biol. Soc. Wash.*, 99(3): 500 (1986): Vernonia streptoclada.
subluteus *(S. Elliot) H. Robinson & B. Kahn in Proc. Biol. Soc. Wash.*, 99(3): 500 (1986): Vernonia sublutea.
swinglei *(H. Humb.) H. Robinson & B. Kahn in Proc. Biol. Soc. Wash.*, 99(3): 500 (1986): Vernonia swinglei.

DISTERIGMA (Ericac.).
balslevii *J.L. Luteyn in Op. Bot.*, 92: 113 (1987)—Ecuador, Colombia.

DIURANTHERA (Anthericac.).
chinglingensis *J.Q. Xing & T.C. Cui in Acta Bot. Bor.-Occid. Sin.*, 7(3): 203 (1987)—China.

DIURIS (Orchidac.).
chrysantha *D.L. Jones & M.A. Clements in Proc. Roy. Soc. Queensl.*, 98: 130 (1987)—Australia (Queensland, New South Wales).
parvipetala *(Dockr.) D.L. Jones & M.A. Clements in Proc. Roy. Soc. Queensl.*, 98: 132 (1987): D. punctata var. parvipetala.

DODECATHEON (Primulac.).
pulchellum
f. margaritaceum *(Fassett) C.F. Reed in Phytologia*, 63(5): 411 (1987): D. amethystinum f. margaritaceum.
f. strictlerae *(Fern.) C.F. Reed in Phytologia*, 63(5): 411 (1987): D. meadia f. strictlerae.

DOLICHORHYNCHUS I.C. Hedge & Kit Tan in Pl. Syst. Evol., 156(3–4): 197 (1987). *CRUCIFERAE.*
arabicus *I.C. Hedge & Kit Tan in Pl. Syst. Evol.*, 156(3–4): 198 (1987)—Saudi Arabia.

DOMBEYA (Sterculiac.).
longicuspis
var. bosseri *L.C. Barnett & L.J. Dorr in Bull. Mus. Nation. Hist. Nat., B, Adansonia*, Ser. 4, 8(4): 366 (1986 publ. 1987)—Madagascar.
moratii *L.C. Barnett & L.J. Dorr in Bull. Mus. Nation. Hist. Nat., B, Adansonia*, Ser. 4, 8(4): 365 (1986 publ. 1987)—Madagascar.

DORONICUM (Compos.).
kuepferi *R. Chacón Aumente in An. Jard. Bot. Madrid*, 43(2): 269 (1986 publ. 1987)—Spain.

DOROTHEANTHUS (Aizoac.).
bellidiformis
subsp. hestermalensis *H.-D. Ihlenfeldt & M. Struck in Beitr. Biol. Pfl.*, 61(3): 445 (1986 publ. 1987)—South Africa (Cape Province).

DRABA (Crucif.).
altaica
var. modesta *(W.W. Smith) W.T. Wang in Acta Bot. Yunnanica*, 9(1): 5 (1987): D. modesta.
daochengensis *W.T. Wang in Acta Bot. Yunnanica*, 9(1): 7 (1987)—China.
dolichotricha *W.T. Wang in Acta Bot. Yunnanica*, 9(1): 5 (1987)—China.
involucrata
var. lasiocarpa *W.T. Wang in Acta Bot. Yunnanica*, 9(1): 6 (1987)—China.
moupinensis
var. calcicola *(O.E. Schulz) W.T. Wang in Acta Bot. Yunnanica*, 9(1): 4 (1987): D. calcicola.
qinghaiensis *L.L. Lu in Acta Phytotax. Sin.*, 25(4): 319 (1987)—China.
staintonii *Jafri ex H. Hara in Enum. Fl. Pl. Nepal*, 2: 42 (1979)—Nepal.
zangbeiensis *L.L. Lu in Acta Phytotax. Sin.*, 25(4): 320 (1987)—Xizang.

DRACOCEPHALUM (Labiat.).
subsect. Aristodracon *A.L. Budantsev in Bot. Zhurn.*, 72(2): 265 (1987).
ser. Botryoidea *A.L. Budantsev in Bot. Zhurn.*, 72(2): 264 (1987).
sect. Confertodracon *A.L. Budantsev in Bot. Zhurn.*, 72(2): 262 (1987).
ser. Cryptodracon *(Benth.) A.L. Budantsev in Bot. Zhurn.*, 72(2): 265 (1987): D. sect. Cryptodracon.
ser. Discoloria *A.L. Budantsev in Bot. Zhurn.*, 72(2): 264 (1987).
ser. Fruticulosa *Schischk. ex A.L. Budantsev in Bot. Zhurn.*, 72(2): 263 (1987).
ser. Kafiristanica *A.L. Budantsev in Bot. Zhurn.*, 72(2): 264 (1987).
ser. Komaroviana *Schischk. ex A.L. Budantsev in Bot. Zhurn.*, 72(2): 266 (1987).
ser. Multicaulia *Schischk. ex A.L. Budantsev in Bot. Zhurn.*, 72(2): 265 (1987).
ser. Nodulosa *Schischk. ex A.L. Budantsev in Bot. Zhurn.*, 72(2): 263 (1987).
ser. Nutantia *Schischk. ex A.L. Budantsev in Bot. Zhurn.*, 72(2): 263 (1987).
sect. Palmata *A.L. Budantsev in Bot. Zhurn.*, 72(2): 262 (1987).
ser. Peregrina *Schischk. ex A.L. Budantsev in Bot. Zhurn.*, 72(2): 265 (1987).
ser. Pinnata *A.L. Budantsev in Bot. Zhurn.*, 72(2): 264 (1987).
ser. Renatiana *A.L. Budantsev in Bot. Zhurn.*, 72(2): 266 (1987).
ser. Setigera *A.L. Budantsev in Bot. Zhurn.*, 72(2): 266 (1987).
heterophyllum
subsp. ovalifolium *A.L. Budantsev in Bot. Zhurn.*, 72(1): 93 (1987)—China, Mongolia, U.S.S.R. (Siberia).
junatovii *A.L. Budantsev in Bot. Zhurn.*, 72(1): 92 (1987)—Mongolia.
origanoides
subsp. bungeanum *(Schischk. & Serg.) A.L. Budantsev in Bot. Zhurn.*, 72(1): 93 (1987): D. bungeanum.

DRACONIA (Compos.).
campestris
f. oligocephala *(Abromeit) K. Sutorý in Čas. Morav. Muz., Vědy Přir.*, 70: 212 (1985): Artemisia campestris c. oligocephala.

DROGUETIA (Urticac.).
hildebrandtii *I. Friis & C.M. Wilmot-Dear in Nordic J. Bot.*, 7(2): 126 (1987)—Madagascar.
iners
subsp. burchellii *(N.E. Br.) I. Friis & C.M. Wilmot-Dear in Nordic J. Bot.*, 7(2): 126 (1987): D. burchellii.
subsp. urticoides *(Wight) I. Friis & C.M. Wilmot-Dear in Nordic J. Bot.*, 7(2): 126 (1987): Forsskaolea urticoides.

DROSERA (Droserac.).
rechingeri *A. Strid in Pl. Syst. Evol.*, 155(1–4): 343 (1987)—Australia (W. Australia).
slackii *M.R. Cheek in Kew Bull.*, 42(3): 738 (1987)—South Africa (Cape Province).

DUBOUZETIA (Elaeocarpac.).
sect. Oligovula *M.J.E. Coode in Kew Bull.*, 42(4): 803 (1987).
sect. Spongosperma *M.J.E. Coode in Kew Bull.*, 42(4): 809 (1987).
australiensis *M.J.E. Coode in Kew Bull.*, 42(4): 805 (1987)—Australia (N. Terr.).

DUCHESNEA (Rosac.).
×hara-kurosawae *N. Naruhashi & M. Sugimoto in J. Phytogeogr. Taxon.*, 34(1): 11 (1986)—Japan. (D. chrysantha × indica).

DUMASIA (Legumin.).
villosa
subsp. bicolor *(Hayata) H. Ohashi & Y. Tateishi in Sci. Rep. Tohoku Univ.*, Ser. 4, 38(4): 310 (1984): Dumasia bicolor.

DUNBARIA (Legumin.).
maymyoensis *P. Satyanarayana & K. Thothathri in Bull. Bot. Surv. India*, 27(1–4): 152 (1985 publ. 1987)—Burma.

DURANTA (Verbenac.).
arida
subsp. serpentina *R.W. Sanders & W.S. Judd in Moscosoa*, 4: 217 (1986)—Hispaniola (Haiti).

DYCKIA (Bromeliac.).
estevesii *W. Rauh in J. Bromeliad Soc.*, 37(3): 120 (1987)—Brazil.
scrutor *L.B. Smith in J. Bromeliad Soc.*, 37(4): 151 (1987)—Brazil.

DYSOSMA (Berberidac.).
furfuracea *S.Y. Bao in Acta Phytotax. Sin.*, 25(2): 155 (1987)—China.

E

EBRACTEOLA (Aizoac.).
fulleri *(L. Bol.) H.F. Glen in Bothalia*, 16(2): 222 (1986): Ruschia fulleri.
wilmaniae *(L. Bol.) H.F. Glen in Bothalia*, 16(2): 222 (1986): Mesembryanthemum wilmaniae.

ECDYSANTHERA (Apocynac.).
lakhimpurensis *S.K. Srivastava & B.N. Mehrotra in Candollea*, 41(2): 381 (1986)—Assam.

ECHEANDIA (Anthericac.).
coalcomanensis *R.W. Cruden in Contrib. Univ. Mich. Herb.*, 16: 130 (1987)—Mexico.
gentryi *R.W. Cruden in Contrib. Univ. Mich. Herb.*, 16: 132 (1987)—Mexico.
imbricata *R.W. Cruden in Contrib. Univ. Mich. Herb.*, 16: 133 (1987)—Mexico.
macvaughii *R.W. Cruden in Contrib. Univ. Mich. Herb.*, 16: 129 (1987), as 'mcvaughii'—Mexico.
occidentalis *R.W. Cruden in Contrib. Univ. Mich. Herb.*, 16: 132 (1987)—Mexico.
parvicapsulata *R.W. Cruden in Contrib. Univ. Mich. Herb.*, 16: 130 (1987)—Mexico.
pihuamensis *R.W. Cruden in Contrib. Univ. Mich. Herb.*, 16: 131 (1987)—Mexico.
robusta *R.W. Cruden in Contrib. Univ. Mich. Herb.*, 16: 131 (1987)—Mexico.
sinaloensis *R.W. Cruden in Contrib. Univ. Mich. Herb.*, 16: 130 (1987)—Mexico.
udipratensis *R.W. Cruden in Contrib. Univ. Mich. Herb.*, 16: 132 (1987)—Mexico.

ECHINOCEREUS (Cactac.).
nicholii *(L. Benson) B.D. Parfitt in Phytologia*, 63(3): 157 (1987): E. engelmannii var. nicholii.

ECHINOCHLOA (Gramin.).
walteri
var. laevigata *(Wieg.) S.R. Hill in Castanea*, 51(4): 277 (1986): E. walteri f. laevigata.

ECHINOCODON A.A. Kolakovskiĭ in Soobshch. Akad. Nauk Gruz. SSR, 121(2): 387 (1986), non D.Y. Hong (1984). *CAMPANULACEAE.*
primulifolium *(Brot.) A.A. Kolakovskiĭ in Soobshch. Akad. Nauk Gruz. SSR*, 121(2): 387 (1986): Campanula primulifolia.

ECHINOCORYNE H. Robinson in Proc. Biol. Soc. Wash., 100(3): 586 (1987). *COMPOSITAE.*
echinocephala *(H. Robinson) H. Robinson in Proc. Biol. Soc. Wash.*, 100(3): 587 (1987): Vernonia echinocephala.
holosericea *(Mart.) H. Robinson in Proc. Biol. Soc. Wash.*, 100(3): 587 (1987): Vernonia holosericea.
pungens *(Gardn.) H. Robinson in Proc. Biol. Soc. Wash.*, 100(3): 587 (1987): Vernonia pungens.
schwenkiaefolia *(Mart.) H. Robinson in Proc. Biol. Soc. Wash.*, 100(3): 587 (1987): Vernonia schwenkiaefolia.
stricta *(Gardn.) H. Robinson in Proc. Biol. Soc. Wash.*, 100(3): 588 (1987): Vernonia stricta.
subulata *(Baker) H. Robinson in Proc. Biol. Soc. Wash.*, 100(3): 588 (1987): Vernonia subulata.

ECHINODORUS (Alismatac.).
bracteatus
subsp. efenestratus *(Fassett) R.R. Haynes & L.B. Holm-Nielsen in Brittonia,* 38(4): 330 (1986): E. bracteatus var. efenestratus.
cordifolius
subsp. fluitans *(Fassett) R.R. Haynes & L.B. Holm-Nielsen in Brittonia,* 38(4): 329 (1986): E. fluitans.
grandiflorus
subsp. aureus *(Fassett) R.R. Haynes & L.B. Holm-Nielsen in Brittonia,* 38(4): 330 (1986): E. grandiflorus var. aureus.
macrophyllus
subsp. scaber *(Rataj) R.R. Haynes & L.B. Holm-Nielsen in Brittonia,* 38(4): 331 (1986): E. scaber.
reticulatus *R.R. Haynes & L.B. Holm-Nielsen in Brittonia,* 38(4): 327 (1986)—Surinam.
subalatus
subsp. andrieuxii *(Hook. & Arn.) R.R. Haynes & L.B. Holm-Nielsen in Brittonia,* 38(4): 327 (1986): Alisma andrieuxii.

ECHINOLAENA (Gramin.).
oplismenoides *(Munro ex Doell) M.T. Stieber in Syst. Bot.,* 12(2): 212 (1987): Ichnanthus oplismenoides.
standleyi *(A.S. Hitchc.) M.T. Stieber in Syst. Bot.,* 12(2): 214 (1987): Ichnanthus standleyi.

ECHINOPS (Compos.).
taeckholmianus *A. Amin in Candollea,* 42(1): 411 (1987), as 'taeckholmiana'—Egypt.

ECHINOPSIS (Cactac.).
candicans *(Gillies ex Salm-Dyck) F.A.C. Weber ex D. Hunt in Bradleya,* 5: 92 (1987): Cereus candicans.
schieliana *(Backeberg) D. Hunt in Bradleya,* 5: 92 (1987): Lobivia schieliana.
tegeleriana *(Backeberg) D. Hunt in Bradleya,* 5: 92 (1987): Lobivia tegeleriana.
thionantha *(Spegazzini) D. Hunt in Bradleya,* 5: 92 (1987): Echinocactus thionanthus.

EHRHARTA (Gramin.).
thunbergii *G.E. Gibbs Russell in Bothalia,* 17(2): 192 (1987), nom. nov.: Aira villosa *L.f.*

ELAEAGNUS (Elaeagnac.).
calcarea *Z.R. Xu in Guihaia,* 5(4): 348 (1985)—China.
lipoensis *Z.R. Xu in Acta Sci. Nat. Univ. Sunyatseni,* 1987(2): 46 (1987)—China.
parvifolia
var. nuda *(Serv.) D. Basu in Bull. Bot. Surv. India,* 27(1–4): 241 (1985 publ. 1987): E. umbellata var. nuda.
var. pedunculata *D. Basu in J. Econ. Taxon. Bot.,* 8(2): 427 (1986), as 'pedunculatus'—Nepal, India.

ELAEAGNUS :–
var. villosa *(Serv.) D. Basu in Bull. Bot. Surv. India,* 27(1–4): 241 (1985 publ. 1987): E. umbellata var. villosa.
s-stylata *Z.R. Xu in Acta Sci. Nat. Univ. Sunyatseni,* 1987(2): 45 (1987)—China.

ELATOSTEMA (Urticac.).
xinningense *W.T. Wang in Guihaia,* 5(4): 323 (1985)—China.
yungshunense *W.T. Wang in Guihaia,* 5(4): 325 (1985)—China.

ELEOCHARIS (Cyperac.).
ser. Websteria *(S.H. Wright) G.C. Tucker in J. Arnold Arbor.,* 68(4): 387 (1987): gen. Websteria.
abnorma *Y.D. Chen in Bull. Bot. Res. North-East. Forest. Inst.,* 7(2): 124 (1987)—China.
callosa *Y.D. Chen in Bull. Bot. Res. North-East. Forest. Inst.,* 7(2): 123 (1987)—China.
confervoides *(Poiret) G.C. Tucker in J. Arnold Arbor.,* 68(4): 388 (1987): Scirpus confervoides.
erhaiensis *Y.D. Chen in Bull. Bot. Res. North-East. Forest. Inst.,* 7(2): 122 (1987)—China.
glabella *Y.D. Chen in Bull. Bot. Res. North-East. Forest. Inst.,* 7(2): 118 (1987)—China.
kuoi *Y.D. Chen in Bull. Bot. Res. North-East. Forest. Inst.,* 7(2): 120 (1987)—China.
paradoxa *Y.D. Chen in Bull. Bot. Res. North-East. Forest. Inst.,* 7(2): 121 (1987)—China.
penchaoi *Y.D. Chen in Bull. Bot. Res. North-East. Forest. Inst.,* 7(2): 119 (1987)—China.
qinghaiensis *Y.D. Chen in Bull. Bot. Res. North-East. Forest. Inst.,* 7(2): 117 (1987)—China.
rugosa *D.A. Simpson in Kew Bull.,* 42(4): 900 (1987)—Brazil.

ELLEANTHUS (Orchidac.).
hirsutis *K. Barringer in Syst. Bot.,* 12(1): 162 (1987)—Peru.
steyermarkii *K. Barringer in Syst. Bot.,* 12(1): 165 (1987)—Ecuador.

×ELYHORDEUM (Gramin.).
macounii *(Vasey) M.E. Barkworth & D.R. Dewey in Great Basin Nat.,* 43(4): 570 (1983): Elymus macounii.

×ELYLEYMUS (Gramin.).
aristatus *(Merrill) M.E. Barkworth & D.R. Dewey in Great Basin Nat.,* 43(4): 570 (1983): Elymus aristatus.

ELYMUS (Gramin.).
acutus *(DC.) M.-A. Thiébaud in Candollea* 42(1): 340 (1987): Triticum acutum.
dentatus
var. elatus *(Hook. f.) G. Singh in J. Econ. Taxon. Bot.,* 8(2): 497 (1986): Agropyron dentatum var. elatum.
var. kashmiricus *(Melderis) G. Singh in J. Econ. Taxon. Bot.,* 8(2): 497 (1986): Agropyron dentatus var. kashmiricum.

ELYMUS :–
var. scabrus *(Nevski) G. Singh in J. Econ. Taxon. Bot.,* 8(2): 497 (1986): Agropyron dentatus var. scabrum.
hispanicus *(Boiss.) S. Talavera in Lagascalia,* 14(1): 170 (1986): Agropyrum panormitanus var. hispanicus.
lanceolatus
subsp. albicans *(Scribner & J.G. Smith) M.E. Barkworth & D.R. Dewey in Great Basin Nat.,* 43(4): 568 (1983): Agropyron albicans.
×pseudorepens *(Scribner & J.G. Smith) M.E. Barkworth & D.R. Dewey in Great Basin Nat.,* 43(4): 568 (1983): Agropyron pseudorepens.
repens
var. aristatus *(Baumgarten) Melderis & D. McClintock in D. McClintock, Suppl. Wild Flow. Guernsey:* 48 (1987): Agropyron repens var. aristatum.
semicostatus
var. thomsonii *(Hook. f.) G. Singh in J. Econ. Taxon. Bot.,* 8(2): 498 (1986): Agropyron semicostatum var. thomsonii.
var. validus *(Melderis) G. Singh in J. Econ. Taxon. Bot.,* 8(2): 498 (1986): Agropyron striatum var. validum.
transhyrcanus
var. togarovii *A. Ataeva in Izv. Akad. Nauk Turkm. SSR, Biol. Nauk,* 1987(3): 54 (1987)—U.S.S.R. (Middle Asia).
violaceus *(Hornem.) J. Feilberg in Medd. Grønland, Biosci.,* 15: 12 (1984): Triticum violaceum.

ELYTRARIA (Acanthac.).
mexicana *P.A. Fryxell & S.D. Koch in Aliso,* 11(4): 541 (1987)—Mexico.

ELYTRIGIA (Gramin.).
×acuta
nothosubsp. obtusiuscula *(Lange) M. Kerguélen in Lejeunia,* 120: 86 (1987): Agropyron × obtusiusculum.
araucana *(L. Parodi) Covas ex J.H. Hunziker & C.C. Xifreda in Darwiniana,* 27(1–4): 561 (1986): Agropyron attenuatum var. araucanum.
attenuata *(Kunth) Covas ex J.H. Hunziker & C.C. Xifreda in Darwiniana,* 27(1–4): 562 (1986): Triticum attenuatum.
breviaristata *(A. Hitchc.) Covas ex J.H. Hunziker & C.C. Xifreda in Darwiniana,* 27(1–4): 562 (1986): Agropyron breviaristatum.
fuegiana *(Speg.) Covas ex J.H. Hunziker & C.C. Xifreda in Darwiniana,* 27(1–4): 562 (1986): Triticum fuegianum.
intermedia
var. kugitangi *A. Ataeva in Izv. Akad. Nauk Turkm. SSR, Biol. Nauk,* 1987(3): 54 (1987)—U.S.S.R. (Middle Asia).
subsp. latronum *(Godron) M. Kerguélen in Lejeunia,* 120: 87 (1987): Triticum latronum.

ELYTRIGIA :–
×laxa *(Fries) M. Kerguélen in Lejeunia,* 120: 88 (1987): Triticum × laxum.
mendocina *(L. Parodi) Covas ex J.H. Hunziker & C.C. Xifreda in Darwiniana,* 27(1–4): 562 (1986): Agropyron mendocinum.
patagonica *(Speg.) Covas ex J.H. Hunziker & C.C. Xifreda in Darwiniana,* 27(1–4): 562 (1986): Triticum fuegianum var. patagonicum.
pulcherrima
var. glabra *A. Ataeva in Izv. Akad. Nauk Turkm. SSR, Biol. Nauk,* 1987(3): 54 (1987)—U.S.S.R. (Middle Asia).
scabrifolia *(Doell) Covas ex J.H. Hunziker & C.C. Xifreda in Darwiniana,* 27(1–4): 562 (1986): Triticum repens var. scabrifolium.
scabriglumis *(Hackel) Covas ex J.H. Hunziker & C.C. Xifreda in Darwiniana,* 27(1–4): 563 (1986): Agropyron repens var. scabriglume.
tilcarensis *(J.H. Hunz.) Covas ex J.H. Hunziker & C.C. Xifreda in Darwiniana,* 27(1–4): 563 (1986): Agropyron tilcarense.
trichophora
var. glabra *A. Ataeva in Izv. Akad. Nauk Turkm. SSR, Biol. Nauk,* 1987(3): 54 (1987)—U.S.S.R. (Middle Asia).

EMBELIA (Myrsinac.).
brassii *H. Sleumer in Blumea,* 32(2): 390 (1987)—New Guinea (Papua New Guinea).
kuborensis *H. Sleumer in Blumea,* 32(2): 392 (1987)—New Guinea (Papua New Guinea).
ottoniana *H. Sleumer in Blumea,* 32(2): 389 (1987)—New Guinea (Papua New Guinea).
pullenii *H. Sleumer in Blumea,* 32(2): 392 (1987)—New Guinea (Papua New Guinea).
tsjeriam-cottam
var. ferruginea *(Wall. ex Cl.) K. Haridasan & R.R. Rao, Forest Fl. Meghalaya,* 2: 559 (1987): E. robusta var. ferruginea.

EMILIA (Compos.).
khaopawtaensis *H. Koyama in Acta Phytotax. Geobot.,* 37(4–6): 114 (1986)—Thailand, Malaya.

ENCYCLIA (Orchidac.).
isochila *(Rchb. f.) D.D. Dod in Moscosoa,* 4: 193 (1986): Epidendrum isochilum.
var. tridens *(Rchb. f.) D.D. Dod in Moscosoa,* 4: 193 (1986): Epidendrum isochilum var. tridens.
papillosa *(Batem.) I. Aguirre-Olavarrieta in Orquídea (México),* 10(2): 387 (1987): Epidendrum papillosum.

EPACRIS (Epacridac.).
apsleyensis *R.K. Crowden in Pap. Proc. Roy. Soc. Tasmania,* 120: 17 (1986)—Tasmania.
grandis *R.K. Crowden in Pap. Proc. Roy. Soc. Tasmania,* 120: 19 (1986)—Tasmania.

EPIDENDRUM (Orchidac.).
appendiculatum *T. Hashimoto in Bull. Nation. Sci. Mus., B (Tokyo),* 13(1): 36 (1987)—Peru.

EPIDENDRUM :–
aristatum *J.D. Ackerman & A.M. Montalvo in Selbyana,* 9(1): 126 (1986)—Ecuador.
brachybotrys *J.D. Ackerman & A.M. Montalvo in Selbyana,* 9(1): 129 (1986)—Panama.
dressleri *E. Hágsater in Orquídea (México),* 10(2): 340 (1987)—Mexico.
greenwoodii *E. Hágsater in Orquídea (México),* 10(2): 338 (1987)—Mexico, Guatemala, Salvador.
macroclinium *E. Hágsater in Orquídea (México),* 10(2): 319 (1987)—Mexico, Belize, Costa Rica, Guatemala, Nicaragua, Panama.
nelsonii *E. Hágsater in Orquídea (México),* 10(2): 342 (1987)—Guatemala.

EPILOBIUM (Onagrac.).
clavatotrichon *(J. Rohlena) A.W. Hill in Index Kewensis,* Supp. 6: 76 (1916–1920 publ. 1926): E. palustre var. clavatotrichon.

EPIMEDIUM (Berberidac.).
reticulatum *C.Y. Wu ex S.Y. Bao in Acta Phytotax. Sin.,* 25(2): 156 (1987)—China.

EPIPACTIS (Orchidac.).
helleborine
subsp. latina *W. Rossi & E. Klein in Orchidee,* 38(2): 93 (1987)—Italy.

EPIPOGIUM (Orchidac.).
sect. Roseum *A. Nageswara Rao in Indian Orchid J.,* 2(1): 15 (1987).

ERAGROSTIS (Gramin.).
gloeophylla *S.M. Phillips in Kew Bull.,* 42(4): 929 (1987)—Somali Republic, Ethiopia.
schweinfurthii
var. kiwuensis *(Jedw.) S.M. Phillips in Kew Bull.,* 42(4): 930 (1987): E. kiwuensis.

ERANTHIS (Ranunculac.).
sect. Shibateranthis *(Nakai) M. Tamura in Acta Phytotax. Geobot.,* 38: 96 (1987): gen. Shibateranthis.

EREMOCAULON T.R. Soderstrom & X. Londoño in Amer. J. Bot., 74(1): 37 (1987). *GRAMINEAE.*
aureofimbriatum *T.R. Soderstrom & X. Londoño in Amer. J. Bot.,* 74(1): 37 (1987)—Brazil.

EREMOPHILA (Myoporac.).
deserti *(Cunn. ex Benth.) R.J. Chinnock in Fl. S. Australia,* 3: 1338 (1986): Myoporum deserti.

ERIA (Orchidac.).
chandrasekharanii *(Bhargavan & Mohanan) C. Sathish Kumar & K.S. Manilal in Taxon,* 35(4): 720 (1986): Porpax chandrasekharanii.
medogensis *S.C. Chen & Z.H. Tsi in Acta Phytotax. Sin.,* 25(5): 329 (1987)—Xizang.

ERIA :–
muscicola
var. ponmudiana *M. Mohanan & A.N. Henry in J. Econ. Taxon. Bot.,* 8(2): 425 (1986)—India.

ERIGERON (Compos.).
awapensis *S.L. Welsh in Great Basin Nat.,* 43(2): 365 (1983)—U.S.A. (Utah).
canaani *S.L. Welsh in Great Basin Nat.,* 43(2): 366 (1983)—U.S.A. (Utah).
carringtoniae *S.L. Welsh in Great Basin Nat.,* 43(2): 366 (1983), as 'carringtonae'—U.S.A. (Utah).
goodrichii *S.L. Welsh in Great Basin Nat.,* 43(2): 366 (1983)—U.S.A. (Utah).
maguirei
var. harrisonii *S.L. Welsh in Great Basin Nat.,* 43(2): 367 (1983)—U.S.A. (Utah).
punjabensis *T. Akhtar & M.N. Chaudhri in Pakistan Syst.,* 3(1): 23 (1987)—Pakistan.
speciosus
var. mollis *(Gray) S.L. Welsh in Great Basin Nat.,* 43(2): 278 (1983): E. glabellus var. mollis.
var. uintahensis *(Cronq.) S.L. Welsh in Great Basin Nat.,* 43(2): 279 (1983): E. uintahensis.
untermanniorum *S.L. Welsh in Great Basin Nat.,* 43(2): 367 (1983), as 'untermannii'—U.S.A. (Utah).
wahwahensis *S.L. Welsh in Great Basin Nat.,* 43(2): 368 (1983)—U.S.A. (Utah).

ERIOCAULON (Eriocaulac.).
bolei *P.V. Bole & M.R. Almeida in J. Bombay Nat. Hist. Soc.,* 83(3): 600 (1986 publ. 1987)—India.
nantoense
var. satsumense *S. Hatusima & T. Sakata in J. Phytogeogr. Taxon.,* 34(2): 64 (1986)—Japan.

ERIOCHLOA (Gramin.).
acuminata
var. minor *(Vasey) R.B. Shaw in Sida,* 12(1): 174 (1987): E. punctata var. minor.
aristata
var. boxiana *(Hitchc.) R.B. Shaw in Sida,* 12(1): 177 (1987): E. boxiana.
setosa
var. ekmanii *(Hitchc.) R.B. Shaw in Sida,* 12(1): 202 (1987): Eriochloa ekmanii.

ERIONEURON (Gramin.).
avenaceum
var. kurtzianum *(L.R. Parodi) A.M. Anton in Kurtziana,* 10: 62 (1977): Tridens avenacea var. kurtziana.
var. longiglume *(L.R. Parodi) A.M. Anton in Kurtziana,* 10: 65 (1977): Tridens avenacea var. longiglumis.
var. pygmaeum *(Hackel) A.M. Anton in Kurtziana,* 10: 63 (1977): Triodia avenacea var. pygmaea.

ERIONEURON :–
pilosum
var. longearistatum *(Kurtz) A.M. Anton in Kurtziana,* 10: 58 (1977): Triodia avenacea var. longearistata.

ERVATAMIA (Apocynac.).
balansae
var. macrophylla *Trân Đính Lý in Feddes Repert.*, 97(7–8): 451 (1986)—Vietnam.

ERYNGIUM (Umbellif.).
rechingeri *S.G. Tamamschian & M.G. Pimenov in Fl. Iran.*, 162: 55 (1987)—Afghanistan.

ERYSIMUM (Crucif.).
sect. Canescentia *V.I. Dorofeev in Bot. Zhurn.*, 72(11): 1538 (1987).
subsect. Canescentia *V.I. Dorofeev in Bot. Zhurn.*, 72(11): 1538 (1987).
sect. Contracta *V.I. Dorofeev in Bot. Zhurn.*, 72(11): 1537 (1987).
subgen. Cuspidaria *(DC.) V.I. Dorofeev in Novosti Sist. Vyssh. Rast.*, 23: 66 (1986): E. sect. Cuspidaria.
subsect. Cylindrica *V.I. Dorofeev in Bot. Zhurn.*, 72(11): 1542 (1987).
subsect. Gelida *V.I. Dorofeev in Bot. Zhurn.*, 72(11): 1540 (1987).
sect. Hieracifolia *V.I. Dorofeev in Bot. Zhurn.*, 72(11): 1539 (1987).
subsect. Iberica *V.I. Dorofeev in Bot. Zhurn.*, 72(11): 1538 (1987).
sect. Pachygonum *(C.A. Mey.) V.I. Dorofeev in Bot. Zhurn.*, 72(11): 1540 (1987): E. subgen. Pachygonum.
subsect. Subangulata *V.I. Dorofeev in Bot. Zhurn.*, 72(11): 1542 (1987).
gelidum
subsp. krynitzkii *(Bordz.) V.I. Dorofeev in Bot. Zhurn.*, 72(11): 1541 (1987): E. krynitzkii.
mayorii *M.J. Álvarez Martínez, H.S. Nava & M.A. Fernández Casado in An. Jard. Bot. Madrid,* 43(1): 174 (1986)—Spain.
substrigosum
subsp. meyerianum *(Rupr.) V.I. Dorofeev in Bot. Zhurn.*, 72(11): 1537 (1987): Erysimastrum meyerianum.

ERYTHROCOCCA (Euphorbiac.).
atrovirens
var. flaccida *(Pax) A. Radcliffe-Smith in Kew Bull.*, 42(2): 395 (1987): Claoxylon flaccidum.
uniflora *M.G. Gilbert in Kew Bull.*, 42(2): 363 (1987)—Ethiopia.

ERYTHRORCHIS (Orchidac.).
cassythoides *(A. Cunn. ex Lindl.) L.A. Garay in Bot. Mus. Leafl. Harvard Univ.*, 30(4): 234 (1986): Dendrobium cassythoides.
ochobiensis *(Hayata) L.A. Garay in Bot. Mus. Leafl. Harvard Univ.*, 30(4): 234 (1986): Galeola ochobiensis.

ERYTHROXYLUM (Erythroxylac.).
andrei *T. Plowman in Fieldiana, Bot., n.s.*, 19: 2 (1987)—Brazil.
caatingae *T. Plowman in Fieldiana, Bot., n.s.*, 19: 5 (1987)—Brazil.
maracasense *T. Plowman in Fieldiana, Bot., n.s.*, 19: 8 (1987)—Brazil.
mattos-silvae *T. Plowman in Fieldiana, Bot., n.s.*, 19: 12 (1987)—Brazil.
membranaceum *T. Plowman in Fieldiana, Bot. n.s.*, 19: 17 (1987)—Brazil.
petrae-caballi *T. Plowman in Fieldiana, Bot., n.s.*, 19: 20 (1987)—Brazil.
santosii *T. Plowman in Fieldiana, Bot., n.s.*, 19: 25 (1987)—Brazil.
splendidum *T. Plowman in Fieldiana, Bot., n.s.*, 19: 25 (1987)—Brazil.
stipulosum *T. Plowman in Fieldiana, Bot., n.s.*, 19: 31 (1987)—Brazil.
tenue *T. Plowman in Fieldiana, Bot., n.s.*, 19: 35 (1987)—Brazil.

ESCHWEILERA (Lecythidac.).
laevicarpa *S.A. Mori in Mem. New York Bot. Gard.*, 44: 32 (1987)—Brazil, French Guiana, Peru.
pedicellata *(Richard) S.A. Mori in Mem. New York Bot. Gard.*, 44: 34 (1987): Lecythis pedicellata.
squamata *S.A. Mori in Mem. New York Bot. Gard.*, 44: 33 (1987)—French Guiana.

EUCALYPTUS (Myrtac.).
deuaensis *D.J. Boland & P.M. Gilmour in Brunonia,* 9(1): 105 (1987)—Australia (New South Wales).
virens *M.I.H. Brooker & A.R. Bean in Brunonia,* 9(2): 223 (1987)—Australia (Queensland).

EUCHARIS (Amaryllidac.).
cyaneosperma *A.W. Meerow in Sida,* 12(1): 39 (1987)—Peru, Bolivia, Brazil.
formosa *A.W. Meerow in Sida,* 12(1): 42 (1987)—Ecuador, Colombia, Peru.
moorei *(Baker) A.W. Meerow in Sida,* 12(1): 35 (1987): E. grandiflora var. moorei.

EUCLEA (Ebenac.).
dewinteri *E. Retief in Bothalia,* 16(2): 228 (1986)—South Africa (Transvaal).

EUGENIA (Myrtac.).
batingabranca *M. Sobral in Napaea, Rev. Bot.*, 1: 23 (1987)—Brazil.
cacuminis *Alain H. Liogier in Phytologia,* 61(6): 356 (1986)—Puerto Rico.
constanzae *Alain H. Liogier in Phytologia,* 61(6): 357 (1986)—Hispaniola (Dominican Republic).
higueyana *Alain H. Liogier in Phytologia,* 61(6): 357 (1986)—Hispaniola (Dominican Republic).
holdridgei *Alain H. Liogier in Phytologia,* 61(6): 358 (1986)—Hispaniola (Haiti).

EUGENIA :–
incertissima *M. Sobral in Candollea,* 42(1): 109 (1987), nom. nov.: E. obtusa *Barb. Rodr. ex Chodat & Hassler.*
jimenezii *Alain H. Liogier in Phytologia,* 61(6): 358 (1986)—Hispaniola (Dominican Republic).
mosenii *(Kausel) M. Sobral in Napaea, Rev. Bot.,* 1: 25 (1987): Myrceugenia mosenii.
padronii *Alain H. Liogier in Phytologia,* 61(6): 359 (1986)—Puerto Rico.
samanensis *Alain H. Liogier in Phytologia,* 61(6): 359 (1986)—Hispaniola (Dominican Republic).
sanjuanensis *P.E. Sánchez Vindas in Phytologia,* 63(5): 402 (1987)—Nicaragua.
sotoesparzae *P.E. Sánchez Vindas in Phytologia,* 61(3): 139 (1986)—Mexico.
uxpanapensis *P.E. Sánchez Vindas & L.M. Ortega Torres in Phytologia,* 63(5): 404 (1987)—Mexico.

EUONYMUS (Celastrac.).
hernandezii *C.L. Lundell in Phytologia,* 63(2): 73 (1987)—Mexico.
lipoensis *Z.R. Xu in Acta Sci. Nat. Univ. Sunyatseni,* 1987(2): 44 (1987)—China.
verrucosoides
var. apertus *(Loes.) C.Y. Wu in Fl. Xizangica,* 3: 121 (1986), without basionym ref.: E. striatus var. apertus.

EUPATORIUM (Compos.).
maculatum
subsp. bruneri *(Gray) G.W. Douglas in Canad. J. Bot.,* 64(11): 2726 (1986): E. bruneri.

EUPHORBIA (Euphorbiac.).
subgen. Lacanthis *(Rafinesque) M.G. Gilbert in Kew Bull.,* 42(1): 238 (1987): gen. Lacanthis.
ser. Lutescentes *(Prokh.) A. Radcliffe-Smith in Kew Mag.,* 4(3): 113 (1987): E. subsect. Lutescentes.
ser. Purpuratae *(Prokh.) A. Radcliffe-Smith in Kew Mag.,* 4(3): 113 (1987): E. subsect. Purpuratae.
atroflora *S. Carter in Kew Bull.,* 42(2): 393 (1987)—Kenya.
baga
var. parvifolia *L.E. Newton in Bot. J. Linn. Soc.,* 94(4): 464 (1987)—Ghana.
biharamulensis *S. Carter in Kew Bull.,* 42(3): 675 (1987)—Tanzania.
borenensis *M.G. Gilbert in Kew Bull.,* 42(2): 391 (1987)—Ethiopia, Kenya.
breviarticulata
var. trunciformis *S. Carter in Kew Bull.,* 42(3): 676 (1987)—Kenya, Ethiopia.
bussei
var. kibwezensis *(N.E. Br.) S. Carter in Kew Bull.,* 42(3): 678 (1987): E. kibwezensis.
bwambensis *S. Carter in Kew Bull.,* 42(3): 681 (1987)—Uganda.

EUPHORBIA :–
caespitosa
var. ventanicola *(Croizat) R. Subils in Kurtziana,* 10: 209 (1977): E. caespitosa subsp. ventanicola.
candelabrum
var. bilocularis *(N.E. Br.) S. Carter in Kew Bull.,* 42(3): 681 (1987): E. bilocularis.
catamarcensis *(Croizat) R. Subils in Kurtziana,* 10: 150 (1977): Chamaesyce catamarcensis.
cataractarum *S. Carter in Kew Bull.,* 42(2): 379 (1987)—Tanzania, Zambia.
collina
var. andina *(Phil.) R. Subils in Kurtziana,* 10: 214 (1977): E. andina.
var. chrysophylla *(Klotzsch & Garcke) R. Subils in Kurtziana,* 10: 215 (1977): Tithymalus chrysophyllus.
var. glaucescens *(Chod. & Wilczek) R. Subils in Kurtziana,* 10: 217 (1977): E. portulacoides var. glaucescens.
var. minor *(Hicken) R. Subils in Kurtziana,* 10: 218 (1977): E. portulacoides var. minor.
var. nahuelhuapina *(Croizat) R. Subils in Kurtziana,* 10: 219 (1977): E. portulacoides subsp. nahuelhuapina.
var. patagonica *(Hieron.) R. Subils in Kurtziana,* 10: 219 (1977): E. patagonica.
var. spathulata *(Chod. & Wilcz.) R. Subils in Kurtziana,* 10: 220 (1977): E. portulacoides var. spathulata.
cryptocaulis *M.G. Gilbert in Kew Bull.,* 42(1): 231 (1987)—Ethiopia.
cuprispina *S. Carter in Kew Bull.,* 42(2): 378 (1987)—Kenya.
dalettiensis *M.G. Gilbert in Kew Bull.,* 42(2): 391 (1987)—Ethiopia.
dauana *S. Carter in Kew Bull.,* 42(2): 376 (1987)—Kenya.
davidii *R. Subils in Kurtziana,* 17: 125 (1984)—Argentina.
decaryi
var. spirosticha *W. Rauh & G. Buchloh in Cact. Succ. J. (U.S.A.),* 59(1): 9 (1987)—Madagascar.
discrepans *S. Carter in Kew Bull.,* 42(2): 380 (1987)—Tanzania.
exilispina *S. Carter in Kew Bull.,* 42(2): 382 (1987)—Tanzania.
fissispina *Bally & S. Carter in Kew Bull.,* 42(2): 375 (1987)—Ethiopia.
greenwayi
subsp. breviaculeata *S. Carter in Kew Bull.,* 42(2): 381 (1987)—Tanzania.
heterochroma
subsp. tsavoensis *S. Carter in Kew Bull.,* 42(2): 387 (1987)—Kenya.
heterospina *S. Carter in Kew Bull.,* 42(2): 389 (1987)—Kenya, Uganda.
subsp. baringoensis *S. Carter in Kew Bull.,* 42(2): 390 (1987)—Kenya.
hieronymi *R. Subils in Kurtziana,* 10: 221 (1977), nom. nov.: E. dioica *Hieronymus.*

EUPHORBIA :–
inaequilatera
var. spanothrix *S. Carter in Kew Bull.,* 42(2): 370 (1987)—Kenya, Tanzania.
laikipiensis *S. Carter in Kew Bull.,* 42(2): 373 (1987)—Kenya.
magnicapsula *S. Carter in Kew Bull.,* 42(3): 678 (1987)—Tanzania, Kenya.
var. lacertosa *S. Carter in Kew Bull.,* 42(3): 679 (1987)—Kenya, Uganda.
monadenioides *M.G. Gilbert in Kew Bull.,* 42(1): 233 (1987)—Ethiopia, Uganda.
nyassae
subsp. mentiens *S. Carter in Kew Bull.,* 42(2): 379 (1987)—Zambia, Tanzania.
nyikae
var. neovolkensii *(Pax) S. Carter in Kew Bull.,* 42(3): 677 (1987): E. neovolkensii.
oranensis *(Croizat) R. Subils in Kurtziana,* 10: 144 (1977): Chamaesyce oranensis.
papillosa
var. erythrorrhiza *(Boiss.) R. Subils in Kurtziana,* 10: 230 (1977): E. erythrorrhiza.
polygalifolia
subsp. vasconcensis *Vivant ex M. Kerguélen in Lejeunia,* 120: 93 (1987)—France.
portucasadiana *(Croizat) R. Subils in Kurtziana,* 10: 163 (1977): Chamaesyce portucasadiana.
scarlatina *S. Carter in Kew Bull.,* 42(2): 392 (1987)—Kenya.
schillingii *A. Radcliffe-Smith in Kew Mag.,* 4(3): 112 (1987)—Nepal.
selousiana *S. Carter in Kew Bull.,* 42(2): 369 (1987)—Tanzania.
similiramea *S. Carter in Kew Bull.,* 42(2): 371 (1987)—Tanzania, Kenya.
taruensis *S. Carter in Kew Bull.,* 42(2): 377 (1987)—Kenya.
tenuispinosa
var. robusta *Bally & S. Carter in Kew Bull.,* 42(2): 377 (1987)—Kenya.

EUPHRASIA (Scrophulariac.).
alpina
subsp. asturica *(Pugsley) H.S. Nava Fernández & M.A. Fernández Casado in An. Jard. Bot. Madrid,* 43(1): 180 (1986): E. asturica.
amphisysepala *W.R. Barker in J. Adelaide Bot. Gard.,* 10(2): 204 (1987)—Tasmania.

EUPHRONIA (Trigoniac.).
acuminatissima *J.A. Steyermark in Ann. Missouri Bot. Gard.,* 74(1): 91 (1987)—Venezuela.

×**EURYGRAECUM** hort. in Orchid Rev., 95(1121): centre page pull-out p. 9 (1987). *ORCHIDACEAE.* (Angraecum × Eurychone).

EUTHRYPTOCHLOA T.A. Cope in Kew Bull., 42(3): 707 (1987). *GRAMINEAE.*
longiligula *T.A. Cope in Kew Bull.,* 42(3): 707 (1987)—China.

EUTREMA (Crucif.).
edwardsii
var. heterophyllum *(W.W. Smith) W.T. Wang in Acta Bot. Yunnanica,* 9(1): 18 (1987): Braya heterophylla.
koreana *(Nak.) K. Hammer in Kulturpflanze,* 34: 98 (1986): Wasabia koreana.

EVODIELLA (Rutac.).
muelleri
f. dinggi *J.F. Veldkamp & R.J. Rouwenhorst in Blumea,* 32(1): 203 (1987), nom. nov.: E. hooglandii *B.L. v.d. Linden.*
f. kanange *J.F. Veldkamp & R.J. Rouwenhorst in Blumea,* 32(1): 204 (1987)—New Guinea (Papua New Guinea).
velutina *J.F. Veldkamp in Blumea,* 32(1): 205 (1987)—New Guinea (Papua New Guinea).

EXODECONUS (Solanac.).
maritima *(Benth.) W.G. D'Arcy in Solanaceae Newsl.,* 2(4): 19 (1986): Thinogeton maritima.

F

FALLOPIA (Polygonac.).
ciliinervis *(Nak.) K. Hammer in Kulturpflanze,* 34: 99 (1986): Pleuropterus ciliinervis.

FARGESIA (Gramin.).
qinlingensis *T.P. Yi & J.X. Shao in J. Bamboo Res.,* 6(1): 42 (1987)—China.

FARSETIA (Crucif.).
aegyptia
subsp. edgeworthii *(Hook. f. & Thoms.) B. Jonsell in Symb. Bot. Upsal.,* 25(3): 43 (1986): F. edgeworthii.
cornus-africani *B. Jonsell in Symb. Bot. Upsal.,* 25(3): 70 (1986)—Somali Republic.
dhofarica *B. Jonsell & A.G. Miller in Symb. Bot. Upsal.,* 25(3): 98 (1986)—Oman, South Yemen.
divaricata *B. Jonsell in Symb. Bot. Upsal.,* 25(3): 49 (1986)—Somali Republic.
emarginata *B. Jonsell in Symb. Bot. Upsal.,* 25(3): 38 (1986)—Somali Republic.
latifolia *B. Jonsell & A.G. Miller in Symb. Bot. Upsal.,* 25(3): 57 (1986)—Oman.
nummularia *B. Jonsell in Symb. Bot. Upsal.,* 25(3): 44 (1986)—Ethiopia, Kenya.
pedicellata *B. Jonsell in Symb. Bot. Upsal.,* 25(3): 51 (1986)—Somali Republic.
spinulosa *B. Jonsell in Symb. Bot. Upsal.,* 25(3): 49 (1986)—Somali Republic.
stenoptera
subsp. boivinii *(Fournier) B. Jonsell in Symb. Bot. Upsal.,* 25(3): 95 (1986): F. boivinii.

FARSETIA :–
tenuisiliqua *B. Jonsell in Symb. Bot. Upsal.*, 25(3): 55 (1986)—Somali Republic.

FERULA (Umbellif.).
behboudiana *(Rech. f. & Esfand.) D.F. Chamberlain in Fl. Iran.*, 162: 395 (1987): Smyrniopsis behboudiana.
dictyocarpa *K.H. Rechinger in Fl. Iran.*, 162: 423 (1987)—Afghanistan.
gabrielii *K.H. Rechinger in Fl. Iran.*, 162: 398 (1987)—Iran.
ghorana *K.H. Rechinger in Fl. Iran.*, 162: 424 (1987)—Afghanistan.
kandaharica *K.H. Rechinger in Fl. Iran.*, 162: 424 (1987)—Afghanistan.
kashanica *K.H. Rechinger in Fl. Iran.*, 162: 400 (1987)—Iran.
lutensis *K.H. Rechinger in Fl. Iran.*, 162: 396 (1987)—Iran.
pachycaulos *K.H. Rechinger in Fl. Iran.*, 162: 398 (1987)—Afghanistan.
persica
var. latisecta *D.F. Chamberlain in Fl. Iran.*, 162: 401 (1987), without latin descr.—Iran.
racemosoumbellata *(Gilli) K.H. Rechinger in Fl. Iran.*, 162: 401 (1987): Johrenia racemoso-umbellata.
rechingeri *D.F. Chamberlain in Fl. Iran.*, 162: 422 (1987)—Afghanistan.
songarica
subsp. mongolica *V.M. Vinogradova & R.V. Kamelin in Novosti Sist. Vŷssh. Rast.*, 23: 101 (1986)—Mongolia.
tabasensis *K.H. Rechinger in Fl. Iran.*, 162: 419 (1987)—Iran.
trachelocarpa *K.H. Rechinger in Fl. Iran.*, 162: 425 (1987)—Afghanistan.
xanthocarpa *K.H. Rechinger in Fl. Iran.*, 162: 394 (1987)—Afghanistan.

FERULAGO (Umbellif.).
sect. Aucheria *M.G. Pimenov & L.P. Tomkovich in Bot. Zhurn.*, 72(7): 968 (1987).
sect. Bernardia *M.G. Pimenov & L.P. Tomkovich in Bot. Zhurn.*, 72(7): 968 (1987).
sect. Fieldianae *M.G. Pimenov & L.P. Tomkovich in Bot. Zhurn.*, 72(7): 971 (1987).
subgen. Galbanifera *M.G. Pimenov & L.P. Tomkovich in Bot. Zhurn.*, 72(7): 968 (1987).
sect. Humiles *M.G. Pimenov & L.P. Tomkovich in Bot. Zhurn.*, 72(7): 968 (1987).
sect. Lophosciadium *(DC.) M.G. Pimenov & L.P. Tomkovich in Bot. Zhurn.*, 72(7): 968 (1987): gen. Lophosciadium.
sect. Uloptera *(Fenzl) M.G. Pimenov & L.P. Tomkovich in Bot. Zhurn.*, 72(7): 971 (1987): gen. Uloptera.
angulata
subsp. carduchorum *(Boiss. & Hausskn.) D.F. Chamberlain in Fl. Iran.*, 162: 432 (1987): F. carduchorum.

FESTUCA (Gramin.).
subgen. Erosiflorae *E.B. Alekseev in Novosti Sist. Vŷssh. Rast.*, 23: 11 (1986).
ampla
subsp. simplex *(Pérez Lara) J.A. Devesa in Lagascalia,* 14(1): 164 (1986): F. scaberrima var. simplex.
atlantica
subsp. oxyphylla *(Litard. & Maire) A.M. Romo in Collect. Bot.*, 16(2): 337 (1986): F. atlantica var. oxyphylla.
atlantigena *(Litar.) A.M. Romo in Collect. Bot.*, 16(2): 338 (1986): F. rubra var. atlantigena.
camerunensis *E.B. Alekseev in Bot. Zhurn.*, 72(9): 1267 (1987)—Cameroun.
cordubensis *J.A. Devesa in Lagascalia,* 14(1): 165 (1986)—Spain.
djurdjurae *(Battandier & Trabut) A.M. Romo in Collect. Bot.*, 16(2): 338 (1986): F. duriuscula f. djurdjurae.
dyris *(Maire & Trabut) A.M. Romo in Collect. Bot.*, 16(2): 338 (1986): F. alpina var. dyris.
elgonensis *E.B. Alekseev in Bot. Zhurn.*, 72(9): 1266 (1987)—Uganda, Sudan.
embergeri *(Litard. & Maire) A.M. Romo in Collect. Bot.*, 16(2): 338 (1986): F. ovina var. embergeri.
fontqueriana *(St.-Yves) A.M. Romo in Collect. Bot.*, 16(2): 338 (1986): F. ovina var. fontqueriana.
gaetula
subsp. embergeri *(Litard.) A.M. Romo in Collect. Bot.*, 16(2): 338 (1986): F. rubra subvar. embergeri.
gamisansii *M. Kerguélen in Lejeunia,* 120: 95 (1987), nom. nov.: F. ovina var. litardierei *St.-Yves.*
iberica *(Hackel) K. Richter, Pl. Europ.*, 1: 99 (1890): F. rubra var. iberica.
karabaghensis *S.G. Musaev in Bot. Zhurn.*, 72(1): 94 (1987), nom. nov.: F. varia var. caucasica *St.-Yves.*
maroccana
subsp. pozzicola *(Litard. & Maire) A.M. Romo in Collect. Bot.*, 16(2): 339 (1986): F. deserti subvar. pozzicola.
maskerensis *(Litard.) A.M. Romo in Collect. Bot.*, 16(2): 339 (1986): F. ovina var. maskerensis.
nevadensis *(Hackel) K. Richter, Pl. Europ.*, 1: 101 (1890): F. rubra subsp. nevadensis.
nigricans *(Hackel) K. Richter, Pl. Europ.*, 1: 99 (1890): F. rubra var. nigricans.
numidica *(Battandier & Trabut) A.M. Romo in Collect. Bot.*, 16(2): 339 (1986): F. plicata subsp. numidica.
oraria *(Dumort.) H.C. van Hall, Fl. Belg. Sept.*, 1: 110 (1825): F. rubra * (subsp.) oraria.
paucispicula *V. de la Fuente García & D. Sánchez-Mata in Candollea,* 41(2): 441 (1986), nom. nov.: F. rubra subsp. fontqueri *Litard.*
pilosella *E.B. Alekseev in Byull. Mosk. Obshch. Ispŷt. Prir., Biol.*, 91(3): 118 (1986)—Yemen.

FESTUCA :–
sauvagei *A.M. Romo in Collect. Bot.*, 16(2): 339 (1986), nom. nov.: F. ovina var. maroccana *St.-Yves.*
sudanensis *E.B. Alekseev in Bot. Zhurn.*, 72(9): 1264 (1987)—Sudan.
trabutii *(St.-Yves) A.M. Romo in Collect. Bot.*, 16(2): 339 (1986): F. ovina subvar. trabutii.
weilleri *(Litard.) A.M. Romo in Collect. Bot.*, 16(2): 340 (1986): F. ovina var. weilleri.
yemenensis *E.B. Alekseev in Byull. Mosk. Obshch. Ispŷt. Prir., Biol.*, 91(3): 118 (1986)—Yemen.
yvesiana *(Litard. & Maire) A.M. Romo in Collect. Bot.*, 16(2): 339 (1986): F. rubra var. yvesiana.
yvesii
subsp. bellottii *(Auquier & Kerguélen) M. Kerguélen in Lejeunia*, 120: 102 (1987): F. durissima subsp. bellottii.

FICUS (Morac.).
guntheri *J.H. Torres-Romero in Caldasia*, 14(67): 207 (1985), as 'guntherii'—Colombia.

FIRMIANA (Sterculiac.).
colorata *(Roxb.) R. Brown Pterocymbium:* 235 (4 June 1844): Sterculia colorata. †

FLEMINGIA (Legumin.).
bhutanica *A.J.C. Grierson in Notes Roy. Bot. Gard. Edinburgh*, 44(2): 257 (1987)—Bhutan.

FLUEGGEA (Euphorbiac.).
neowawraea *W.J. Hayden in Brittonia*, 39(2): 268 (1987), nom. nov.: Neowawraea phyllanthoides *Rock.*

FORSYTHIA (Oleac.).
mira *M.C. Chang in Invest. Stud. Nat., Mus. Hist. Nat. Shanghaiense*, 7: 16 (1987)—China.

FREZIERA (Theac.).
carinata *A.L. Weitzman in J. Arnold Arbor.*, 68(3): 325 (1987)—Venezuela.
echinata *A.L. Weitzman in J. Arnold Arbor.*, 68(3): 328 (1987)—Colombia.
minima *A.L. Weitzman in J. Arnold Arbor.*, 68(3): 331 (1987)—Ecuador.
stuebelii *(Hieron.) A.L. Weitzman in J. Arnold Arbor.*, 68(3): 333 (1987): Taonabo stuebelii.

FRITILLARIA (Liliac.).
sect. Alpinea *Y.K. Yang & J.K. Wu in J. Wuhan Bot. Res.*, 5(2): 144 (1987).
ser. Anhuiensae *Y.K. Yang & J.Z. Shao in J. Wuhan Bot. Res.*, 5(2): 145 (1987).
sect. Davida *Y.K. Yang & J.K. Wu in J. Wuhan Bot. Res.*, 5(2): 145 (1987).
ser. Delavayae *Y.K. Yang & J.K. Wu in J. Wuhan Bot. Res.*, 5(2): 145 (1987).

FRITILLARIA :–
ser. Fuscae *Y.K. Yang & J.K. Wu in J. Wuhan Bot. Res.*, 5(2): 145 (1987).
ser. Imperialae *Y.K. Yang & J.K. Wu in J. Wuhan Bot. Res.*, 5(2): 144 (1987).
sect. Maximowicza *Y.K. Yang & J.K. Wu in J. Wuhan Bot. Res.*, 5(2): 146 (1987).
sect. Rosulia *Y.K. Yang & J.Z. Shao in J. Wuhan Bot. Res.*, 5(2): 145 (1987).
ser. Wuyangensae *Y.K. Yang & Z.Y. Gao in J. Wuhan Bot. Res.*, 5(2): 146 (1987).
ser. Zhufenense *Y.K. Yang & J.K. Wu in J. Wuhan Bot. Res.*, 5(2): 144 (1987).
anhuiensis
f. jinzhaiensis *Y.K. Yang & J.Z. Shao in J. Wuhan Bot. Res.*, 5(2): 143 (1987)—China.
borealixingjiangensis *Y.K. Yang, S.X. Zhang & G.J. Liu in J. Wuhan Bot. Res.*, 5(2): 128 (1987), as 'boreali-xingjiangensis'—Xinjiang.
chekiangensis *(Hsiao & K.C. Hsia) Y.K. Yang, Z.H. Lin & C. Lin in J. Wuhan Bot. Res.*, 5(2): 143 (1987): F. thunbergii var. chekiangensis.
chuanbeiensis *Y.K. Yang, D.H. Jiang & Y.H. Yang in J. Wuhan Bot. Res.*, 5(2): 137 (1987)—China.
var. huyabeimu *Y.K. Yang & D.H. Jiang in J. Wuhan Bot. Res.*, 5(2): 138 (1987)—China.
fujiangensis *Y.K. Yang, D.H. Jiang & Y.H. Yu in J. Wuhan Bot. Res.*, 5(2): 139 (1987)—China.
halabulanica *X.Z. Duan & X.J. Zheng in Acta Phytotax. Sin.*, 25(1): 57 (1987)—Xinjiang.
heboksarensis *X.Z. Duan & X.J. Zheng in Acta Phytotax. Sin.*, 25(1): 62 (1987)—Xinjiang.
himalaica *Y.K. Yang, Y.Q. Ye et al. in J. Wuhan Bot. Res.*, 5(2): 126 (1987)—Xizang.
huangshanensis
f. tonglingensis *(S.C. Chen & S.F. Yin) Y.K. Yang & Y.H. Zhang in J. Wuhan Bot. Res.*, 5(2): 141 (1987): F. monantha var. tonglingensis.
ionica
var. ochridana *K. Micevski in Acta Bot. Croat.*, 37: 212 (1978)—Jugoslavia.
karelinii
var. albiflora *X.Z. Duan & X.J. Zheng in Acta Phytotax. Sin.*, 25(1): 63 (1987)—Xinjiang.
lanzhouensis *Y.K. Yang, P.P. Ling & G. Yao in J. Wuhan Bot. Res.*, 5(2): 132 (1987)—China.
lhiinzeensis *Y.K. Yang, Y.Q. Ye et al. in J. Wuhan Bot. Res.*, 5(2): 126 (1987)—Xizang.
lishiensis *Y.K. Yang & J.K. Wu in J. Wuhan Bot. Res.*, 5(2): 134 (1987)—China.
var. yichengensis *Y.K. Yang & P.P. Ling in J. Wuhan Bot. Res.*, 5(2): 135 (1987)—China.
meleagroides
var. flavovirens *X.Z. Duan & X.J. Zheng in Acta Phytotax. Sin.*, 25(1): 58 (1987)—Xinjiang.

FRITILLARIA :–
monantha
var. ningguoica *Y.K. Yang & M.M. Fang in J. Wuhan Bot. Res.*, 5(2): 143 (1987)—China.
przewalskii
var. gannanica *Y.K. Yang & J.Z. Ren in J. Wuhan Bot. Res.*, 5(2): 132 (1987)—China.
rhodocanakis
subsp. argolica *E. Zaharof in Willdenowia*, 16(2): 348 (1987)—Greece.
rixii *E. Zaharof in Nordic J. Bot.*, 6(6): 727 (1986)—Greece.
shaanxiica *Y.K. Yang, S.X. Zhang & D.K. Zhang in J. Wuhan Bot. Res.*, 5(2): 133 (1987)—China.
shennongjiaensis *Y.K. Yang & Z. Zheng in J. Wuhan Bot. Res.*, 5(2): 136 (1987), with two types cited—China.
var. zhengbaensis *Y.K. Yang & J.X. Yang in J. Wuhan Bot. Res.*, 5(2): 137 (1987)—China.
shuchengensis *Y.K. Yang, D.Q. Wang & J.Z. Shao in J. Wuhan Bot. Res.*, 5(2): 140 (1987)—China.
tachengensis *X.Z. Duan & X.J. Zheng in Acta Phytotax. Sin.*, 25(1): 61 (1987)—Xinjiang.
var. nivea *Y.K. Yang & S.X. Zhang in J. Wuhan Bot. Res.*, 5(2): 129 (1987)—Xinjiang.
taipaiensis
var. fengxianensis *Y.K. Yang & J.K. Wu in J. Wuhan Bot. Res.*, 5(2): 134 (1987)—China.
f. platyphylla *Y.K. Yang & S.X. Zhang in J. Wuhan Bot. Res.*, 5(2): 134 (1987)—China.
var. yuxiensis *Y.K. Yang, Z.Y. Gao & C.S. Zhou in J. Wuhan Bot. Res.*, 5(2): 135 (1987)—China.
tianshanica *Y.K. Yang & L.R. Hsu in J. Wuhan Bot. Res.*, 5(2): 127 (1987)—Xinjiang.
tortifolia *X.Z. Duan & X.J. Zheng in Acta Phytotax. Sin.*, 25(1): 59 (1987)—Xinjiang.
var. citrina *X.Z. Duan & X.J. Zheng in Acta Phytotax. Sin.*, 25(1): 60 (1987)—Xinjiang.
unibracteata
var. ganziensis *Y.K. Yang & J.K. Wu in J. Wuhan Bot. Res.*, 5(2): 137 (1987)—China.
verticillata
var. jimunaica *X.Z. Duan & X.J. Zheng in Acta Phytotax. Sin.*, 25(1): 61 (1987)—Xinjiang.
wanjiangensis *Y.K. Yang, J.Z. Shao & Y.H. Zhang in J. Wuhan Bot. Res.*, 5(2): 141 (1987)—China.
wenxianensis *Y.K. Yang & J.K. Wu in J. Wuhan Bot. Res.*, 5(2): 129 (1987)—China.
xiaobeimu *Y.K. Yang, J.Z. Shao & M.M. Fang in J. Wuhan Bot. Res.*, 5(2): 142 (1987)—China.

FRITILLARIA :–
xibeiensis *Y.K. Yang, C.X. Feng & H.Z. Yang in J. Wuhan Bot. Res.*, 5(2): 130 (1987)—China.
yuminensis
var. varians *Y.K. Yang & G.J. Liu in J. Wuhan Bot. Res.*, 5(2): 129 (1987)—Xinjiang.

FUIRENA (Cyperac.).
obcordata *P.L. Forbes in S. Afr. J. Bot.*, 53(2): 185 (1987)—South Africa (Natal, Transvaal), Mozambique, Namibia. †

FUSISPERMUM (Violac.).
laxiflorum *W.H.A. Hekking in Proc. Kon. Nederl. Akad. Wetensch., C*, 87(2): 124 (1984)—Panama.

G

GALACTIA (Legumin.).
densiflora *M.T. Germán R. & M. Sousa S. in Iselya*, 2(3): 109 (1985)—Mexico.
microphylla *(Chapman) Rogers ex D. Isely in Brittonia*, 38(4): 354 (1986): G. floridana var. microphylla.

GALEARIS (Orchidac.).
spectabilis
f. albiflora *(Ulke) C.F. Reed in Phytologia*, 63(5): 410 (1987): Orchis spectabilis f. albiflora.

GALEOPSIS (Labiat.).
ladanum
subsp. cebennensis *(Br.-Bl.) M. Kerguélen in Lejeunia*, 120: 103 (1987): G. intermedia subsp. cebennensis.

GALIUM (Rubiac.).
asperuloides
subsp. hoffmeisteri *(Klotzsch) H. Hara in Enum. Fl. Pl. Nepal*, 2: 201 (1979): Asperula hoffmeisteri.
bourgeanum
subsp. maroccanum *(Ball) J.M. Montserrat in Treb. Inst. Bot. Barcelona*, 11: 24 (1987): G. bourgeanum var. maroccanum.
campylotrichum *S. Nazimuddin & F. Ehrendorfer in Pl. Syst. Evol.*, 155(1–4): 73 (1987)—Pakistan.
densiflorum
var. saurense *(Litv.) N.N. Tsvelev in Novosti Sist. Vȳssh. Rast.*, 23: 158 (1986): G. saurense.
elegans
f. vestitum *(D. Don) H. Hara in Enum. Fl. Pl. Nepal*, 2: 201 (1979): G. vestitum.

GALIUM :–
gymnopetalum *F. Ehrendorfer & E. Schönbeck-Temesy in Pl. Syst. Evol.*, 155(1–4): 78 (1987)—India.
lahulense *F. Ehrendorfer & E. Schönbeck-Temesy in Pl. Syst. Evol.*, 155(1–4): 79 (1987)—India.
mahadivense *G. Singh in G. Singh & P. Kachroo, Forest Fl. Srinagar:* 125 (1987), as 'mahadivensis'—India.
nepalense *F. Ehrendorfer & E. Schönbeck-Temesy in Pl. Syst. Evol.*, 155(1–4): 82 (1987)—Nepal.
pseudocruciata *(J. Rohlena) A.W. Hill in Index Kewensis,* Supp. 6: 88 (1916–1920 publ. 1926): G. vernum var. pseudocruciata.
saipalense *F. Ehrendorfer & E. Schönbeck-Temesy in Pl. Syst. Evol.*, 155(1–4): 82 (1987)—Nepal.
subfalcatum *S. Nazimuddin & F. Ehrendorfer in Pl. Syst. Evol.*, 155(1–4): 71 (1987)—Pakistan.
tetraphyllum *S. Nazimuddin & F. Ehrendorfer in Pl. Syst. Evol.*, 155(1–4): 73 (1987)—Pakistan.

GAMOCHAETA (Compos.).
coarctata *(Willd.) M. Kerguélen in Lejeunia,* 120: 104 (1987): Gnaphalium coarctatum.

GARCINIA (Guttif.).
barkeriana *(Urb. & Ekm.) Alain H. Liogier in Phytologia,* 61(6): 353 (1986): Rheedia barkeriana.
darwiniana *K.R. Keshava Murthy, S.N. Yoganarasimhan & K. Vasudevan Nair in Current Sci.*, 56(9): 425 (1987)V—India. N1
hessii *(Britt.) Alain H. Liogier in Phytologia,* 61(6): 353 (1986): Rheedia hessii.
portoricensis *(Urb.) Alain H. Liogier in Phytologia,* 61(6): 353 (1986): Rheedia portoricensis.
verticillata *(Urb.) Alain H. Liogier in Phytologia,* 61(6): 353 (1986): Rheedia verticillata.

GASTERIA (Aloeac.).
vlokii *E.J. van Jaarsveld in Cact. Succ. J. (U.S.A.),* 59(4): 170 (1987)—South Africa (Cape Province).

×**GASTISOCALPA** hort. in Orchid Rev., 95(1127): centre page pull-out p. 9 (1987). *ORCHIDACEAE.* (Gastrochilus × Luisia × Pomatocalpa).

GASTONIA (Araliac.).
crassa *(Hemsley) F. Friedmann in Bull. Mus. Nation. Hist. Nat., B, Adansonia,* Ser. 4, 8(3): 246 (1986): Indokingia crassa.
lionnetii *F. Friedmann in Bull. Mus. Nation. Hist. Nat., B, Adansonia,* Ser. 4, 8(3): 253 (1986)—Seychelles Is.
sechellarum
var. contracta *F. Friedmann in Bull. Mus. Nation. Hist. Nat., B, Adansonia,* Ser. 4, 8(3): 251 (1986)—Seychelles Is.

GASTONIA :–
var. curiosae *F. Friedmann in Bull. Mus. Nation. Hist. Nat., B, Adansonia,* Ser. 4, 8(3): 251 (1986)—Seychelles Is.

GASTROLOBIUM (Legumin.).
graniticum *(S. Moore) M.D. Crisp in C.H. Stirton (ed.), Advances in Legume System.*, 3: 130 (1987): Oxylobium graniticum.
heterophyllum *(Turcz.) M.D. Crisp in C.H. Stirton (ed.), Advances in Legume System.*, 3: 130 (1987): Chorizema heterophyllum.
parviflorum *(Benth.) M.D. Crisp in C.H. Stirton (ed.), Advances in Legume System.*, 3: 130 (1987): Oxylobium parviflorum.
racemosum *(Turcz.) M.D. Crisp in C.H. Stirton (ed.), Advances in Legume System.*, 3: 130 (1987): Mirbelia racemosa.
rigidum *(C. Gardner) M.D. Crisp in C.H. Stirton (ed.), Advances in Legume System.*, 3: 130 (1987): Oxylobium rigidum.
spectabile *(Endl.) M.D. Crisp in C.H. Stirton (ed.), Advances in Legume System.*, 3: 130 (1987): Oxylobium spectabile.
tetragonophyllum *(E. Pritzel) M.D. Crisp in C.H. Stirton (ed.), Advances in Legume System.*, 3: 130 (1987): Oxylobium tetragonophyllum.

GAUDICHAUDIA (Malpighiac.).
chasei *W.R. Anderson in Contrib. Univ. Mich. Herb.*, 16: 68 (1987)—Mexico.
cycloptera *(DC.) W.R. Anderson in Contrib. Univ. Mich. Herb.*, 16: 69 (1987): Hiraea ? cycloptera.
krusei *W.R. Anderson in Contrib. Univ. Mich. Herb.*, 16: 69 (1987)—Mexico.
macvaughii *W.R. Anderson in Contrib. Univ. Mich. Herb.*, 16: 72 (1987), as 'mcvaughii'—Mexico.

GAULTHERIA (Ericac.).
wardii
var. serrulata *C.Y. Wu & T.Z. Hsu in Fl. Xizangica,* 3: 699 (1986)—Xizang.

GENISTA (Legumin.).
nothosect. Spades *F. Llamas in Stud. Bot. (Salamanca),* 5: 209 (1986). (G. sect. Erinacoides × sect. Spartioides).
sandrasica *P. Hartvig & A. Strid in Bot. Jahrb.*, 108(2–3): 328 (1987)—Turkey.

GENTIANA (Gentianac.).
ser. Angulosae *R.I. Gagnidze in Zam. Sist. Geogr. Rast.*, 41: 51 (1986).
ser. Aquaticae *Grossh. ex R.I. Gagnidze in Zam. Sist. Geogr. Rast.*, 41: 51 (1986).
ser. Asclepiadeae *Grossh. ex R.I. Gagnidze in Zam. Sist. Geogr. Rast.*, 41: 50 (1986).
sect. Gelida *R.I. Gagnidze in Zam. Sist. Geogr. Rast.*, 41: 51 (1986), as comb. nov. but basionym invalid.
ser. Nivales *R.I. Gagnidze in Zam. Sist. Geogr. Rast.*, 41: 52 (1986).
ser. Prostratae *Grossh. ex R.I. Gagnidze in Zam. Sist. Geogr. Rast.*, 41: 51 (1986).

GENTIANA :–
ser. Triflorae *Grossh. ex R.I. Gagnidze in Zam. Sist. Geogr. Rast.*, 41: 50 (1986).
acaulis
var. minor *(R. Nègre) M. Kerguélen in Lejeunia,* 120: 104 (1987): G. kochiana var. minor.
angulosa
var. pontica *(Soltok.) R.I. Gagnidze in Zam. Sist. Geogr. Rast.*, 41: 52 (1986): G. pontica.
crenulatotruncata *(Marq.) T.N. He in Fl. Xizangica,* 3: 962 (1986): G. prostrata var. crenulato-truncata.
harwanensis *G. Singh in G. Singh & P. Kachroo, Forest Fl. Srinagar:* 126 (1987)—India.
malyschevii *(V. Zuev) I.A. Gubanov & R.V. Kamelin in Byull. Mosk. Obshch. Ispўt. Prir., Biol.*, 92(1): 122 (1987): Gentianella malyschevii.
waltonii
var. lhasaensis *(Hsiao & K.C. Hsia) T.N. He in Fl. Xizangica,* 3: 917 (1986): G. lhasaensis.

GENTIANELLA (Gentianac.).
chlorantha *J.S. Pringle in Sida,* 11(4): 357 (1986)—Peru.
cuspidata *(Griseb.) J.S. Pringle in Sida,* 11(4): 368 (1986): Gentiana cuspidata.
dianthoides *(H.B.K.) Fabris ex J.S. Pringle in Sida,* 11(4): 368 (1986): Gentiana dianthoides.
dielsiana *(Gilg) J.S. Pringle in Sida,* 11(4): 368 (1986): Gentiana dielsiana.
dolichopoda *(Gilg) J.S. Pringle in Sida,* 11(4): 368 (1986): Gentiana dolichopoda.
helianthemoides *(Gilg) J.S. Pringle in Sida,* 11(4): 368 (1986): Gentiana helianthemoides.
gilgiana *(Reimers) Fabris ex J.S. Pringle in Sida,* 11(4): 368 (1986): Gentiana gilgiana.
mendocina *(Gilg) J.S. Pringle in Sida,* 11(4): 368 (1986): Gentiana mendocina.
persquarrosa *(Reimers) J.S. Pringle in Sida,* 11(4): 368 (1986): Gentiana persquarrosa.
radicata *(Griseb.) J.S. Pringle in Sida,* 11(4): 368 (1986): Gentiana radicata.
riojae *(Gilg) Fabris ex J.S. Pringle in Sida,* 11(4): 368 (1986): Gentiana riojae.
vaginalis *(Griseb.) J.S. Pringle in Sida,* 11(4): 368 (1986): Gentiana vaginalis.

GERANIUM (Geraniac.).
camaense *C.C. Huang in Fl. Xizangica,* 3: 10 (1986), without type—Xizang.

GERBERA (Compos.).
connata *Y.C. Tseng in Acta Bot. Austro Sinica,* 3: 7 (1986)—Xizang.
latiligulata *Y.C. Tseng in Acta Bot. Austro Sinica,* 3: 11 (1986)—China.
lijiangensis *Y.C. Tseng in Acta Bot. Austro Sinica,* 3: 6 (1986)—China, Xizang.

GERBERA :–
macrocephala *Y.C. Tseng in Acta Bot. Austro Sinica,* 3: 12 (1986)—China.
pterodonta *Y.C. Tseng in Acta Bot. Austro Sinica,* 3: 9 (1986)—China.
saxatilis *Chang ex Y.C. Tseng in Acta Bot. Austro Sinica,* 3: 10 (1986)—China.

GEUM (Rosac.).
subgen. Constrictohirtistylis *G. Panigrahi & K.M. Purohit in Bull. Bot. Surv. India,* 26(3–4): 238 (1984 publ. 1986).
×catlingii *J.-P. Bernard & R. Gauthier in Nat. Canad.,* 113(4): 321 (1986)—Canada (Quebec, Ontario). (G. canadense × urbanum).
×macneillii *J.-P. Bernard & R. Gauthier in Nat. Canad.,* 113(4): 321 (1986)—Canada (Quebec). (G. laciniatum × urbanum).

GIGANTOCHLOA (Gramin.).
achmadii *E.A. Widjaja in Reinwardtia,* 10(3): 373 (1987)—Sumatra.
atroviolacea *E.A. Widjaja in Reinwardtia,* 10(3): 323 (1987)—Java.
manggong *E.A. Widjaja in Reinwardtia,* 10(3): 365 (1987)—Java, Lesser Sunda Is., Burma.
pruriens *E.A. Widjaja in Reinwardtia,* 10(3): 369 (1987)—Sumatra.
pseudoarundinacea *(Steud.) E.A. Widjaja in Reinwardtia,* 10(3): 305 (1987): Bambusa pseudoarundinacea.
vietnamica *Nguyen to Quyen in Bot. Zhurn.,* 72(6): 829 (1987)—Vietnam.
vinhphuica *Nguyen to Quyen in Bot. Zhurn.,* 72(6): 830 (1987)—Vietnam.

GIRARDINIA (Urticac.).
septentrionalis *I.A. Grudzinskaya in Bot. Zhurn.,* 72(7): 960 (1987)—U.S.S.R. (Soviet Far East), China, Korea.

GLEDITSIA (Legumin.).
medogensis *Z.C. Ni in Acta Phytotax. Sin.,* 25(3): 231 (1987)—Xizang.

GLOEOSPERMUM (Violac.).
pauciflorum *W.H.A. Hekking in Proc. Kon. Nederl. Akad. Wetensch., C,* 89(4): 399 (1986)—Panama.

GLOSSOSTIPULA D.H. Lorence in Candollea, 41(2): 454 (1986). *RUBIACEAE.*
blepharophylla *(Standley) D.H. Lorence in Candollea,* 41(2): 457 (1986): Randia blepharophylla.
concinna *(Standley) D.H. Lorence in Candollea,* 41(2): 458 (1986): Randia concinna.

GLYCINE (Legumin.).
arenaria *M.D. Tindale in Brunonia,* 9(2): 187 (1987)—Australia (W. Australia, N. Terr.).
curvata *M.D. Tindale in Brunonia,* 9(1): 99 (1987)—Australia (Queensland).

GLYCINE :–
microphylla *(Benth.) M.D. Tindale in Brunonia,* 9(2): 181 (1987): Leptolobium microphyllum.

GLYCOSMIS (Rutac.).
medogensis *D.D. Tao ex C.C. Huang in Fl. Xizangica,* 3: 30 (1986)—Xizang.
oligantha *C.C. Huang in Guihaia,* 7(2): 122 (1987)—China.

GNAPHALIUM (Compos.).
microcephalum
subsp. thermale *(E. Nels.) G.W. Douglas in Canad. J. Bot.,* 64(11): 2726 (1986): G. thermale.

GNEPHOSIS (Compos.).
drummondii *(A. Gray) P.S. Short in Muelleria,* 6(5): 317 (1987): Chrysocoryne drummondii.
multiflora *(P. Short) P.S. Short in Muelleria,* 6(5): 317 (1987): Chrysocoryne multiflora.
tridens *(P. Short) P.S. Short in Muelleria,* 6(5): 318 (1987): Chrysocoryne tridens.
trifida *(P. Short) P.S. Short in Muelleria,* 6(5): 318 (1987): Chrysocoryne trifida.
uniflora *(Turcz.) P.S. Short in Muelleria,* 6(5): 318 (1987): Chrysocoryne uniflora.

GNIDIA (Thymelaeac.).
hirsuta *(L.) M. Thulin in Taxon,* 35(4): 724 (1986): Lobelia hirsuta.

GOLDFUSSIA (Acanthac.).
echinata *(Nees) K. Haridasan & R.R. Rao, Forest Fl. Meghalaya,* 2: 663 (1987): Strobilanthes echinata.
medogensis *H.W. Li in Fl. Xizangica,* 4: 413 (1985)—Xizang.

GOMPHOLOBIUM (Legumin.).
asperulum *(S. Moore) M.D. Crisp in C.H. Stirton (ed.), Advances in Legume System.,* 3: 129 (1987): Burtonia asperula.
confertum *(DC.) M.D. Crisp in C.H. Stirton (ed.), Advances in Legume System.,* 3: 129 (1987): Burtonia conferta.
gompholobioides *(F. Muell.) M.D. Crisp in C.H. Stirton (ed.), Advances in Legume System.,* 3: 129 (1987): Burtonia gompholobioides.
simplicifolium *(F. Muell. & Tate) M.D. Crisp in C.H. Stirton (ed.), Advances in Legume System.,* 3: 129 (1987): Burtonia simplicifolia.
villosum *(Meissner) M.D. Crisp in C.H. Stirton (ed.), Advances in Legume System.,* 3: 129 (1987): Burtonia villosa.

GONGYLOLEPIS (Compos.).
terramarae *J.A. Steyermark in Ann. Missouri Bot. Gard.,* 74(1): 115 (1987)—Venezuela.

GONGYLOSCIADIUM K.H. Rechinger in Fl. Iran., 162: 308 (1987). *UMBELLIFERAE.*

GONGYLOSCIADIUM :–
falcarioides *(Bornm. & Wolff) K.H. Rechinger in Fl. Iran.,* 162: 308 (1987): Pimpinella falcarioides.

GONIOCHILUS M.W. Chase in Contrib. Univ. Mich. Herb., 16: 124 (1987). *ORCHIDACEAE.*
leochilinus *(Reichb. f.) M.W. Chase in Contrib. Univ. Mich. Herb.,* 16: 125 (1987): Rodriguezia leochilina.

GONOLOBUS (Asclepiadac.).
colombianus *G. Morillo in An. Jard. Bot. Madrid,* 43(2): 235 (1986 publ. 1987)—Colombia.
heterophyllus *(Hemsley) G. Morillo in An. Jard. Bot. Madrid,* 43(2): 237 (1986 publ. 1987): Fischeria heterophylla.
stipitatus *G. Morillo in An. Jard. Bot. Madrid,* 43(2): 237 (1986 publ. 1987)—Colombia.

GOODALLIA T. Bowdich, Exc. Madeira (ed. S. Bowdich): 61 (1825); cf. D.J. Mabberley in Bot Macaronesica, 6: 59 (1980). *CRASSULACEAE.*

GOODENIA (Goodeniac.).
benthamiana *R.C. Carolin in Fl. S. Australia,* 3: 1392 (1986), nom. nov.: G. amplexans var. parvifolia *Benth.*
blackiana *R.C. Carolin in Fl. S. Australia,* 3: 1393 (1986)—Australia (Victoria, S. Australia).
brunnea *R.C. Carolin in Fl. S. Australia,* 3: 1393 (1986)—Australia (S. Australia).
quasilibera *R.C. Carolin in Fl. S. Australia,* 3: 1403 (1986)—Australia (W. Australia, S. Australia).
saccata *R.C. Carolin in Fl. S. Australia,* 3: 1403 (1986)—Australia (S. Australia).

GOODYERA (Orchidac.).
hispaniolae *D.D. Dod in Moscosoa,* 4: 183 (1986)—Hispaniola (Dominican Republic, Haiti).

GOSSYPIUM (Malvac.).
anomalum
subsp. senarense *(Wawra & Peyr.) K. Vollesen in Kew Bull.,* 42(2): 339 (1987): G. senarense.
bricchettii *(Ulbrich) K. Vollesen in Kew Bull.,* 42(2): 349 (1987): Cienfugosia bricchettii.
herbaceum
subsp. africanum *(Watt) K. Vollesen in Kew Bull.,* 42(2): 343 (1987): G. obtusifolium var. africanum.
schwendimanii *P.A. Fryxell & S.D. Koch in Aliso,* 11(4): 546 (1987)—Mexico.
trifurcatum *K. Vollesen in Kew Bull.,* 42(2): 342 (1987)—Somali Republic.

GOVENIA (Orchidac.).
bella *E.W. Greenwood in Orquídea (México),* 10(2): 230 (1987)—Mexico.

GRAPHARDISIA (Myrsinac.).
murphyae *C.L. Lundell in Phytologia,* 63(2): 77 (1987)—Costa Rica.

GRAPTOPETALUM (Crassulac.).
pentandrum
subsp. superbum *M. Kimnach in Cact. Succ. J. (U.S.A.),* 59(4): 142 (1987)—Cultivation ex Mexico.

GRAYIA (Chenopodiac.).
brandegei
var. plummeri *H.C. Stutz & S.C. Sanderson in Madroño,* 34(2): 148 (1987)—U.S.A. (Utah, Colorado, Wyoming, New Mexico).

GRENACHERIA (Myrsinac.).
royenii *H. Sleumer in Blumea,* 32(2): 395 (1987)—New Guinea (W. Irian).

GRINDELIA (Compos.).
cabrerae *L.A. Espinar in Kurtziana,* 9: 81 (1976)—Argentina.

GUINDILIA (Sapindac.).
cristata *(Radlk.) A.T. Hunziker in Kurtziana,* 9: 100 (1976): Valenzuelia cristata.
dissecta *(Covas & Burk.) A.T. Hunziker in Kurtziana,* 9: 100 (1976): Valenzuelia dissecta.

GUTIERREZIA (Compos.).
petradoria *(Welsh & Goodrich) S.L. Welsh in Great Basin Nat.,* 43(2): 288 (1983): Xanthocephalum petradoria.
pomariensis *(Welsh) S.L. Welsh in Great Basin Nat.,* 43(2): 288 (1983): G. sarothrae var. pomariensis.

GUZMANIA (Bromeliac.).
alborosea *H.E. Luther in Selbyana,* 9(1): 187 (1986)—Ecuador.
andreettae *W. Rauh in Abh. Akad. Wiss. Lit. Mainz, Math.-Naturwiss. Klasse, Trop. Subtrop. Pflanzenwelt,* 58: 46 (1986)—Ecuador.
terrestris *L.B. Smith & J.A. Steyermark in Ann. Missouri Bot. Gard.,* 73(4): 715 (1986 publ. 1987)—Venezuela.
thyrsoidea *W. Rauh in Abh. Akad. Wiss. Lit. Mainz, Math.-Naturwiss. Klasse, Trop. Subtrop. Pflanzenwelt,* 58: 50 (1986)—Costa Rica.

GYMNANTHEMUM (Compos.).
coloratum *(Willd.) H. Robinson & B. Kahn in Proc. Biol. Soc. Wash.,* 99(3): 501 (1986): Eupatorium coloratum.

GYMNANTHERA (Asclepiadac.).
fruticosa *K.L. Wilson in J. Adelaide Bot. Gard.,* 10(1): 113 (1987)—Australia (N. Terr.).

GYMNEMA (Asclepiadac.).
dunnii *(Maiden & Betche) P.I. Forster in Austrobaileya,* 2(4): 401 (1987): Marsdenia rostrata var. dunnii.

GYMNOCALYCIUM (Cactac.).
achirasense *Till & Schatzl ex H. Till in Kakt. And. Sukk.,* 38(8): 191 (1987)—Argentina.
bayrianum *Till ex H. Till in Kakt. And. Sukk.,* 38(8): 191 (1987)—Argentina.
borthii *Koop ex H. Till in Kakt. And. Sukk.,* 38(8): 191 (1987)—Argentina.
bruchii
var. brigittae *J. Piltz in Succulenta,* 66(10): 213 (1987)—Argentina.
horridispinum *Frank ex H. Till in Kakt. And. Sukk.,* 38(8): 191 (1987)—Argentina.
intertextum *Backeberg ex H. Till in Kakt. And. Sukk.,* 38(8): 191 (1987)—Argentina.
pugionacanthum *Backeberg ex H. Till in Kakt. And. Sukk.,* 38(8): 191 (1987)—Argentina.
schroederianum
subsp. bayense *R. Kiesling in Cact. Succ. J. (U.S.A.),* 59(2): 48 (1987)—Argentina.
subsp. paucicostatum *R. Kiesling in Cact. Succ. J. (U.S.A.),* 59(2): 49 (1987)—Argentina.
striglianum *Jeggle ex H. Till in Kakt. And. Sukk.,* 38(8): 191 (1987)—Argentina.

GYMNOPETALUM G.A.W. Arnott in Madras J. Lit. Sci., 12: 52 (1840). *CUCURBITACEAE.*
ceylanicum *G.A.W. Arnott in Madras J. Lit. Sci.,* 12: 52 (1840), nom. illegit.—E. Indies.
wightii *G.A.W. Arnott in Madras J. Lit. Sci.,* 12: 52 (1840)—E. Indies.

GYNOPACHIS (Rubiac.).
beccariana *(Baillon) D.D. Tirvengadum in Bull. Mus. Nation. Hist. Nat., B, Adansonia,* Ser. 4, 8(3): 284 (1986): Randia beccariana.
binata *(King & Gamble) D.D. Tirvengadum in Bull. Mus. Nation. Hist. Nat., B, Adansonia,* Ser. 4, 8(3): 283 (1986): Randia binata.
impressinervis *(King & Gamble) D.D. Tirvengadum in Bull. Mus. Nation. Hist. Nat., B, Adansonia,* Ser. 4, 8(3): 283 (1986): Randia impressinervis.
jambosoides *(Valeton) D.D. Tirvengadum in Bull. Mus. Nation. Hist. Nat., B, Adansonia,* Ser. 4, 8(3): 284 (1986): Randia jambosoides.
pulcherrima *(Merr.) D.D. Tirvengadum in Bull. Mus. Nation. Hist. Nat., B, Adansonia,* Ser. 4, 8(3): 285 (1986): Randia pulcherrima.
umbellata *(Elm.) D.D. Tirvengadum in Bull. Mus. Nation. Hist. Nat., B, Adansonia,* Ser. 4, 8(3): 283 (1986): Randia umbellata.

GYPSACANTHUS E.J. Lott, V. Jaramillo L. & J. Rzedowski in Bol. Soc. Bot. Méx., 46: 47 (1984 publ. 1986). *ACANTHACEAE.*

GYPSACANTHUS :–
nelsonii *E.J. Lott, V. Jaramillo L. & J. Rzedowski in Bol. Soc. Bot. Méx.*, 46: 47 (1984 publ. 1986)—Mexico.

GYPSOPHILA (Caryophyllac.).
makranica *(Rech.f.) S.A. Ghazanfar in Fl. Pakistan,* 175: 94 (1986): Saponaria makranica.

GYRINOPS (Thymelaeac.).
vidalii *Pham-hoang Ho in Bull. Mus. Nation. Hist. Nat., B, Adansonia,* Ser. 4, 8(3): 242 (1986)—Laos.

H

HAAGEOCEREUS (Cactac.).
weberbaueri *(Schumann ex Vaupel) D. Hunt in Bradleya,* 5: 92 (1987): Cereus weberbaueri.

HABENARIA (Orchidac.).
bougainvillae *J. Renz in Pl. Syst. Evol.*, 155(1–4): 326 (1987)—Bougainville Is.
brownelliana *P.M. Catling in Brittonia,* 39(3): 402 (1987)—Belize, Mexico.
elongata
var. leptophylla *J. Renz in Pl. Syst. Evol.*, 155(1–4): 330 (1987)—New Guinea (Papua New Guinea).
ensigera *J. Renz in Pl. Syst. Evol.*, 155(1–4): 319 (1987)—New Guinea (W. Irian).
indica *C. Sathish Kumar & K.S. Manilal in Taxon,* 35(4): 719 (1986), nom. nov.: H. affinis *Wight*.
ixtlanensis *E.W. Greenwood in Orquídea (México),* 10(2): 297 (1987)—Mexico.
rechingeri *J. Renz in Pl. Syst. Evol.*, 155(1–4): 324 (1987)—New Guinea (W. Irian).
trichoglossa *J. Renz in Pl. Syst. Evol.*, 155(1–4): 322 (1987)—New Guinea (Papua New Guinea).

HABRACANTHUS (Acanthac.).
harlingii *D.C. Wasshausen in Op. Bot.*, 92: 272 (1987)—Ecuador.

HAEMODORUM (Haemodorac.).
discolor *T.D. Macfarlane in Fl. Australia,* 45: 464, 138 (1987)—Australia (W. Australia).
gracile *T.D. Macfarlane in Fl. Australia,* 45: 464, 147 (1987)—Australia (W. Australia).
loratum *T.D. Macfarlane in Fl. Australia,* 45: 464, 140 (1987)—Australia (W. Australia).
venosum *T.D. Macfarlane in Fl. Australia,* 45: 464, 138 (1987)—Australia (W. Australia).

HAGENBACHIA (Anthericac.).
columbiana *R.W. Cruden in Nordic J. Bot.*, 7(3): 259 (1987)—Colombia.

HAGENBACHIA :–
hassleriana *(Baker) R.W. Cruden in Nordic J. Bot.*, 7(3): 259 (1987): Chlorophytum hasslerianum.

HAKEA (Proteac.).
cygna *B. Lamont in Bot. J. Linn. Soc.*, 94(4): 439 (1987)—Australia (W. Australia).
subsp. needlei *B. Lamont in Bot. J. Linn. Soc.*, 94(4): 440 (1987)—Australia (W. Australia).
erecta *B. Lamont in Bot. J. Linn. Soc.*, 94(4): 440 (1987)—Australia (W. Australia).

HALOGETON (Chenopodiac.).
kashmirianus *C. Grey-Wilson & B.M. Wadhwa in Kew Bull.*, 42(2): 473 (1987)—India, Xizang.

HAMAMELIS (Hamamelidac.).
mollis
var. oblongifolia *M.B. Deng & G. Yao in Bull. Nanjing Bot. Gard. Mem. Sun Yat Sen,* 1984–1985: 125 (1986)—China.

HANSTEINIA (Acanthac.).
blepharorhachis *(Lindau) L.H. Durkee in Fieldiana, Bot.*, n.s., 18: 43 (1986): Kolobochilus blepharorhachis.
sessilifolia *(Oerst.) L.H. Durkee in Fieldiana, Bot.*, n.s., 18: 44 (1986): Glockeria sessilifolia.

HAPLOCHOREMA (Zingiberac.).
magnum *R.M. Smith in Notes Roy. Bot. Gard. Edinburgh,* 44(2): 214 (1987)—Borneo (Sarawak).
pauciflorum *R.M. Smith in Notes Roy. Bot. Gard. Edinburgh,* 44(2): 213 (1987)—Borneo (Sarawak).
var. bullatum *R.M. Smith in Notes Roy. Bot. Gard. Edinburgh,* 44(2): 213 (1987)—Borneo (Sarawak).

HAPLOPAPPUS (Compos.).
armerioides
var. gramineus *S.L. Welsh & F.J. Smith in Great Basin Nat.*, 43(2): 371 (1983)—U.S.A. (Utah).
compactus *(Hall) L.C. Anderson in Great Basin Nat.*, 43(2): 358 (1983): H. bloomeri subsp. compactus.
crispus *L.C. Anderson in Great Basin Nat.*, 43(2): 359 (1983)—U.S.A. (Utah).
racemosus
var. prionophyllus *(Greene) S.L. Welsh in Great Basin Nat.*, 43(2): 294 (1983): Pyrrocoma prionophylla.
var. sessiliflorus *(Greene) S.L. Welsh in Great Basin Nat.*, 43(2): 294 (1983): Pyrrocoma sessiliflora.
thiniicola *J. Rzedowski & E. Ezcurra in Cienc. Interamer.*, 26(1–2): 16 (1986)—Mexico.
watsonii
var. rydbergii *(Blake) S.L. Welsh in Great Basin Nat.*, 43(2): 295 (1983): H. rydbergii.

HAPLOPAPPUS :–
zionis *L.C. Anderson in Great Basin Nat.*, 43(2): 360 (1983)—U.S.A. (Utah).

HARRISIA (Cactac.).
hahniana *(Back.) M. Kimnach & P.C. Hutchison in Cact. Succ. J. (U.S.A.)*, 59(2): 59 (1987): Mediocactus hahnianus.
tetracantha *(Labouret) D. Hunt in Bradleya*, 5: 92 (1987): Cereus tetracanthus.

HATIORA (Cactac.).
subgen. Rhipsalidopsis *(Britton & Rose) W. Barthlott in Bradleya*, 5: 100 (1987): gen. Rhipsalidopsis.
gaertneri *(Regel) W. Barthlott in Bradleya*, 5: 100 (1987): Epiphyllum russellianum var. gaertneri.
×graeseri *(Werdermann) W. Barthlott in Bradleya*, 5: 100 (1987): Rhipsaphyllopsis graeseri.
herminiae *(Campos-Porto & Castellanos) Backeb. ex W. Barthlott in Bradleya*, 5: 100 (1987): Hariota herminiae.
rosea *(Lagerheim) W. Barthlott in Bradleya*, 5: 100 (1987): Rhipsalis rosea.

HEBE (Scrophulariac.).
sect. Glaucae *M.J. Heads in Bot. Soc. Otago Newsl.*, 5: 11 (1987).
ser. Occlusae *M.J. Heads in Bot. Soc. Otago Newsl.*, 5: 11 (1987).
sect. Subdistichae *M.J. Heads in Bot. Soc. Otago Newsl.*, 5: 11 (1987).
crawii *M.J. Heads in Bot. Soc. Otago Newsl.*, 5: 11 (1987)—New Zealand.

HECHTIA (Bromeliac.).
fragilis *K. Burt-Utley & J.F. Utley in Brittonia*, 39(1): 40 (1987)—Mexico.
lyman-smithii *K. Burt-Utley & J.F. Utley in Brittonia*, 39(1): 37 (1987)—Mexico.

HEDEOMA (Labiat.).
hyssopifolium
var. chihuahuensis *J. Henrickson in Sida*, 11(4): 413 (1986)—Mexico.

HEDERA (Araliac.).
helix
subsp. hibernica *(Kirchner) D. McClintock, Suppl. Wild Flow. Guernsey:* 23 (1987): H. helix var. hibernica.

HEDINIA (Crucif.).
ser. Hediniopsis *(Botsch. & Petrovsky) B.A. Yurtsev, A.A. Korobkov & S.A. Balandin in Bot. Zhurn.*, 72(8): 1035 (1987): gen. Hediniopsis.
czukotica *(Botsch. & Petrovsky) B.A. Yurtsev, A.A. Korobkov & S.A. Balandin in Bot. Zhurn.*, 72(8): 1035 (1987): Hediniopsis czukotica.

HEDINIOPSIS V.P. Bochantsev & V.V. Petrovskiĭ in Bot. Zhurn., 71(11): 1548 (1986). *CRUCIFERAE.*

HEDINIOPSIS :–
czukotica *V.P. Bochantsev & V.V. Petrovskiĭ in Bot. Zhurn.*, 71(11): 1549 (1986)—U.S.S.R. (Soviet Far East).

HEDYOTIS (Rubiac.).
corymbosa
var. linearis *(DC.) D.B. Deb & R. Dutta in J. Bombay Nat. Hist. Soc.*, 83(3): 692 (1986 publ. 1987): Oldenlandia linearis.
nigricans
var. floridana *(Standley) R.P. Wunderlin in Sida*, 11(4): 400 (1986): Houstonia floridana.

HEINEKENIA (Legumin.).
eremitica *(Santos) G. Kunkel, Kanarischen Ins. und ihre Pflanzenwelt:* 153 (1987): Lotus eremiticus.

HELDREICHIA (Crucif.).
atalayi *Kit Tan in Pl. Syst. Evol.*, 154(1–2): 113 (1986)—Turkey.

HELIANTHEMUM (Cistac.).
×carmen-joanae *J. Mansanet & I. Mateu in Mediterranea*, 8: 85 (1985)—Spain. (H. lavandulifolium × pilosum var. racemosum).
polygonoides *M. Peinado, J.M. Martínez-Parras, F. Alcaraz & I. Espuelas in Candollea*, 42(1): 361 (1987)—Spain.
urrielense *(Laínz) H.S. Nava Fernández & M.A. Fernández-Casado in An. Jard. Bot. Madrid*, 43(1): 28 (1986): H. nummularium subsp. urrielense.

HELIANTHUS (Compos.).
agrestis
f. almae *E.R. Wolde in Phytologia*, 63(4): 310 (1987)—U.S.A. (Florida).
f. oswaldii *E.R. Wolde in Phytologia*, 63(4): 310 (1987)—U.S.A. (Florida).

HELICHRYSUM (Compos.).
ambiguum
subsp. vinaceum *L. Haegi in Fl. S. Australia*, 3: 1535 (1986)—Australia (S. Australia).
bampsianum *S. Lisowski in Bull. Jard. Bot. Nation. Belg.*, 56(3–4): 352 (1986)—Zaïre, Tanzania.
devredii *S. Lisowski in Bull. Jard. Bot. Nation. Belg.*, 56(3–4): 350 (1986)—Zaïre.
duvigneaudii *S. Lisowski in Bull. Jard. Bot. Nation. Belg.*, 56(3–4): 364 (1986)—Zaïre.
eremaeum *L. Haegi in Fl. S. Australia*, 3: 1535 (1986)—Australia (S. Australia, W. Australia, N. Terr.).
jeffreyanum *S. Lisowski in Bull. Jard. Bot. Nation. Belg.*, 56(3–4): 356 (1986)—Zaïre.
kalandanum *S. Lisowski in Bull. Jard. Bot. Nation. Belg.*, 56(3–4): 353 (1986)—Zaïre.
lejolyanum *S. Lisowski in Bull. Jard. Bot. Nation. Belg.*, 56(3–4): 362 (1986)—Zaïre.
malaisseanum *S. Lisowski in Bull. Jard. Bot. Nation. Belg.*, 56(3–4): 357 (1986)—Zaïre.
mauritianum *A.J. Scott in Kew Bull.*, 42(4): 854 (1987)—Mauritius.

HELICHRYSUM :–
polhillianum *S. Lisowski in Bull. Jard. Bot. Nation. Belg.*, 56(3–4): 360 (1986)—Zaïre.
robbrechtianum *S. Lisowski in Bull. Jard. Bot. Nation. Belg.*, 56(3–4): 349 (1986)—Zaïre.
symoensianum *S. Lisowski in Bull. Jard. Bot. Nation. Belg.*, 56(3–4): 361 (1986)—Zaïre.
theresae *S. Lisowski in Bull. Jard. Bot. Nation. Belg.*, 56(3–4): 358 (1986)—Zaïre.
zairense *S. Lisowski in Bull. Jard. Bot. Nation. Belg.*, 56(3–4): 366 (1986)—Zaïre.

HELICONIA (Heliconiac.).
atropurpurea *G.S. Daniels & F.G. Stiles in Brenesia,* 15, Supl.: 15 (1979)—Costa Rica, Panama.
barryana *W.J. Kress in Selbyana,* 9(1): 165 (1986)—Panama.
bella *W.J. Kress in Selbyana,* 9(1): 158 (1986)—Panama.
calatheaphyllus *G.S. Daniels & F.G. Stiles in Brenesia,* 15, Supl.: 17 (1979)—Costa Rica.
colgantea *R.R. Smith ex G.S. Daniels & F.G. Stiles in Brenesia,* 15, Supl.: 18 (1979)—Costa Rica.
deflexa *G.S. Daniels & F.G. Stiles in Brenesia,* 15, Supl.: 19 (1979)—Costa Rica, Panama.
golfodulcensis *G.S. Daniels & F.G. Stiles in Brenesia,* 15, Supl.: 20 (1979)—Costa Rica.
gracilis *G.S. Daniels & F.G. Stiles in Brenesia,* 15, Supl.: 21 (1979)—Costa Rica.
ignescens *G.S. Daniels & F.G. Stiles in Brenesia,* 15, Supl.: 22 (1979)—Costa Rica.
irrasa
var. albicosta *G.S. Daniels & F.G. Stiles in Brenesia,* 15, Supl.: 27 (1979)—Costa Rica.
var. glabra *G.S. Daniels & F.G. Stiles in Brenesia,* 15, Supl.: 25 (1979)—Costa Rica.
subsp. undulata *G.S. Daniels & F.G. Stiles in Brenesia,* 15, Supl.: 25 (1979)—Costa Rica, Nicaragua.
var. undulata *G.S. Daniels & F.G. Stiles in Brenesia,* 15, Supl.: 25 (1979)—Costa Rica, Nicaragua.
lankesteri
var. rubra *G.S. Daniels & F.G. Stiles in Brenesia,* 15, Supl.: 28 (1979)—Costa Rica.
lennartiana *W.J. Kress in Selbyana,* 9(1): 161 (1986)—Panama.
lindsayana *W.J. Kress in Selbyana,* 9(1): 158 (1986)—Panama.
lophocarpa *G.S. Daniels & F.G. Stiles in Brenesia,* 15, Supl.: 31 (1979)—Costa Rica, Panama.
lutea *W.J. Kress in Selbyana,* 9(1): 163 (1986)—Panama.
mathiasiae *G.S. Daniels & F.G. Stiles in Brenesia,* 15, Supl.: 33 (1979), as 'mathiasii'—Costa Rica, Nicaragua.
monteverdensis *G.S. Daniels & F.G. Stiles in Brenesia,* 15, Supl.: 34 (1979)—Costa Rica.

HELICONIA :–
pogonantha
var. holerythra *G.S. Daniels & F.G. Stiles in Brenesia,* 15, Supl.: 39 (1979)—Costa Rica, Panama.
var. pubescens *G.S. Daniels & F.G. Stiles in Brenesia,* 15, Supl.: 37 (1979)—Costa Rica.
ramonensis *G.S. Daniels & F.G. Stiles in Brenesia,* 15, Supl.: 39 (1979)—Costa Rica.
sarapiquensis *G.S. Daniels & F.G. Stiles in Brenesia,* 15, Supl.: 41 (1979)—Costa Rica, Panama.
secunda
var. viridiflora *G.S. Daniels & F.G. Stiles in Brenesia,* 15, Supl.: 42 (1979)—Costa Rica.
talamancana *G.S. Daniels & F.G. Stiles in Brenesia,* 15, Supl.: 42 (1979)—Costa Rica.
thomasiana *W.J. Kress in Selbyana,* 9(1): 156 (1986)—Panama.
trichocarpa *G.S. Daniels & F.G. Stiles in Brenesia,* 15, Supl.: 44 (1979)—Costa Rica.
trichocarpa
var. caducispatha *G.S. Daniels & F.G. Stiles in Brenesia,* 15, Supl.: 44 (1979)—Costa Rica.
umbrophila *G.S. Daniels & F.G. Stiles in Brenesia,* 15, Supl.: 45 (1979)—Costa Rica.
wilsonii *G.S. Daniels & F.G. Stiles in Brenesia,* 15, Supl.: 46 (1979)—Costa Rica.

HELICTERES (Sterculiac.).
vegae *C.L. Cristóbal in Bonplandia,* 6(1): 74 (1987)—Mexico.

HELICTOTRICHON (Gramin.).
adsurgens *(Schur ex Simonkai) H.J. Conert in G. Hegi, Illustr. Fl. Mitteleur.*, ed. 3, 1(3: Lief. 4): 249 (1987): Avena adsurgens.
subsp. ausserdorferi *(Ascherson & Graebner) H.J. Conert in G. Hegi, Illustr. Fl. Mitteleur.*, ed. 3, 1(3: Lief. 4): 250 (1987): Avena alpina C. ausserdorferi.
praeustum
subsp. pseudoviolaceum *(Kerner ex Dalla Torre) H.J. Conert in G. Hegi, Illustr. Fl. Mitteleur.*, ed. 3, 1(3: Lief. 4): 255 (1987): Avena pseudoviolacea.
quadridentulum *(Doell) S.A. Renvoize in Kew Bull.*, 42(4): 921 (1987): Avena quadridentula.

HELIOPSIS (Compos.).
novogaliciana *B.L. Turner in Phytologia,* 63(1): 1 (1987)—Mexico.
sinaloensis *B.L. Turner in Phytologia,* 63(1): 2 (1987)—Mexico.

HELIPTERUM (Compos.).
stoveae *D.A. Cooke in Fl. S. Australia,* 3: 1546 (1986)—Australia (S. Australia).

HELLADIA M. Král in Preslia, 59(4): 309 (1987). *CRASSULACEAE.*
cretica *(C. Presl) M. Král in Preslia,* 59(4): 309 (1987): Sedum creticum.

HELLADIA :-
tristriata *(Boiss.) M. Král in Preslia,* 59(4): 309 (1987): Sedum tristriatum.
tymphaea *(Quézel & Contandr.) M. Král in Preslia,* 59(4): 309 (1987): Sedum tymphaeum.

HELLENOCARUM (Umbellif.).
multiflorum
subsp. strictum *(Griseb.) Kit Tan in Pl. Syst. Evol.,* 154(1–2): 122 (1986): Bunium strictum.
pisidicum *Kit Tan in Pl. Syst. Evol.,* 154(1–2): 121 (1986)—Turkey.

×**HELPILIA** hort. in Orchid Rev., 95(1121): centre page pull-out p. 9 (1987). *ORCHIDACEAE.* (Helcia × Trichopilia).

HEMEROCALLIS (Hemerocallidac.).
fulva
var. aurantiaca *(Baker) M. Hotta in Acta Phytotax. Geobot.,* 37(1–3): 21 (1986): H. aurantiaca.
lilioasphodelus
var. yezoensis *(Hara) M. Hotta in Acta Phytotax. Geobot.,* 37(1–3): 19 (1986): H. yezoensis.
major *(Baker) M. Hotta in Acta Phytotax. Geobot.,* 37(1–3): 22 (1986): H. aurantiaca var. major.
middendorffii
var. exaltata *(Stout) M. Hotta in Acta Phytotax. Geobot.,* 37(1–3): 19 (1986): H. exaltata.

HEMIBOEA (Gesneriac.).
henryi
var. guandongensis *Z.Y. Li in Acta Phytotax. Sin.,* 25(3): 225 (1987)—China.
subcapitata
var. sordidopuberula *Z.Y. Li in Acta Phytotax. Sin.,* 25(3): 224 (1987)—China.

HEMILOPHIA (Crucif.).
pulchella
var. rockii *(O.E. Schulz) W.T. Wang in Acta Bot. Yunnanica,* 9(1): 1 (1987): H. rockii.

HENSMANIA (Anthericac.).
chapmanii *G.J. Keighery in Fl. Australia,* 45: 486, 251 (1987)—Australia (W. Australia).
stoniella *G.J. Keighery in Fl. Australia,* 45: 486, 251 (1987)—Australia (W. Australia).

HERACLEUM (Umbellif.).
dissectum
var. voroschilovii *(Gorovoi) M.G. Pimenov in Sosud. Rast. Sovet. Dal'nego Vostoka,* 2: 269 (1987): H. voroschilovii.
rechingeri *I. Mandenova in Fl. Iran.,* 162: 496 (1987)—Iran.

HERITIERA (Sterculiac.).
kanikensis *N.C. Majumdar & L.K. Banerjee in Bull. Bot. Surv. India,* 27(1–4): 150 (1985 publ. 1987)—India.

HERMINIUM (Orchidac.).
kumaunensis *S. Deva & H.B. Naithani, Orchid Fl. N.W. Himalaya:* 159 (1986)—India.

HERNIARIA (Illecebrac.).
sect. Austroamericanae *M.N. Chaudhri in Pakistan Syst.,* 3(1): 2 (1987).
austroamericana *M.N. Chaudhri & R. Rutishauser in Pakistan Syst.,* 3(1): 2 (1987)—Argentina, Bolivia.
pauciflora *(J. Rohlena) A.W. Hill in Index Kewensis,* Supp. 6: 99 (1916–1920 publ. 1926): H. hirsuta var. pauciflora.

HETERANTHERA (Pontederiac.).
subgen. Zosterella *(J.K. Small) T.J. Rosatti in J. Arnold Arbor.,* 68(1): 59 (1987): gen. Zosterella.

HETEROPLEXIS (Compos.).
microcephala *Y.L. Chen in Guihaia,* 5(4): 339 (1985)—China.

HETEROPTERIS (Malpighiac.).
alata *(W. Anderson) W.R. Anderson in Contrib. Univ. Mich. Herb.,* 16: 75 (1987): H. beecheyana var. alata.
aliciae *W.R. Anderson in Contrib. Univ. Mich. Herb.,* 16: 75 (1987)—Brazil.
fluminensis *(Grisebach) W.R. Anderson in Contrib. Univ. Mich. Herb.,* 16: 77 (1987): Hiraea fluminensis.
huberi *W.R. Anderson in Contrib. Univ. Mich. Herb.,* 16: 78 (1987)—Venezuela.
prunifolia *(H.B.K.) W.R. Anderson in Contrib. Univ. Mich. Herb.,* 16: 80 (1987): Hiraea ? prunifolia.
standleyana *W.R. Anderson in Contrib. Univ. Mich. Herb.,* 16: 80 (1987), nom. nov.: Banisteria rosea *Standley.*

HETEROSTIGMA (Pandanac.).
heudelotianum *C. Gaudichaud-Beaupré, Bot. Voy. Bonite, Atlas,* t. 25, f. 15–31 (1841)—Locality not stated. †

HETEROTHECA (Compos.).
barbata *(Rydb.) J.C. Semple in Brittonia,* 39(3): 379 (1987): Chrysopsis barbata.
chihuahuana *(B. Turner & Sundberg) B.L. Turner in Phytologia,* 63(2): 128 (1987): Osbertia chihuahuana.
fulcrata
var. arizonica *J.C. Semple in Brittonia,* 39(3): 380 (1987)—U.S.A. (Arizona).
var. senilis *(Wooton & Standley) J.C. Semple in Brittonia,* 39(3): 380 (1987): Chrysopsis senilis.
horrida
subsp. cinerascens *(S.F. Blake) J.C. Semple in Brittonia,* 39(3): 381 (1987): Chrysopsis viscida subsp. cinerascens.
oregona
var. compacta *(Keck) J.C. Semple in Brittonia,* 39(3): 382 (1987): Chrysopsis oregona var. compacta.

HETEROTHECA :–
var. rudis *(Greene) J.C. Semple in Brittonia,* 39(3): 382 (1987): Chrysopsis rudis.
var. scaberrima *(A. Gray) J.C. Semple in Brittonia,* 39(3): 383 (1987)—Chrysopsis oregona var. scaberrima.
pumila *(Greene) J.C. Semple in Brittonia,* 39(3): 383 (1987): Chrysopsis pumila.
thiniicola *(Rzed. & Ezcurra) B.L. Turner in Phytologia,* 63(2): 128 (1987): Haplopappus thiniicola.
vandevenderorum *B.L. Turner in Phytologia,* 63(2): 127 (1987)—Mexico.
villosa
var. pedunculata *(Greene) V. Harms ex J.C. Semple in Brittonia,* 39(3): 383 (1987): Chrysopsis pedunculata.
zionensis *J.C. Semple in Brittonia,* 39(3): 384 (1987)—U.S.A. (Utah).

HETEROTROPA (Aristolochiac.).
kumageana *(Masamune) F. Maekawa ex T. Yahara in J. Fac. Sci. Univ. Tokyo, Bot.,* 14(2): 81 (1987): Asarum kumageanum.

HETEROZEUXINE T. Hashimoto in Ann. Tsukuba Bot. Gard., 5: 21 (1986). *ORCHIDACEAE.*
nervosa *(Wall. ex Lindl.) T. Hashimoto in Ann. Tsukuba Bot. Gard.,* 5: 21 (1986): Monochilus nervosus.
odorata *(Fukuyama) T. Hashimoto in Ann. Tsukuba Bot. Gard.,* 5: 21 (1986): Zeuxine odorata.

HEXASTYLIS (Aristolochiac.).
shuttleworthii
var. harperi *L.L. Gaddy in Sida,* 12(1): 54 (1987)—U.S.A. (Georgia, Alabama).

HIBISCUS (Malvac.).
peralbus *P.A. Fryxell in Syst. Bot.,* 12(2): 277 (1987)—Australia (W. Australia).
zygomorphus *P.A. Fryxell & S.D. Koch in Aliso,* 11(4): 549 (1987)—Mexico.

HIERACIUM (Compos.).
scouleri
var. albertinum *(Farr) G.W. Douglas & G.A. Allen in Canad. J. Bot.,* 64(11): 2726 (1986): H. albertinum.
westii *P.D. Sell in Watsonia,* 16(3): 283 (1987)—Great Britain.

HIEROCHLOE (Gramin.).
hirta
subsp. praetermissa *G. Wiemarck in Symb. Bot. Upsal.,* 27(2): 176 (1986 publ. 1987)—Sweden, Germany, Switzerland, U.S.S.R.

HILLIA (Rubiac.).
bonoi *J.A. Steyermark in Ernstia,* 42: 5 (1987)—Venezuela.

HILLIARDIA B. Nordenstam in Op. Bot., 92: 147 (1987). *COMPOSITAE.*

HILLIARDIA :–
zuurbergensis *(Oliv.) B. Nordenstam in Op. Bot.,* 92: 149 (1987): Matricaria zuurbergensis.

HILLIELLA (O.E. Schulz) Y.H. Zhang & H.W. Li in Acta Bot. Yunnanica, 8(4): 401 (1986): Cochlearia sect. Hilliella (Crucif.).
alatipes *(Hand.-Mazz.) Y.H. Zhang & H.W. Li in Acta Bot. Yunnanica,* 8(4): 402 (1986): Cochlearia alatipes.
var. macranthus *Y.H. Zhang in Acta Bot. Yunnanica,* 8(4): 403 (1986)—China.
var. micranthus *Y.H. Zhang in Acta Bot. Yunnanica,* 8(4): 403 (1986)—China.
changhuaensis *Y.H. Zhang in Acta Bot. Yunnanica,* 9(2): 155 (1987)—China.
formosana *(Hayata) Y.H. Zhang & H.W. Li in Acta Bot. Yunnanica,* 8(4): 403 (1986): Cochlearia formosana.
fumarioides *(Dunn) Y.H. Zhang & H.W. Li in Acta Bot. Yunnanica,* 8(4): 403 (1986): Cochlearia fumarioides.
guangdongensis *Y.H. Zhang in Acta Bot. Yunnanica,* 9(2): 157 (1987)—China.
hui *(O.E. Schulz) Y.H. Zhang & H.W. Li in Acta Bot. Yunnanica,* 8(4): 404 (1986): Cochlearia hui.
hunanensis *Y.H. Zhang in Acta Bot. Yunnanica,* 9(2): 160 (1987)—China.
lichuanensis *Y.H. Zhang in Acta Bot. Yunnanica,* 9(2): 158 (1987)—China.
longistyla *Y.H. Zhang in Acta Bot. Yunnanica,* 9(2): 153 (1987)—China.
paradoxa *(Hance) Y.H. Zhang & H.W. Li in Acta Bot. Yunnanica,* 8(4): 404 (1986): Cardamine paradoxa.
rivulorum *(Dunn) Y.H. Zhang & H.W. Li in Acta Bot. Yunnanica,* 8(4): 404 (1986): Nasturtium rivulorum.
sinuata *(Kuan) Y.H. Zhang & H.W. Li in Acta Bot. Yunnanica,* 8(4): 404 (1986): Cochlearia sinuata.
var. qianwuensis *Y.H. Zhang in Acta Bot. Yunnanica,* 8(4): 405 (1986)—China.
warburgii *(O.E. Schulz) Y.H. Zhang & H.W. Li in Acta Bot. Yunnanica,* 8(4): 405 (1986): Cochlearia warburgii.

HIPPEASTRUM (Amaryllidac.).
wilsoniae *L.J. Doran & F.W. Meyer in Herbertia,* 43(1): 10 (1987)—Bolivia.

HIPPOCREPIS (Legumin.).
emerus
subsp. emeroides *(Boiss. & Spruner) W. Greuter & Burdet in Willdenowia,* 16(2): 443 (1987): Coronilla emeroides.

HIPTAGE (Malpighiac.).
calcicola *P. Sirirugsa in Nordic J. Bot.,* 7(3): 277 (1987)—Thailand.
gracilis *P. Sirirugsa in Nordic J. Bot.,* 7(3): 279 (1987)—Thailand.
monopteryx *P. Sirirugsa in Nordic J. Bot.,* 7(3): 279 (1987)—Thailand.

HODGSONIA (Cucurbitac.).
macrocarpa
var. capniocarpa *(Ridl.) Tsai ex A.M. Lu & Z.Y. Zhang in Fl. Reipubl. Popul. Sin.,* 73(1): 259 (1986): Hodgsonia capniocarpa.

HOLBOELLIA (Lardizabalac.).
khasiana *T.K. Paul & M.P. Nayar in Bull. Bot. Surv. India,* 27(1–4): 105 (1985 publ. 1987)—India.

HOLCOGLOSSUM (Orchidac.).
sect. Kimballianum *E.A. Christenson in Notes Roy. Bot. Gard. Edinburgh,* 44(2): 255 (1987).
amesianum *(Reichb. f.) E.A. Christenson in Notes Roy. Bot. Gard. Edinburgh,* 44(2): 255 (1987): Vanda amesiana.
saprophyticum *(Gagnep.) E.A. Christenson in Notes Roy. Bot. Gard. Edinburgh,* 44(2): 255 (1987): Vanda saprophytica.
sinicum *E.A. Christenson in Notes Roy. Bot. Gard. Edinburgh,* 44(2): 255 (1987)—China.
subulifolium *(Reichb. f.) E.A. Christenson in Notes Roy. Bot. Gard. Edinburgh,* 44(2): 255 (1987): Vanda subulifolia.

HOLCOLEMMA (Gramin.).
dispar *W.D. Clayton in Kew Bull.,* 42(2): 402 (1987), nom. nov.: Panicum inaequale *F. Muell.*

HOMOPHOLIS (Gramin.).
proluta *(F. Muell.) R.D. Webster, Australian Paniceae (Poac.):* 96 (1987): Panicum prolutum.

HOPEA (Dipterocarpac.).
ponga
var. cauveriana *K.R. Keshava Murthy & S.N. Yoganarasimhan in Current Sci.,* 56(11): 544 (1987)—India.

HORDEUM (Gramin.).
vulgare
var. abdulbasirovii *D.S. Omarov in Trudy̆ Prikl. Bot. Genet. Selek.,* 95: 57 (1985), as 'abdulbasirovi'—Cultivation.
var. bachteevii *D.S. Omarov in Trudy̆ Prikl. Bot. Genet. Selek.,* 95: 57 (1985), as 'bachteevi'—Cultivation.
var. lukyanovae *D.S. Omarov in Trudy̆ Prikl. Bot. Genet. Selek.,* 95: 57 (1985)—Cultivation.
var. lvovii *D.S. Omarov in Trudy̆ Prikl. Bot. Genet. Selek.,* 95: 57 (1985), as 'lvovi'—Cultivation.
var. valentinae *D.S. Omarov in Trudy̆ Prikl. Bot. Genet. Selek.,* 95: 58 (1985)—Cultivation.

HORNSTEDTIA (Zingiberac.).
gracilis *R.M. Smith in Notes Roy. Bot. Gard. Edinburgh,* 44(2): 233 (1987)—Borneo (Sabah).

HORSFIELDIA (Myristicac.).
coryandra *W.J.J.O. de Wilde in Blumea,* 32(2): 464 (1987)—New Guinea (Papua New Guinea).
discolor *W.J.J.O. de Wilde in Blumea,* 32(2): 469 (1987)—Borneo (Kalimantan, ?Sarawak, ?Brunei, ?Sabah).
elongata *W.J.J.O. de Wilde in Blumea,* 32(2): 465 (1987)—Malaya.
urceolata *W.J.J.O. de Wilde in Blumea,* 32(2): 462 (1987)—New Guinea (Papua New Guinea).

HOUSTONIA (Rubiac.).
canadensis
var. setiscaphia *(L.G. Carr) C.F. Reed in Phytologia,* 63(5): 412 (1987): H. setiscaphia.

HOYA (Asclepiadac.).
obscura *Elmer ex C.M. Burton in Hoyan,* 8(1): 15 (1986)—Philippine Is.
revoluta *Wight ex J.D. Hooker, Fl. Brit. India,* 4(1): 55 (1883); *cf. C.M. Burton in Hoyan,* 9(2): 33 (1987)— Malaya, Singapore. †

HUERNIA (Asclepiadac.).
formosa *L.C. Leach in Excelsa,* 12: 95 (1986)—Somali Republic.
lavrani *L.C. Leach in Excelsa,* 12: 94 (1986)—Somali Republic.

HYACINTHOIDES (Hyacinthac.).
subgen. Somera *(Salisb.) F. Speta in Phyton (Austria),* 26(2): 288 (1987): gen. Somera.
mauritanica *(Schousb.) F. Speta in Phyton (Austria),* 26(2): 288 (1987): Scilla mauritanica.
reverchonii *(Degen & Hervier) F. Speta in Phyton (Austria),* 26(2): 288 (1987): Scilla reverchonii.

HYDRANGEA (Hydrangeac.).
verticillata *W.H. Gao in Acta Phytotax. Sin.,* 25(5): 410 (1987)—China.
zhewanensis *P.S. Hsu & X.P. Zhang in Invest. Stud. Nat., Mus. Hist. Nat. Shanghaiense,* 7: 12 (1987)—China.

HYDROCOTYLE (Umbellif.).
craibii *H. Eichler in Feddes Repert.,* 98(3–4): 146 (1987), nom. nov.: H. javanica var. chinensis *Dunn ex Shan & Liou.*

HYGROCHILUS (Orchidac.).
parishii
var. marriottiana *(Rchb. f.) U.C. Pradhan in Indian Orchid J.,* 2(1): 11 (1987): Vanda parishii var. marriottiana.
var. purpurea *(N.E. Brown) U.C. Pradhan in Indian Orchid J.,* 2(1): 11 (1987): Vanda parishii var. purpurea.

HYMENOCARPOS (Legumin.).
hispanicus *P. Lassen in Willdenowia,* 16(2): 443 (1987), nom. nov.: Anthyllis lotoides *L. (1753).*

HYMENOLAENA (Umbellif.).
polyphylla *K.H. Rechinger in Fl. Iran.*, 162: 176 (1987)—Afghanistan.

HYMENORCHIS (Orchidac.).
serrulata *(Hallé) N. Hallé in Bull. Mus. Nation. Hist. Nat., B, Adansonia,* Ser. 4, 8(3): 224 (1986): Saccolabium serrulatum.

HYMENOXYS (Compos.).
lapidicola *S.L. Welsh & E. Neese in Great Basin Nat.*, 43(2): 373 (1983)—U.S.A. (Utah).

HYPERICUM (Guttif.).
sect. Knifa *(Adanson) N.K.B. Robson in Bull. Brit. Mus. (Nat. Hist.), Bot.*, 16(1): 2 (1987): gen. Knifa.
sect. Trigynobrathys *(Y. Kimura) N.K.B. Robson in Bull. Brit. Mus. (Nat. Hist.), Bot.*, 16(1): 3 (1987): Sarothra sect. Trigynobrathys.
acostanum *Steyerm. ex N.K.B. Robson in Bull. Brit. Mus. (Nat. Hist.), Bot.*, 16(1): 27 (1987)—Ecuador.
bolivaricum *N.K.B. Robson in Bull. Brit. Mus. (Nat. Hist.), Bot.*, 16(1): 72 (1987)—Colombia.
brasiliense
var. linoides *(St. Hil.) C. Rodríguez Jiménez in Mem. Soc. Cienc. Nat. La Salle,* 33(94–95): 126 (1973): H. linoides.
caracasanum
subsp. turumiquirense *(Steyerm.) N.K.B. Robson in Bull. Brit. Mus. (Nat. Hist.), Bot.*, 16(1): 90 (1987): H. caracasanum var. turumiquirense.
cassiopiforme *N.K.B. Robson in Bull. Brit. Mus. (Nat. Hist.), Bot.*, 16(1): 59 (1987)—Peru.
costaricense *N.K.B. Robson in Bull. Brit. Mus. (Nat. Hist.), Bot.*, 16(1): 70 (1987)—Costa Rica, Panama, Colombia.
cymobrathys *N.K.B. Robson in Bull. Brit. Mus. (Nat. Hist.), Bot.*, 16(1): 103 (1987)—Colombia.
denticulatum
var. virgatum *(Lam.) C. Rodríguez Jiménez in Mem. Soc. Cienc. Nat. La Salle,* 33(94–95): 100 (1973): H. virgatum.
empetrifolium
subsp. oliganthum *(Rech. f.) I. Hagemann in Pl. Syst. Evol.*, 155(1–4): 186 (1987): H. empetrifolium var. oliganthum.
subsp. tortuosum *(Rech. f.) I. Hagemann in Pl. Syst. Evol.*, 155(1–4): 186 (1987): H. empetrifolium var. tortuosum.
gleasonii *N.K.B. Robson in Bull. Brit. Mus. (Nat. Hist.), Bot.*, 16(1): 55 (1987)—Colombia.
horizontale *N.K.B. Robson in Bull. Brit. Mus. (Nat. Hist.), Bot.*, 16(1): 81 (1987)—Colombia.

HYPERICUM :–
irazuense *Kuntze ex N.K.B. Robson in Bull. Brit. Mus. (Nat. Hist.), Bot.*, 16(1): 30 (1987)—Costa Rica, Panama.
jaramilloi *N.K.B. Robson in Bull. Brit. Mus. (Nat. Hist.), Bot.*, 16(1): 86 (1987)—Colombia, Costa Rica.
lancioides
subsp. congestiflorum *(Triana & Planchon) N.K.B. Robson in Bull. Brit. Mus. (Nat. Hist.), Bot.*, 16(1): 101 (1987): H. struthiolifolium var. congestiflorum.
llanganaticum *N.K.B. Robson in Bull. Brit. Mus. (Nat. Hist.), Bot.*, 16(1): 95 (1987)—Ecuador.
loxense
subsp. aequatoriale *(R. Keller) N.K.B. Robson in Bull. Brit. Mus. (Nat. Hist.), Bot.*, 16(1): 44 (1987): H. loxense var. aequatoriale.
magdalenicum *N.K.B. Robson in Bull. Brit. Mus. (Nat. Hist.), Bot.*, 16(1): 63 (1987)—Colombia, Venezuela.
maguirei *N.K.B. Robson in Bull. Brit. Mus. (Nat. Hist.), Bot.*, 16(1): 52 (1987)—Ecuador.
marahuacanum *N.K.B. Robson in Bull. Brit. Mus. (Nat. Hist.), Bot.*, 16(1): 77 (1987)—Venezuela, Colombia.
subsp. chimantaicum *N.K.B. Robson in Bull. Brit. Mus. (Nat. Hist.), Bot.*, 16(1): 80 (1987)—Venezuela.
subsp. strictissimum *N.K.B. Robson in Bull. Brit. Mus. (Nat. Hist.), Bot.*, 16(1): 79 (1987)—Colombia, Venezuela.
martense *N.K.B. Robson in Bull. Brit. Mus. (Nat. Hist.), Bot.*, 16(1): 51 (1987)—Colombia.
papillosum *N.K.B. Robson in Bull. Brit. Mus. (Nat. Hist.), Bot.*, 16(1): 34 (1987)—Colombia.
parallelum *N.K.B. Robson in Bull. Brit. Mus. (Nat. Hist.), Bot.*, 16(1): 76 (1987)—Colombia.
pauciflorum
subsp. involutum *(Labill.) C. Rodríguez Jiménez in Mem. Soc. Cienc. Nat. La Salle,* 33(94–95): 105 (1973): Ascyrum involutum.
phellos
subsp. oroqueanum *N.K.B. Robson in Bull. Brit. Mus. (Nat. Hist.), Bot.*, 16(1): 25 (1987)—Colombia.
subsp. platyphyllum *(Gleason) N.K.B. Robson in Bull. Brit. Mus. (Nat. Hist.), Bot.*, 16(1): 25 (1987): H. platyphyllum.
pleiostylum *C. Rodríguez Jiménez in Mem. Soc. Cienc. Nat. La Salle,* 33(94–95): 118 (1973)—Brazil.
prietoi *N.K.B. Robson in Bull. Brit. Mus. (Nat. Hist.), Bot.*, 16(1): 58 (1987)—Ecuador.
recurvum *N.K.B. Robson in Bull. Brit. Mus. (Nat. Hist.), Bot.*, 16(1): 68 (1987)—Peru.
selaginella *N.K.B. Robson in Bull. Brit. Mus. (Nat. Hist.), Bot.*, 16(1): 102 (1987)—Colombia.

HYPERICUM :–
simonsii *N.K.B. Robson in Bull. Brit. Mus. (Nat. Hist.), Bot.*, 16(1): 34 (1987)—Colombia.
sprucei *N.K.B. Robson in Bull. Brit. Mus. (Nat. Hist.), Bot.*, 16(1): 65 (1987)—Ecuador, Peru.
strictum
subsp. compactum *(Triana & Planchon) N.K.B. Robson in Bull. Brit. Mus. (Nat. Hist.), Bot.*, 16(1): 85 (1987): H. struthiolifolium γ compactum.
valleanum *N.K.B. Robson in Bull. Brit. Mus. (Nat. Hist.), Bot.*, 16(1): 64 (1987)—Colombia.
wurdackii *N.K.B. Robson in Bull. Brit. Mus. (Nat. Hist.), Bot.*, 16(1): 69 (1987)—Peru.

HYPERTHELIA (Gramin.).
kottoensis *B.-M. Descoings & M. Mazade in Bull. Soc. Bot. France, Lett. Bot.*, 134(2): 203 (1987)—Central African Republic.

HYPOXIS (Hypoxidac.).
arillacea *R.J.F. Henderson in Fl. Australia,* 45: 487, 181 (1987)—Australia (Queensland).
exilis *R.J.F. Henderson in Fl. Australia,* 45: 487, 182 (1987)—Australia (Victoria, New South Wales).
gardneri *R.J.F. Henderson in Fl. Australia,* 45: 487, 190 (1987)—Australia (W. Australia).
glabella
var. leptantha *(Benth.) R.J.F. Henderson in Fl. Australia,* 45: 488 (1987): H. leptantha.
gracilis *J.G.C. Lehmann in Sem. Hort. Bot. Hamb.*: 17 (1826)—Locality not stated. †
hygrometrica
var. splendida *R.J.F. Henderson in Fl. Australia,* 45: 488, 184 (1987)—Australia (Victoria, New South Wales).
var. villosisepala *R.J.F. Henderson in Fl. Australia,* 45: 488, 184 (1987)—Australia (Victoria, Queensland, New South Wales), Tasmania.
nervosa *R.J.F. Henderson in Fl. Australia,* 45: 489, 181 (1987)—Australia (W. Australia, N. Terr.).
occidentalis
var. quadriloba *(F. Muell.) R.J.F. Henderson in Fl. Australia,* 45: 489 (1987): H. glabella var. quadriloba.
pratensis
var. tuberculata *R.J.F. Henderson in Fl. Australia,* 45: 489, 185 (1987)—Australia (Queensland, New South Wales).
vaginata
var. brevistigmata *R.J.F. Henderson in Fl. Australia,* 45: 490, 189 (1987)—Australia (Victoria, New South Wales), Tasmania.

HYPTIS (Labiat.).
fasciculata
subsp. fastigiata *(Bentham) R.M. Harley in Fl. Illustr. Catar.*, 1 (Labiadas: gen. 1): 14 (1985): H. fastigiata.

I

ICACINA (Icacinac.).
oliviformis
var. pubescens *(Boutique) J.M. Fay in Kew Bull.*, 42(2): 464 (1987): I. senegalensis var. pubescens.

ICHNANTHUS (Gramin.).
pallens
var. majus *(Nees) M.T. Stieber in Syst. Bot.*, 12(2): 207 (1987): Panicum pallens var. majus.

ICHNOCARPUS (Apocynac.).
baillonii *(Pierre ex Spire) Trân Đính Lý in Feddes Repert.*, 97(9–10): 675 (1986): Micrechites baillonii.

ILEX (Aquifoliac.).
sect. Stigmatophorae *(Loes.) T. Yamazaki in J. Jap. Bot.*, 62(6): 191 (1987): I. subsect. Stigmatophorae.
dabieshanensis *K. Yao & M.B. Deng in Acta Phytotax. Sin.*, 25(4): 324 (1987)—China.
liesneri *J.A. Steyermark in Ann. Missouri Bot. Gard.*, 74(1): 94 (1987)—Venezuela.
maximowicziana
var. kanehirae *(Yamamoto) T. Yamazaki in J. Jap. Bot.*, 62(6): 191 (1987): I. crenata var. kanehirai.
mertensii
var. beecheyi *(Loes.) T. Yamazaki in J. Jap. Bot.*, 62(2): 64 (1987): I. integra var. beecheyi.
rugosa
var. hondoensis *T. Yamazaki in J. Jap. Bot.*, 62(5): 144 (1987)—Japan.
tadiandamolensis *K.R. Keshava Murthy, S.N. Yoganarasimhan & K. Vasudevan Nair in Current Sci.*, 56(6): 270 (1987), as 'tadiandamolense'—India.

IMPATIENS (Balsaminac.).
alliensii *(Zoll.) T. Shimizu in Acta Phytotax. Geobot.*, 38: 53 (1987): Semeiocardium alliensii.
hainanensis *Y.L. Chen in Guihaia,* 7(1): 9 (1987)—Hainan.

INDIGOFERA (Legumin.).
bancroftii *P.G. Wilson in J. Adelaide Bot. Gard.*, 10(1): 125 (1987)—Australia (Queensland).
basedowii
subsp. longibractea *(J.M. Black) P.G. Wilson in J. Adelaide Bot. Gard.*, 10(1): 121 (1987): I. longibractea.
gifbergensis *C.H. Stirton & J.K. Jarvie in Bothalia,* 17(1): 2 (1987)—South Africa (Cape Province).
hebepetala
f. glabra *(Ali) H. Ohashi in Enum. Fl. Pl. Nepal,* 2: 123 (1979): I. hebepetala var. glabra.

INDIGOFERA :–
helmsii *P.G. Wilson in J. Adelaide Bot. Gard.*, 10(1): 119 (1987), nom. nov.: I. uncinata *Ewart & Kerr.*
ionii *J.K. Jarvie & C.H. Stirton in Bothalia,* 17(1): 3 (1987)—South Africa (Cape Province).
kasinii *K. Boonyamalik in Thai Forest Bull., Bot.*, 16: 207 (1986)—Thailand.
rubroglandulosa *G. Germishuizen in Bothalia,* 17(1): 33 (1987)—South Africa (Natal, Cape Province).
thesioides *J.K. Jarvie & C.H. Stirton in Bothalia,* 16(2): 230 (1986)—South Africa (Cape Province).
verruculosa *P.G. Wilson in J. Adelaide Bot. Gard.*, 10(1): 123 (1987)—Australia (N. Terr.).

INULA (Compos.).
scaberrima *(J. Rohlena) A.W. Hill in Index Kewensis,* Supp. 6: 107 (1916–1920 publ. 1926): I. squarrosa var. scaberrima.
sechmenii *P. Hartvig & A. Strid in Bot. Jahrb.*, 108(2–3): 307 (1987)—Turkey.

IOCHROMA (Solanac.).
confertiflorum *(Miers) A.T. Hunziker in Kurtziana,* 15: 100 (1982): Acnistus confertiflorus.
umbellatum *(Ruiz & Pavon) A.T. Hunziker in Kurtziana,* 15: 101 (1982): Lycium umbellatum.
umbellatum *(R. & P.) W.G. D'Arcy in Solanaceae Newsl.*, 2(4): 21 (1986), isonym see A.T. Hunziker *in Kurtziana,* 15: 101 (1982): Lycium umbellatum.

IPOMOEA (Convolvulac.).
bampsiana *J. Lejoly & S. Lisowski in Bull. Jard. Bot. Nation. Belg.*, 56(3–4): 373 (1986)—Zaïre.
barteri
var. longisepala *J. Lejoly & S. Lisowski in Bull. Jard. Bot. Nation. Belg.*, 56(3–4): 380 (1986)—Zaïre.
deccana
var. lobata *(Clarke) S.C. Johri in Bull. Bot. Surv. India,* 26(3–4): 237 (1984 publ. 1986): I. bracteata var. lobata.
duvigneaudii *J. Lejoly & S. Lisowski in Bull. Jard. Bot. Nation. Belg.*, 56(3–4): 374 (1986)—Zaïre.
hirtifolia *R.C. Fang & S.H. Huang in Fl. Xizangica,* 4: 28 (1985)—Xizang.
lapathifolia
subsp. shabensis *J. Lejoly & S. Lisowski in Bull. Jard. Bot. Nation. Belg.*, 56(3–4): 378 (1986)—Zaïre.
linosepala
subsp. kundelungensis *J. Lejoly & S. Lisowski in Bull. Jard. Bot. Nation. Belg.*, 56(3–4): 379 (1986)—Zaïre.
subsp. upembensis *J. Lejoly & S. Lisowski in Bull. Jard. Bot. Nation. Belg.*, 56(3–4): 380 (1986)—Zaïre.

IPOMOEA :–
marginata *(Desr.) H. Manitz in Feddes Repert.*, 85(9–10): 638 (1974): Convolvulus marginatus.
marginata *(Desr.) B. Verdcourt in Kew Bull.*, 42(3): 658 (1987), isonym see *H. Manitz in Feddes Repert.*, 85(9–10): 638 (1974): Convolvulus marginatus.
miquihuanensis *J.A. McDonald in Brittonia,* 39(1): 110 (1987)—Mexico.
reflexisepala *J. Lejoly & S. Lisowski in Bull. Jard. Bot. Nation. Belg.*, 56(3–4): 375 (1986)—Zaïre.
robynsiana *J. Lejoly & S. Lisowski in Bull. Jard. Bot. Nation. Belg.*, 56(3–4): 376 (1986)—Zaïre.
verdcourtiana *J. Lejoly & S. Lisowski in Bull. Jard. Bot. Nation. Belg.*, 56(3–4): 377 (1986)—Zaïre.
zimmermanii *J.A. McDonald in Brittonia,* 39(1): 108 (1987)—Mexico.

IPOMOPSIS (Polemoniac.).
aggregata
subsp. carmenensis *J. Henrickson in Aliso,* 11(4): 591 (1987)—Mexico.
subsp. weberi *V. Grant & D.H. Wilken in Bot. Gaz.*, 147(3): 367 (1986)—U.S.A. (Colorado, Idaho).
pringlei *(Gray) J. Henrickson in Aliso,* 11(4): 594 (1987): Gilia pringlei.
wendtii *J. Henrickson in Aliso,* 11(4): 589 (1987)—Mexico.

IRESINE (Amaranthac.).
diffusa
var. spiculigera *(Seubert) U. Eliasson in Fl. Ecuador,* 28(44): 124 (1987): I. spiculigera.
leptoclada *(Hook. f.) J. Henrickson & S. Sundberg in Aliso,* 11(3): 360 (1986): Dicraurus leptocladus.
pedicellata *U. Eliasson in Fl. Ecuador,* 28(44): 116 (1987)—Ecuador.

IRIS (Iridac.).
ser. Acutilobae *B.D. Gavrilenko in Zam. Sist. Geogr. Rast.*, 41: 74 (1986).
ser. Ibericae *B.D. Gavrilenko in Zam. Sist. Geogr. Rast.*, 41: 74 (1986).
ser. Paradoxae *B.D. Gavrilenko in Zam. Sist. Geogr. Rast.*, 41: 75 (1986).
paradoxa
subsp. choschab *(Baker) B.D. Gavrilenko in Zam. Sist. Geogr. Rast.*, 41: 75 (1986), without basionym ref.: I. paradoxa var. choschab.
vorobievii *N.S. Pavlova in Sosud. Rast. Sovet. Dal'nego Vostoka,* 2: 424 (1987)—U.S.S.R. (Soviet Far East).

IRLBACHIA (Gentianac.).
poeppigii *(Grisebach) L. Cobb & P.J.M. Maas in Proc. Kon. Nederl. Akad. Wetensch.*, C, 86(2): 131 (1983): Pagaea poeppigii.

IRLBACHIA :–
pratensis *(H.B.K.) L. Cobb & P.J.M. Maas in Proc. Kon. Nederl. Akad. Wetensch., C,* 86(2): 132 (1983): Lisianthus pratensis.

ISACHNE (Gramin.).
goiasensis *S.A. Renvoize in Kew Bull.,* 42(4): 928 (1987)—Brazil.

ISIDOREA (Rubiac.).
gonavensis *A. Aiello & A. Borhidi in Acta Bot. Hung.,* 32(1–4): 221 (1986)—Hispaniola (Haiti).
veris *Ekman ex A. Aiello & A. Borhidi in Acta Bot. Hung.,* 32(1–4): 222 (1986)—Hispaniola (Dominican Republic).

ISOTROPIS (Legumin.).
foliosa *M.D. Crisp in Austrobaileya,* 2(4): 412 (1987)—Australia (Queensland).

ITEA (Escalloniac.).
quizhouensis *H.T. Chang & Y.K. Li in Acta Sci. Nat. Univ. Sunyatseni,* 1987(2): 43 (1987)—China.

IXERIS (Compos.).
parva *(Kitamura) T. Yahara in J. Fac. Sci. Univ. Tokyo, Bot.,* 14(2): 115 (1987): I. dentata f. parva.
yakuinsularis *T. Yahara in J. Fac. Sci. Univ. Tokyo, Bot.,* 14(2): 115 (1987)—Japan.

IXIOLAENA (Compos.).
chloroleuca *L. Haegi in Fl. S. Australia,* 3: 1551 (1986)—Australia (S. Australia, N. Terr., Queensland, New South Wales, Victoria).
pluriseta *L. Haegi in Fl. S. Australia,* 3: 1552 (1986)—Australia (S. Australia).

IXORA (Rubiac.).
baileyana *D.M. Bridson & L.G. Adams in Kew Bull.,* 42(1): 214 (1987), nom. nov.: Lasianthus graciliflorus *Bailey.*

J

JABOROSA (Solanac.).
cabrerae *G. Barboza in Kurtziana,* 18: 89 (1986)—Argentina.

JAGERA (Sapindac.).
javanica
subsp. australiana *P.W. Leenhouts in Blumea,* 32(1): 225 (1987)—Australia (Queensland).

JANUSIA (Malpighiac.).
christianeae *W.R. Anderson in Contrib. Univ. Mich. Herb.,* 16: 80 (1987)—Brazil.

JANUSIA :–
hexandra *(Vell.) W.R. Anderson in Contrib. Univ. Mich. Herb.,* 16: 83 (1987): Banisteria hexandra.
occhionii *W.R. Anderson in Contrib. Univ. Mich. Herb.,* 16: 84 (1987)—Brazil, Paraguay.
prancei *W.R. Anderson in Contrib. Univ. Mich. Herb.,* 16: 87 (1987)—Brazil.

JASIONE (Campanulac.).
corymbosa
var. glabra *(Durieu ex Boiss. & Reuter) J. Parnell in Watsonia,* 16(3): 265 (1987): J. glabra.
crispa
subsp. lanuginella *(Litard. & Maire) J. Lambinon & J. Lewalle in Soc. Ech. Pl. Vasc. Eur. Bassin Médit. Bull.,* 21: 59 (1986): J. amethystina var. lanuginella.
heldreichii
var. papillosa *J. Parnell in Watsonia,* 16(3): 266 (1987)—Turkey.
montana
var. imbricans *J. Parnell in Watsonia,* 16(3): 264 (1987)—Spain, France.

JATROPHA (Euphorbiac.).
subsect. Dichtares *A. Radcliffe-Smith in Kew Bull.,* 42(1): 116 (1987).
subsect. Furciferae *A. Radcliffe-Smith in Kew Bull.,* 42(1): 115 (1987).
subsect. Parvifoliae *A. Radcliffe-Smith in Kew Bull.,* 42(1): 119 (1987).
sect. Robecchiana *A. Radcliffe-Smith in Kew Bull.,* 42(1): 120 (1987).
subsect. Tropaeolifoliae *A. Radcliffe-Smith in Kew Bull.,* 42(1): 115 (1987).
dichtar
var. gracilior *A. Radcliffe-Smith in Kew Bull.,* 42(1): 116 (1987)—Somali Republic.
elbae *J. Jiménez Ramírez in Cact. Suc. Mex.,* 31(1): 3 (1986)—Mexico.
rosea *A. Radcliffe-Smith in Kew Bull.,* 42(1): 118 (1987)—Somali Republic.
stevensii *G.L. Webster in Ann. Missouri Bot. Gard.,* 74(1): 117 (1987)—Nicaragua.
stuhlmannii
subsp. somalensis *A. Radcliffe-Smith in Kew Bull.,* 42(1): 107 (1987)—Somali Republic.

JOHNSONIA (Anthericac.).
inconspicua *G.J. Keighery in Fl. Australia,* 45: 490, 246 (1987)—Australia (W. Australia).

JOHRENIA (Umbellif.).
aromatica *K.H. Rechinger in Fl. Iran.,* 162: 376 (1987)—Iran, Iraq.
golestanica *K.H. Rechinger in Fl. Iran.,* 162: 376 (1987)—Iran.

JOHRENIOPSIS M.G. Pimenov in Fl. Iran., 162: 454 (1987). *UMBELLIFERAE.*

JOHRENIOPSIS :–
oligactis *(Rech. f. & H. Riedl) M.G. Pimenov in Fl. Iran.*, 162: 456 (1987): Peucedanum oligactis.
scoparia *(Boiss.) M.G. Pimenov in Fl. Iran.*, 162: 455 (1987): Johrenia scoparia.
seseloides *(C.A. Mey.) M.G. Pimenov in Fl. Iran.*, 162: 455 (1987): Ferula seseloides.
stricticaulis *(Rech. f.) M.G. Pimenov in Fl. Iran.*, 162: 456 (1987): Peucedanum stricticaule.

JUNCUS (Juncac.).
alpinoarticulatus
subsp. americanus *(Farwell) L. Hämet-Ahti in Ann. Bot. Fenn.*, 23(4): 277 (1986): J. alpinus var. americanus.
subsp. fuscescens *(Fern.) L. Hämet-Ahti in Ann. Bot. Fenn.*, 23(4): 280 (1986): J. alpinus var. fuscescens.
×donanae *M.C. Fernández Carvajal in Lagascalia*, 14(1): 117 (1986) as 'donyanae'—Spain. (J. acutiflorus subsp. rugosus × striatus).
hemiendytus
var. abjectus *(F.J. Hermann) B. Ertter in Mem. New York Bot. Gard.*, 39: 76 (1986): J. abjectus.
leiospermus
var. ahartii *B. Ertter in Mem. New York Bot. Gard.*, 39: 49 (1986)—U.S.A. (California).
luciensis *B. Ertter in Mem. New York Bot. Gard.*, 39: 58 (1986)—U.S.A. (California).
tiehmii *B. Ertter in Mem. New York Bot. Gard.*, 39: 60 (1986)—U.S.A. (California, Idaho, Nevada, Oregon), Mexico.

JUSTICIA (Acanthac.).
colorata *(Nees) D.C. Wasshausen in Op. Bot.*, 92: 274 (1987): Sericographis colorata.
orosiensis *L.H. Durkee in Fieldiana, Bot.*, n.s., 18: 56 (1986)—Costa Rica.

K

KADENIA T.V. Lavrova & V.N. Tikhomirov in Byull. Mosk. Obshch. Ispȳt. Prir., Biol., 91(2): 93 (1986). *UMBELLIFERAE.*
dubia *(Schkuhr) T.V. Lavrova & V.N. Tikhomirov in Byull. Mosk. Obshch. Ispȳt. Prir., Biol.*, 91(2): 93 (1986): Seseli dubium.
salina *(Turcz.) T.V. Lavrova & V.N. Tikhomirov in Byull. Mosk. Obshch. Ispȳt. Prir., Biol.*, 91(2): 93 (1986): Cnidium salinum.

KALAKIA (Umbellif.).
marginata *(Boiss.) R. Alava in Fl. Iran.*, 162: 472 (1987): Cymbocarpum marginatum.

KALAKIA :–
f. leiocarpa *(Bornm. & Gauba) R. Alava in Fl. Iran.*, 162: 473 (1987): Ducrosia stenocarpa f. leiocarpa.

KANAHIA (Asclepiadac.).
carlsbergiana *D. Field, I. Friis & M.G. Gilbert in Nordic J. Bot.*, 6(6): 791 (1986)—Ethiopia.

KARATAVIA M.G. Pimenov & T.V. Lavrova in Bot. Zhurn., 72(1): 36 (1987). *UMBELLIFERAE.*
kultiassovii *(Korov.) M.G. Pimenov & T.V. Lavrova in Bot. Zhurn.*, 72(1): 36 (1987): Selinum kultiassovii.

KARIVIA G.A.W. Arnott in Madras J. Lit. Sci., 12: 50 (1840). *CUCURBITACEAE.*

KEISKEA (Labiat.).
elsholtzioides
f. purpurea *X.H. Guo in Bull. Bot. Res. North-East. Forest. Inst.*, 7(2): 137 (1987)—China.

KIBATALIA (Apocynac.).
villosa *Rudjiman in Agric. Univ. Wageningen Pap.*, 86(5): 82 (1986 publ. 1987)—Borneo (Kalimantan, Sarawak), Malaya.

KICKXIA (Scrophulariac.).
aegyptiaca
subsp. fruticosa *(Desf.) M. Qaiser in Fl. Libya*, 88: 65 (1982): Linaria fruticosa.

KIELMEYERA (Guttif.).
abdita *N. Saddi in Kew Bull.*, 42(1): 221 (1987)—Brazil.
bifaria *N. Saddi in Kew Bull.*, 42(1): 223 (1987)—Brazil.
coriacea
var. glabripes *N. Saddi in Bradea*, 4(35): 283 (1986)—Brazil.
var. guiaensis *N. Saddi in Bradea*, 4(35): 283 (1986)—Brazil.
var. intermedia *N. Saddi in Bradea*, 4(35): 283 (1986)—Brazil.
var. pseudotomentosa *N. Saddi in Bradea*, 4(35): 283 (1986)—Brazil.
subsp. tomentosa *(Camb.) N. Saddi in Bradea*, 4(35): 283 (1986): K. tomentosa.
corymbosa
var. glaziouana *N. Saddi in Bradea*, 4(35): 284 (1986), as 'glaziouiana'—Brazil.
var. oligantha *N. Saddi in Bradea*, 4(35): 284 (1986)—Brazil.
lathrophyton *N. Saddi in Kew Bull.*, 42(1): 225 (1987)—Brazil.
petiolaris
var. cipoensis *N. Saddi in Bradea*, 4(35): 284 (1986)—Brazil.
pumila
var. wawrana *N. Saddi in Bradea*, 4(35): 284 (1986), as 'wawraiana'—Brazil.
regalis *N. Saddi in Kew Bull.*, 42(1): 227 (1987)—Brazil.

KIELMEYERA :–
rubriflora
var. affinis *N. Saddi in Bradea,* 4(35): 285 (1986)—Brazil.
var. major *N. Saddi in Bradea,* 4(35): 285 (1986)—Brazil.
variabilis
var. robusta *N. Saddi in Bradea,* 4(35): 285 (1986)—Brazil.
var. stenophylla *N. Saddi in Bradea,* 4(35): 286 (1986)—Brazil.

KINGDON-WARDIA (Gentianac.).
racemosa *(Griseb.) T.N. He in Fl. Xizangica,* 3: 982 (1986): Ophelia racemosa.

KLEINIA (Compos.).
balsamica *(Dalz. & Gibs.) P. Halliday in Kew Bull.,* 42(2): 442 (1987): Notonia balsamica.

KNAUTIA (Dipsacac.).
lebrunii *J. Prudhomme in Bull. Mens. Soc. Linn. Lyon,* 56(4): 113 (1987)—France.

KNEMA (Myristicac.).
glauca
var. riparia *W.J.J.O. de Wilde in Blumea,* 32(1): 126 (1987)—Borneo (Sarawak).
lateritica
var. subtilis *W.J.J.O. de Wilde in Blumea,* 32(1): 118 (1987)—Philippine Is.
mogeana *W.J.J.O. de Wilde in Blumea,* 32(1): 136 (1987)—Borneo (Kalimantan, Sarawak, Sabah).
riangensis *W.J.J.O. de Wilde in Blumea,* 32(1): 138 (1987)—Borneo (Kalimantan, ?Sarawak).
stenophylla
subsp. longipedicellata *(Sinclair) W.J.J.O. de Wilde in Blumea,* 32(1): 130 (1987): K. cinerea f. longipedicellata.
stylosa *(de Wilde) W.J.J.O. de Wilde in Blumea,* 32(1): 119 (1987): K. hirtella var. stylosa.
subhirtella *W.J.J.O. de Wilde in Blumea,* 32(1): 131 (1987)—Borneo (Sarawak, Sabah).
viridis *W.J.J.O. de Wilde in Blumea,* 32(1): 121 (1987)—Borneo (Sarawak).

KNIPHOFIA (Asphodelac.).
albomontana *H. Baijnath in S. Afr. J. Bot.,* 53(4): 307 (1987)—South Africa (Natal).
angustifolia *(Bak.) L.E. Codd in Bothalia,* 16(2): 231 (1986): K. natalensis var. angustifolia.
bruceae *(Codd) L.E. Codd in Bothalia,* 17(2): 185 (1987): K. praecox subsp. bruceae.
ichopensis
var. aciformis *L.E. Codd in Bothalia,* 16(2): 231 (1986)—South Africa (Natal).

KOANOPHYLLON (Compos.).
guerreroana *B.L. Turner in Phytologia,* 63(3): 203 (1987)—Mexico.
rzedowskii *B.L. Turner in Phytologia,* 63(3): 203 (1987)—Mexico.

KOANOPHYLLON :–
sinaloensis *B.L. Turner in Phytologia,* 63(3): 202 (1987)—Mexico.

KOBRESIA (Cyperac.).
bistaminata *W.Z. Di & M.J. Zhong in Acta Bot. Bor.-Occid. Sin.,* 6(4): 275 (1986)—China.
esbirajbhandarii *K.R. Rajbhandari & H. Ohba in J. Jap. Bot.,* 62(9): 272 (1987)—Nepal.
harae *K.R. Rajbhandari & H. Ohba in J. Jap. Bot.,* 62(7): 193 (1987), as 'harai'—Nepal.
mallae *K.R. Rajbhandari & H. Ohba in J. Jap. Bot.,* 62(9): 270 (1987)—Nepal.

KOEHNERIA S.A. Graham, H. Tobe & P. Baas in Ann. Missouri Bot. Gard., 73(4): 805 (1986 publ. 1987). *LYTHRACEAE.*
madagascariensis *(Baker) S.A. Graham, H. Tobe & P. Baas in Ann. Missouri Bot. Gard.,* 73(4): 806 (1986 publ. 1987): Lagerstroemia madagascariensis.

KOELERIA (Gramin.).
arenaria *(Dumort.) B.D. Jackson in Index Kewensis,* 2: 10 (1753–1885 publ. 1894): K. cristata * (subsp.) arenaria.
arenaria *(Dumortier) H.J. Conert in G. Hegi, Illustr. Fl. Mitteleur.,* ed. 3, 1(3: Lief. 4): 270 (1987), isonym: K. cristata subsp. * K. arenaria.
cristata
subsp. helvetica *(Domin) H.J. Conert in G. Hegi, Illustr. Fl. Mitteleur.,* ed. 3, 1(3: Lief. 4): 267 (1987): K. gracilis subsp. helvetica.
dactylis *Chaubard in Bory & Chaubard, Nouv. Fl. Pélop.:* 6 (1838)—Locality not stated.
mannagettae *(Domin) D. Prain in Index Kewensis,* Suppl. 4: (1906–10 publ. 1913): K. alpigena * mannagettae.
narbonnensis *(Domin) D. Prain in Index Kewensis,* Suppl. 4: 135 (1906–10 publ. 1913): K. gracilis subsp. narbonnensis.
pontarlieri *(Domin) D. Prain in Index Kewensis,* Suppl. 4: 135 (1906–10 publ. 1913): K. albescens subsp. pontarlieri.
schroeteriana *(Domin) D. Prain in Index Kewensis,* Suppl. 4: 135 (1906–10 publ. 1913): K. eriostachya subsp. schroeteriana.

KORSHINSKYA (Umbellif.).
rapifera *(Gilli) Podlech & K.H. Rechinger in Fl. Iran.,* 162: 261 (1987): Bupleurum rapiferum.

KOSTELETZKYA (Malvac.).
flavicentrum *P.A. Fryxell & S.D. Koch in Aliso,* 11(4): 549 (1987)—Mexico.

KRAMERIA (Krameriac.).
bahiana *B.B. Simpson in Brittonia,* 39(2): 198 (1987)—Brazil.

KUNTHERIA J.G. Conran & H.T. Clifford in Fl. Australia, 45: 490, 416 (1987). *COLCHICACEAE.*

KUNTHERIA :–
pedunculata *(F. Muell.) J.G. Conran & H.T. Clifford in Fl. Australia,* 45: 491 (1987): Schelhammera pedunculata.

KYLLINGA (Cyperac.).
brevifolioides *(Delahoussaye & Thieret) G.C. Tucker in J. Arnold Arbor.,* 68(4): 408 (1987): Cyperus brevifolioides.

L

LACHENALIA (Hyacinthac.).
dehoopensis *W.F. Barker in S. Afr. J. Bot.,* 53(2): 167 (1987)—South Africa (Cape Province).
kliprandensis *W.F. Barker in S. Afr. J. Bot.,* 53(2): 168 (1987)—South Africa (Cape Province).
leomontana *W.F. Barker in S. Afr. J. Bot.,* 53(2): 169 (1987)—South Africa (Cape Province).
macgregoriorum *W.F. Barker in J. S. Afr. Bot.,* 45(2): 199 (1979); error 'macgregori' corrected by G.D. Duncan *in Flow. Pl. Afr.,* 49(3–4): t. 1951 (1987)— South Africa (Cape Province). †
minima *W.F. Barker in S. Afr. J. Bot.,* 53(2): 166 (1987)—South Africa (Cape Province).
namibiensis *W.F. Barker in S. Afr. J. Bot.,* 53(2): 170 (1987)—Namibia.

LACTUCA (Compos.).
kashmiriana *S.K. Mamgain & R.R. Rao in J. Bombay Nat. Hist. Soc.,* 83(2): 406 (1986)—India.
lahulensis *S.K. Mamgain & R.R. Rao in Bull. Bot. Surv. India,* 27(1–4): 120 (1985 publ. 1987)—India.
pakistanica *T. Akhtar & M.N. Chaudhri in Pakistan Syst.,* 3(1): 25 (1987)—Pakistan.

LAFOENSIA (Lythrac.).
vandelliana
subsp. replicata *(Pohl) A. Lourteig in Mem. Soc. Cienc. Nat. La Salle,* 45(123): 134 (1985): L. replicata.

LAGENANDRA (Arac.).
gomezii *(Schott) J. Bogner & N. Jacobsen in Aqua Planta,* 1987(2): 49 (1987): Cryptocoryne gomezii.

LAGERSTROEMIA (Lythrac.).
subcostata
var. fauriei *(Koehne) Hatusima ex T. Yahara in J. Fac. Sci. Univ. Tokyo, Bot.,* 14(2): 98 (1987): L. fauriei.

LAGYNIAS (Rubiac.).
subgen. Bembea *B. Verdcourt in Kew Bull.,* 42(1): 141 (1987).

LAGYNIAS :–
rufescens *(E.A. Bruce) B. Verdcourt in Kew Bull.,* 42(1): 143 (1987): Ancylanthos rufescens.
subsp. angustiloba *B. Verdcourt in Kew Bull.,* 42(1): 143 (1987)—Tanzania.

LAMIOPHLOMIS (Labiat.).
rotata
var. subglabra *C.Y. Wu in Fl. Xizangica,* 4: 159 (1985)—Xizang.

LAMIUM (Labiat.).
ser. Alba *Gorschk. ex Sh.I. Kutateladze in Fl. Gruzii, ed. 2,* 11: 237 (1987).
ser. Maculata *Gorschk. ex Sh.I. Kutateladze in Fl. Gruzii, ed. 2,* 11: 238 (1987).
ser. Tomentosa *Gorschk. ex Sh.I. Kutateladze in Fl. Gruzii, ed. 2,* 11: 237 (1987).
garganicum
subsp. longiflorum *(Ten.) M. Kerguélen in Lejeunia,* 120: 116 (1987): L. longiflorum.
gilongensis *H.W. Li in Fl. Xizangica,* 4: 163 (1985)—Xizang.

LAMPRANTHUS (Aizoac.).
deltoides
var. caulescens *(Mill.) G.D. Rowley in Excelsa,* 12: 34 (1986): Mesembryanthemum caulescens.

LANTANA (Verbenac.).
depressa
var. floridana *(Moldenke) R.W. Sanders in Syst. Bot.,* 12(1): 55 (1987): L. bahamensis var. floridana.
var. sanibelensis *R.W. Sanders in Syst. Bot.,* 12(1): 55 (1987)—U.S.A. (Florida).
hodgei *R.W. Sanders in J. Arnold Arbor.,* 68(3): 343 (1987)—Windward Is.
trifolia
f. congolensis *(Moldenke) R.B. Fernandes in Bol. Soc. Brot.,* 59: 254 (1986): L. mearnsii var. congolensis.

LAPPULA (Boraginac.).
mogoltavica *M. Pop. ex A.P. Chukavina in Izv. Akad. Nauk Tadzh. SSR, Biol. Nauk,* 3(92): 92 (1983)—U.S.S.R. (Middle Asia).
nevskii *L.M. Raenko in Bot. Zhurn.,* 72(1): 99 (1987)—U.S.S.R. (Middle Asia).

LATHYRUS (Legumin.).
laevigatus
subsp. emodi *(Wall. ex Fritsch) H. Ohashi in Enum. Fl. Pl. Nepal,* 2: 123 (1979): Orobus emodi.
subsp. hispanicus *(Rouy) M. Kerguélen in Lejeunia,* 120: 116 (1987): L. linnaei proles hispanicus.
roseus
subsp. hermonis *A. Liston in Bot. Jahrb.,* 109(1): 64 (1987)—Israel/Syria border, Turkey.

LATHYRUS :-
rotundifolius
subsp. litvinovii *(Iljin) J. Jäger in Wiss. Zeitschr. Martin-Luther-Univ. Halle-Wittemberg, Math.-Nat.,* 30(5): 89 (1981): L. litvinovii.
vestitus
subsp. alefeldii *(White) S.L. Broich in Syst. Bot.,* 12(1): 151 (1987): L. alefeldii.
subsp. laetiflorus *(Greene) S.L. Broich in Syst. Bot.,* 12(1): 151 (1987): L. laetiflorus.
subsp. laevicarpus *S.L. Broich in Syst. Bot.,* 12(1): 151 (1987)—U.S.A. (California).

LAUMONIERA H.P. Nooteboom in Blumea, 32(2): 383 (1987). *SIMAROUBACEAE.*
bruceadelpha *H.P. Nooteboom in Blumea,* 32(2): 383 (1987)—Sumatra.

LAUNAEA (Compos.).
thalassica *N. Kilian, C. Brochmann & ø.H. Rustan in Willdenowia,* 16(2): 491 (1987)—Cape Verde Is.

LAXMANNIA (Anthericac.).
arida *G.J. Keighery in Fl. Australia,* 45: 491, 258 (1987)—Australia (W. Australia, N. Terr.).
grandiflora
subsp. stirlingensis *G.J. Keighery in Fl. Australia,* 45: 491, 261 (1987)—Australia (W. Australia).
jamesii *G.J. Keighery in Fl. Australia,* 45: 491, 258 (1987)—Australia (W. Australia).
omnifertilis *G.J. Keighery in Fl. Australia,* 45: 491, 260 (1987)—Australia (W. Australia).
orientalis *G.J. Keighery in Fl. Australia,* 45: 492, 257 (1987)—Australia (Victoria, S. Australia), Tasmania.
ramosa
subsp. deflexa *G.J. Keighery in Fl. Australia,* 45: 492, 257 (1987)—Australia (W. Australia).
sessiliflora
subsp. australis *G.J. Keighery in Fl. Australia,* 45: 492, 256 (1987)—Australia (W. Australia).
subsp. drummondii *G.J. Keighery in Fl. Australia,* 45: 492, 256 (1987)—Australia (W. Australia).

LEAVENWORTHIA (Crucif.).
texana *W.F. Mahler in Sida,* 12(1): 239 (1987)—U.S.A. (Texas).

LECHENAULTIA (Goodeniac.).
sect. Patentes *D.A. Morrison in Brunonia,* 9(1): 6 (1987).
brevifolia *D.A. Morrison in Brunonia,* 9(1): 18 (1987)—Australia (W. Australia).
lutescens *D.A. Morrison & R.C. Carolin in Brunonia,* 9(1): 15 (1987)—Australia (N. Terr., W. Australia).
papillata *D.A. Morrison in Brunonia,* 9(1): 12 (1987)—Australia (W. Australia).

LECYTHIS (Lecythidac.).
confertiflora *(A.C. Smith) S.A. Mori in Mem. New York Bot. Gard.,* 44: 31 (1987): Eschweilera confertiflora.
persistens
subsp. aurantiaca *S.A. Mori in Mem. New York Bot. Gard.,* 44: 32 (1987)—French Guiana.

LEFEBVREA (Umbellif.).
droopii *C.C. Townsend in Kew Bull.,* 42(3): 589 (1987)—Tanzania.

LEMNA (Lemnac.).
sect. Biformes *E. Landolt in Veröff. Geobot. Inst. Rübel,* 71: 445 (1986).

LEONOHEBE M.J. Heads in Bot. Soc. Otago Newsl., 5: 4 (1987). *SCROPHULARIACEAE.*
sect. Apiti *M.J. Heads in Bot. Soc. Otago Newsl.,* 5: 7 (1987).
sect. Aromaticae *M.J. Heads in Bot. Soc. Otago Newsl.,* 5: 8 (1987).
sect. Buxifoliatae *M.J. Heads in Bot. Soc. Otago Newsl.,* 5: 10 (1987).
sect. Connatae *M.J. Heads in Bot. Soc. Otago Newsl.,* 5: 6 (1987).
sect. Densifoliae *M.J. Heads in Bot. Soc. Otago Newsl.,* 5: 4 (1987).
sect. Flagriformes *M.J. Heads in Bot. Soc. Otago Newsl.,* 5: 8 (1987).
sect. Salicornioides *M.J. Heads in Bot. Soc. Otago Newsl.,* 5: 7 (1987).
annulata *(Petrie) M.J. Heads in Bot. Soc. Otago Newsl.,* 5: 7 (1987): Veronica armstrongii var. annulata.
armstrongii *(Armstrong) M.J. Heads in Bot. Soc. Otago Newsl.,* 5: 7 (1987): Veronica armstrongii.
benthamii *(Hook. f.) M.J. Heads in Bot. Soc. Otago Newsl.,* 5: 7 (1987): Veronica benthami.
cheesemanii *(Buchanan) M.J. Heads in Bot. Soc. Otago Newsl.,* 5: 5 (1987): Mitrasacme cheesemanii.
ciliolata *(Hook. f.) M.J. Heads in Bot. Soc. Otago Newsl.,* 5: 5 (1987): Logania ciliolata.
coarctata *(Cheesem.) M.J. Heads in Bot. Soc. Otago Newsl.,* 5: 8 (1987): Veronica coarctata.
cupressoides *(Hook. f.) M.J. Heads in Bot. Soc. Otago Newsl.,* 5: 8 (1987): Veronica cupressoides.
densifolia *(F. v. Muell.) M.J. Heads in Bot. Soc. Otago Newsl.,* 5: 4 (1987): Paederota densifolia.
epacridea *(Hook. f.) M.J. Heads in Bot. Soc. Otago Newsl.,* 5: 6 (1987): Veronica epacridea.
haastii *(Hook. f.) M.J. Heads in Bot. Soc. Otago Newsl.,* 5: 6 (1987): Veronica haastii.
var. humilis *(Simpson) M.J. Heads in Bot. Soc. Otago Newsl.,* 5: 6 (1987): Hebe macrocalyx var. humilis.

LEONOHEBE :–
var. macrocalyx *(Armstrong) M.J. Heads in Bot. Soc. Otago Newsl.,* 5: 6 (1987): Veronica macrocalyx.
hectorii *(Hook. f.) M.J. Heads in Bot. Soc. Otago Newsl.,* 5: 8 (1987): Veronica hectori.
var. demissa *(Simpson) M.J. Heads in Bot. Soc. Otago Newsl.,* 5: 8 (1987): Hebe demissa.
imbricata *(Petrie) M.J. Heads in Bot. Soc. Otago Newsl.,* 5: 9 (1987): Veronica imbricata.
laingii *(Ckne.) M.J. Heads in Bot. Soc. Otago Newsl.,* 5: 8 (1987): Veronica laingii.
lycopodioides *(Hook. f.) M.J. Heads in Bot. Soc. Otago Newsl.,* 5: 9 (1987): Veronica lycopodioides.
var. patula *(Simpson & Thomson) M.J. Heads in Bot. Soc. Otago Newsl.,* 5: 9 (1987): Hebe lycopodioides var. patula.
masoniae *(Moore) M.J. Heads in Bot. Soc. Otago Newsl.,* 5: 10 (1987): Hebe pauciramosa var. masonae.
var. rotundata *M.J. Heads in Bot. Soc. Otago Newsl.,* 5: 11 (1987)—New Zealand.
mooreae *M.J. Heads in Bot. Soc. Otago Newsl.,* 5: 10 (1987)—New Zealand.
var. telmata *M.J. Heads in Bot. Soc. Otago Newsl.,* 5: 10 (1987)—New Zealand.
ochracea *(Ashwin) M.J. Heads in Bot. Soc. Otago Newsl.,* 5: 7 (1987): Hebe ochracea.
odora *(Hook. f.) M.J. Heads in Bot. Soc. Otago Newsl.,* 5: 10 (1987): Veronica odora.
pauciflora *(Simpson & Thomson) M.J. Heads in Bot. Soc. Otago Newsl.,* 5: 10 (1987): Hebe pauciflora.
pauciramosa *(Ckne. & Allan) M.J. Heads in Bot. Soc. Otago Newsl.,* 5: 10 (1987): Hebe buxifolia var. pauciramosa.
petriei *(Buchanan) M.J. Heads in Bot. Soc. Otago Newsl.,* 5: 6 (1987): Mitrasacme petriei.
var. murrellii *(Simpson & Thomson) M.J. Heads in Bot. Soc. Otago Newsl.,* 5: 6 (1987): Hebe murrellii.
poppelwellii *(Ckne.) M.J. Heads in Bot. Soc. Otago Newsl.,* 5: 9 (1987): Veronica poppelwellii.
propinqua *(Cheesem.) M.J. Heads in Bot. Soc. Otago Newsl.,* 5: 9 (1987): Veronica propinqua.
ramosissima *(Simpson & Thomson) M.J. Heads in Bot. Soc. Otago Newsl.,* 5: 7 (1987): Hebe ramosissima.
salicornioides *(Hook. f.) M.J. Heads in Bot. Soc. Otago Newsl.,* 5: 7 (1987): Veronica salicornioides.
subsimilis *(Col.) M.J. Heads in Bot. Soc. Otago Newsl.,* 5: 9 (1987): Veronica subsimilis.
var. astonii *(Petrie) M.J. Heads in Bot. Soc. Otago Newsl.,* 5: 9 (1987): Veronica astonii.
subulata *(Simpson) M.J. Heads in Bot. Soc. Otago Newsl.,* 5: 9 (1987): Hebe subulata.

LEONOHEBE :–
tetragona *(Hook. f.) M.J. Heads in Bot. Soc. Otago Newsl.,* 5: 9 (1987): Veronica tetragona.
tetrasticha *(Hook. f.) M.J. Heads in Bot. Soc. Otago Newsl.,* 5: 5 (1987): Veronica tetrasticha.
tumida *(Kirk) M.J. Heads in Bot. Soc. Otago Newsl.,* 5: 5 (1987): Veronica tumida.
uniflora *(Kirk) M.J. Heads in Bot. Soc. Otago Newsl.,* 5: 5 (1987): Veronica uniflora.

LEONTODON (Compos.).
taraxacoides
subsp. hispidus *(Roth) M. Kerguélen in Lejeunia,* 120: 119 (1987): Colobium hispidum.

LEONURUS (Labiat.).
kudrjaschevii *R.V. Kamelin & M. Tulyaganova in Bot. Zhurn.,* 72(6): 813 (1987)—U.S.S.R. (Middle Asia).

LEPANTHES (Orchidac.).
actias-luna *C.A. Luer & A.C. Hirtz in Lindleyana,* 2(1): 30 (1987)—Ecuador.
aeora *C.A. Luer & A.C. Hirtz in Lindleyana,* 2(1): 32 (1987)—Ecuador.
avicularia *C.A. Luer & A.C. Hirtz in Lindleyana,* 2(1): 32 (1987)—Ecuador.
binaria *C.A. Luer & A.C. Hirtz in Orchidee,* 38(6): 291 (1987)—Ecuador, Colombia.
bituberculata *C.A. Luer & Hirtz in Lindleyana,* 2(3): 125 (1987)—Ecuador.
callisto *C.A. Luer & A.C. Hirtz in Orchidee,* 38(6): 287 (1987)—Ecuador.
caloura *C.A. Luer & A.C. Hirtz in Orchidee,* 38(6): 285 (1987)—Ecuador.
celox *C.A. Luer & A.C. Hirtz in Lindleyana,* 2(1): 36 (1987)—Ecuador.
chrysina *C.A. Luer & Hirtz in Lindleyana,* 2(3): 128 (1987)—Ecuador.
condorensis *C.A. Luer & A.C. Hirtz in Orchidee,* 38(6): 283 (1987)—Ecuador.
conjuncta *C.A. Luer & Hirtz in Lindleyana,* 2(3): 128 (1987)—Ecuador.
ctenophora *C.A. Luer & Hirtz in Lindleyana,* 2(3): 129 (1987)—Ecuador.
deformis *C.A. Luer & A.C. Hirtz in Orchidee,* 38(1): 38 (1987)—Ecuador.
disjuncta *C.A. Luer & Hirtz in Lindleyana,* 2(3): 129 (1987)—Ecuador.
ejecta *C.A. Luer & A.C. Hirtz in Orchidee,* 38(1): 36 (1987)—Ecuador.
elaminata *C.A. Luer & Hirtz in Lindleyana,* 2(3): 132 (1987)—Ecuador.
embreei *C.A. Luer & Hirtz in Lindleyana,* 2(3): 132 (1987)—Ecuador.
eruca *C.A. Luer & Hirtz in Lindleyana,* 2(3): 135 (1987)—Ecuador.
hirsutula *C.A. Luer & Hirtz in Lindleyana,* 2(3): 135 (1987)—Ecuador.
hubeinii *C.A. Luer in Orchidee,* 38(2): 58 (1987)—Colombia.
implexa *C.A. Luer & A.C. Hirtz in Lindleyana,* 2(1): 36 (1987)—Ecuador.

LEPANTHES :-
jackinpyxa *C.A. Luer & A.C. Hirtz in Lindleyana*, 2(1): 38 (1987)—Ecuador.
lucifer *C.A. Luer & A.C. Hirtz in Amer. Orchid Soc. Bull.*, 56(10): 1016 (1987)—Ecuador.
meganthera *C.A. Luer & A.C. Hirtz in Lindleyana*, 2(1): 38 (1987)—Ecuador.
mephistopheles *C.A. Luer & A.C. Hirtz in Orchidee*, 38(1): 34 (1987)—Ecuador.
nambijae *C.A. Luer & Hirtz in Lindleyana*, 2(3): 138 (1987)—Ecuador.
necopina *C.A. Luer & A.C. Hirtz in Lindleyana*, 2(1): 42 (1987)—Ecuador.
pachyrhiza *C.A. Luer & Hirtz in Lindleyana*, 2(3): 138 (1987)—Ecuador.
rauhii *C.A. Luer in Orchidee*, 38(2): 59 (1987)—Panama.
satyrica *C.A. Luer & Hirtz in Lindleyana*, 2(3): 141 (1987)—Ecuador.
seegeri *C.A. Luer in Orchidee*, 38(2): 60 (1987)—Panama.
sobrina *C.A. Luer & Hirtz in Lindleyana*, 2(3): 141 (1987)—Ecuador.
sororcula *C.A. Luer & Hirtz in Lindleyana*, 2(3): 144 (1987)—Ecuador.
stegastes *C.A. Luer & Hirtz in Lindleyana*, 2(3): 144 (1987)—Ecuador.
surrogata *C.A. Luer & A.C. Hirtz in Lindleyana*, 2(1): 42 (1987)—Ecuador.
tectorum *C.A. Luer & Hirtz in Lindleyana*, 2(3): 147 (1987)—Ecuador.
thylax *C.A. Luer & A.C. Hirtz in Lindleyana*, 2(1): 45 (1987)—Ecuador.
titanica *C.A. Luer & Hirtz in Lindleyana*, 2(3): 147 (1987)—Ecuador.
tungurahuae *C.A. Luer & Hirtz in Lindleyana*, 2(3): 151 (1987)—Ecuador.
via-incarum *C.A. Luer & Hirtz in Lindleyana*, 2(3): 151 (1987)—Ecuador.
vibrissa *C.A. Luer & Hirtz in Lindleyana*, 2(3): 155 (1987)—Ecuador.
xenos *C.A. Luer & A.C. Hirtz in Orchidee*, 38(1): 32 (1987)—Ecuador.

LEPANTHOPSIS (Orchidac.).
aristata *D.D. Dod in Moscosoa*, 4: 143 (1986)—Hispaniola (Haiti).
atrosetifera *D.D. Dod in Moscosoa*, 4: 145 (1986)—Hispaniola (Haiti).
cucullata *D.D. Dod in Moscosoa*, 4: 145 (1986)—Hispaniola (Haiti).
haitiensis *D.D. Dod in Moscosoa*, 4: 148 (1986)—Hispaniola (Haiti).
lingulata *D.D. Dod in Moscosoa*, 4: 150 (1986)—Hispaniola (Haiti).
ornipteridion *D.D. Dod in Moscosoa*, 4: 152 (1986)—Hispaniola (Haiti).
paryskii *D.D. Dod in Moscosoa*, 4: 152 (1986)—Hispaniola (Haiti).
quisqueyana *D.D. Dod in Moscosoa*, 4: 155 (1986)—Hispaniola (Dominican Republic, Haiti).

LEPIDIUM (Crucif.).
solomonii *I.A. Al-Shehbaz in Ann. Missouri Bot. Gard.*, 73(4): 830 (1986 publ. 1987)—Bolivia.

LEPIDONIA (Compos.).
callilepis *(Gleason) H. Robinson & V.A. Funk in Bot. Jahrb.*, 108(2–3): 225 (1987): Vernonia callilepis.
corae *(Standl. & Steyerm.) H. Robinson & V.A. Funk in Bot. Jahrb.*, 108(2–3): 225 (1987): Vernonia corae.
jonesii *(B. Turner) H. Robinson & V.A. Funk in Bot. Jahrb.*, 108(2–3): 225 (1987): Vernonia jonesii.
lankesteri *(Blake) H. Robinson & V.A. Funk in Bot. Jahrb.*, 108(2–3): 225 (1987): Vernonia lankesteri.
mexicana *(Less.) H. Robinson & V.A. Funk in Bot. Jahrb.*, 108(2–3): 225 (1987): Vernonia mexicana.
salvinae *(Hemsl.) H. Robinson & V.A. Funk in Bot. Jahrb.*, 108(2–3): 225 (1987): Vernonia salvinae.

LEPILAENA (Zannichelliac.).
patentifolia *E.L. Robertson in Fl. S. Australia*, 4: 1736 (1986)—Australia (S. Australia, Victoria), Tasmania.

LEPISMIUM (Cactac.).
subgen. Acanthorhipsalis *(Schumann) W. Barthlott in Bradleya*, 5: 99 (1987): Rhipsalis subgen. Acanthorhipsalis.
subgen. Lymanbensonia *(Kimnach) W. Barthlott in Bradleya*, 5: 99 (1987): gen. Lymanbensonia.
subgen. Ophiorhipsalis *(Schumann) W. Barthlott in Bradleya*, 5: 99 (1987): Rhipsalis subgen. Ophiorhipsalis.
subgen. Pfeiffera *(Salm-Dyck) W. Barthlott in Bradleya*, 5: 99 (1987): gen. Pfeiffera.
aculeatum *(F.A.C. Weber) W. Barthlott in Bradleya*, 5: 99 (1987): Rhipsalis aculeata.
brevispinum *W. Barthlott in Bradleya*, 5: 99 (1987)—Peru.
crenatum *(Britton) W. Barthlott in Bradleya*, 5: 99 (1987): Hariota crenata.
houlletianum *(Lemaire) W. Barthlott in Bradleya*, 5: 99 (1987): Rhipsalis houlletiana.
var. regnellii *(Lindberg) W. Barthlott in Bradleya*, 5: 99 (1987): Rhipsalis regnellii.
ianthothele *(Monville) W. Barthlott in Bradleya*, 5: 99 (1987): Cereus ianthothele.
incachacanum *(Cardenas) W. Barthlott in Bradleya*, 5: 99 (1987): Rhipsalis incachacana.
lorentzianum *(Grisebach) W. Barthlott in Bradleya*, 5: 99 (1987): Rhipsalis lorentziana.
lumbricoides *(Lemaire) W. Barthlott in Bradleya*, 5: 99 (1987): Cereus lumbricoides.

LEPISMIUM :-
micranthum *(Vaupel) W. Barthlott in Bradleya,* 5: 99 (1987): Cereus micranthus.
miyagawae *(Barthlott & Rauh) W. Barthlott in Bradleya,* 5: 99 (1987): Pfeiffera miyagawae.
monacanthum *(Grisebach) W. Barthlott in Bradleya,* 5: 99 (1987): Rhipsalis monacantha.
paranganiense *(Cardenas) W. Barthlott in Bradleya,* 5: 99 (1987): Acanthorhipsalis paranganiensis.
warmingianum *(Schumann) W. Barthlott in Bradleya,* 5: 99 (1987): Rhipsalis warmingiana.

×**LEPTODENDRUM** hort. in Orchid Rev., 95(1127): centre page pull-out p. 9 (1987). *ORCHIDACEAE.* (Epidendrum × Leptotes).

LEPTONYCHIA (Sterculiac.).
banahaensis
var. papuana *J.F. Veldkamp & R.C.H. Flipphi in Blumea,* 32(2): 450 (1987)—New Guinea (W. Irian, Papua New Guinea), Moluccas.
parviflora *(Ridley) J.F. Veldkamp & R.C.H. Flipphi in Blumea,* 32(2): 455 (1987): Leptonychiopsis parviflora.

LEPTORHYNCHOS (Compos.).
scabrus *(Benth.) L. Haegi in Fl. S. Australia,* 3: 1558 (1986): Rhytidanthe scabra.

LEPTOSEMA (Legumin.).
aculeatum *M.D. Crisp in J. Adelaide Bot. Gard.,* 10(1): 131 (1987)—Australia (W. Australia).
daviesioides *(Turcz.) M.D. Crisp in J. Adelaide Bot. Gard.,* 10(1): 131 (1987): Kaleniczenkia daviesioides.

LERCHEA (Rubiac.).
beccariana *(Bakh. f.) B. Axelius in Blumea,* 32(1): 101 (1987): L. paniculata var. beccariana.
corymbosa *B. Axelius in Blumea,* 32(1): 97 (1987)—Sumatra.
parviflora *B. Axelius in Blumea,* 32(1): 105 (1987)—Sumatra.

LESPEDEZA (Legumin.).
davurica
var. beipiaoensis *P.H. Huang & J.S. Ma in Acta Phytotax. Sin.,* 25(5): 369 (1987)—China.
hirta
var. curtissii *(Clewell) D. Isely in Brittonia,* 38(4): 354 (1986): L. hirta subsp. curtissii.

LEUCAENA (Legumin.).
leucocephala
subsp. glabrata *(Rose) S. Zárate Pedroche in Phytologia,* 63(4): 305 (1987): L. glabrata.

LEUCHERIA (Compos.).
lithospermifolia
subsp. integrifolia *(Phil.) J. Grau & Zinnecker in Bot. Jahrb.,* 108(2–3): 231 (1987): Chabraea integrifolia.
viscida *(Bertero ex Colla) J. Grau & Zinnecker in Bot. Jahrb.,* 108(2–3): 231 (1987): Chabraea viscida.

LEUCOJUM (Amaryllidac.).
trichophyllum
var. broteri *(Jordan & Fourr.) B. Valdés in Lagascalia,* 14(1): 176 (1986): Acis broteri.

LEUCOTHOE (Ericac.).
griffithiana
var. sessilifolia *C.Y. Wu & T.Z. Hsu in Fl. Xizangica,* 3: 684 (1986)—Xizang.

LEUTEA M.G. Pimenov in Fl. Iran., 162: 445 (1987). *UMBELLIFERAE.*
cupularis *(Boiss.) M.G. Pimenov in Fl. Iran.,* 162: 446 (1987): Peucedanum cupulare.
gracillima *M.G. Pimenov in Fl. Iran.,* 162: 448 (1987)—Iran.
nematoloba *(Rech. f.) M.G. Pimenov in Fl. Iran.,* 162: 447 (1987): Peucedanum nematolobum.
petiolaris *(DC.) M.G. Pimenov in Fl. Iran.,* 162: 449 (1987): Ferula petiolaris.
polyscias *(Boiss.) M.G. Pimenov in Fl. Iran.,* 162: 446 (1987): Peucedanum polyscias.
rechingeri *(Leute) M.G. Pimenov in Fl. Iran.,* 162: 448 (1987): Peucedanum rechingeri.

LEYCESTERIA (Caprifoliac.).
glaucophylla
var. thibetica *(H.J. Wang) J.F. Huang in Fl. Xizangica,* 4: 509 (1985): L. thibetica.

LEYMUS (Gramin.).
salinus
subsp. salmonis *(C. Hitchc.) R.J. Atkins in Great Basin Nat.,* 43(4): 569 (1983): Elymus ambiguus var. salmonis.

LICANIA (Chrysobalanac.).
fusicarpa *(Kosterm.) G.T. Prance in Brittonia,* 39(3): 366 (1987): Hunga fusicarpa.

LIGULARIA (Compos.).
discoidea *S.W. Liu in Fl. Xizangica,* 4: 832 (1985)—Xizang.
myriocephala *Ling in Fl. Xizangica,* 4: 831 (1985)—Xizang.
nyingchiensis *S.W. Liu in Fl. Xizangica,* 4: 834 (1985)—Xizang.
rumicifolia *(Drumm.) S.W. Liu in Fl. Xizangica,* 4: 832 (1985): Senecio rumicifolius.
tenuicaulis
var. purpuracea *S.W. Liu in Fl. Xizangica,* 4: 829 (1985)—Xizang.

LIGUSTRUM (Oleac.).
amamianum
var. rotundifolium *(Blume) B.M. Miao in Invest. Stud. Nat., Mus. Hist. Nat. Shanghaiense,* 6: 72 (1986): L. japonicum var. rotundifolium.
delavayanum
subsp. morrisonense *(Kanehira & Sasaki) B.M. Miao in Invest. Stud. Nat., Mus. Hist. Nat. Shanghaiense,* 6: 94 (1986): L. morrisonense.
longitubum *Hsu ex M.C. Chang & B.M. Miao in Invest. Stud. Nat., Mus. Hist. Nat. Shanghaiense,* 6: 106 (1986)—China.
sinense
var. concavum *M.C. Chang in Invest. Stud. Nat., Mus. Hist. Nat. Shanghaiense,* 6: 81 (1986)—China.
var. luodianense *M.C. Chang in Invest. Stud. Nat., Mus. Hist. Nat. Shanghaiense,* 6: 77 (1986)—China.
var. rugosulum *(W.W. Smith) M.C. Chang in Invest. Stud. Nat., Mus. Hist. Nat. Shanghaiense,* 6: 78 (1986): L. rugosulum.

LIMNAS (Gramin.).
malyschevii *O.D. Nikiforova in Bot. Zhurn.,* 72(3): 391 (1987)—U.S.S.R. (Siberia, Soviet Far East).

LIMNOPHYTON (Alismatac.).
australiense *H.I. Aston in Muelleria,* 6(5): 311 (1987)—Australia (Queensland).

LIMONIUM (Plumbaginac.).
algusae *(Brullo) W. Greuter in Willdenowia,* 16(2): 449 (1987): L. oleifolium subsp. algusae.
battandieri *W. Greuter & Burdet in Willdenowia,* 16(2): 449 (1987)—Algeria.
baudinii *I.A. Linchevskiĭ in Novosti Sist. Vȳssh. Rast.,* 23: 107 (1986)—Tasmania.
brevipetiolatum *R. Artelari & M. Erben in Mitt. Bot. Staatssamml. München,* 22: 507 (1986)—Greece.
gougetianum
subsp. multiceps *(Pomel) Quézel & Santa ex W. Greuter & Burdet in Willdenowia,* 16(2): 450 (1987): Statice multiceps.
gymnesicum *M. Erben in Mitt. Bot. Staatssamml. München,* 22: 204 (1986), nom. nov.: L. gougetianum subsp. balearicum *Pignatti.*
letourneuxii *(Batt.) W. Greuter & Burdet in Willdenowia,* 16(2): 450 (1987): Statice letourneuxii.
opulentum *(Lojac.) W. Greuter in Willdenowia,* 16(2): 450 (1987): Statice opulenta.
pescadense *W. Greuter & Burdet in Willdenowia,* 16(2): 450 (1987)—Algeria.
pseudodictyocladum *W. Greuter & Burdet in Willdenowia,* 16(2): 450 (1987)—Locality not stated.

LIMONIUM :–
retirameum *W. Greuter & Burdet in Willdenowia,* 16(2): 451 (1987)—Sardinia.
secundirameum *(Lojac.) W. Greuter & Burdet in Willdenowia,* 16(2): 451 (1987): Statice secundiramea.
sinisicum *M. Erben in Mitt. Bot. Staatssamml. München,* 22: 210 (1986)—Sardinia.
solanderi *I.A. Linchevskiĭ in Novosti Sist. Vȳssh. Rast.,* 23: 105 (1986)—Australia (Queensland).
ursanum *M. Erben in Mitt. Bot. Staatssamml. München,* 22: 214 (1986)—Sardinia.
wiedmannii *M. Erben in Mitt. Bot. Staatssamml. München,* 22: 206 (1986)—Balearic Is.

LIMOSELLA (Scrophulariac.).
granitica *W.R. Barker in Fl. S. Australia,* 3: 1284 (1986)—Australia (S. Australia).

LINARIA (Scrophulariac.).
aguillonensis *(García Martínez) X.R. García Martínez & F.J. Silva-Pando in An. Jard. Bot. Madrid,* 43(1): 179 (1986): L. thymifolia subsp. aguillonensis.
laxiflora
subsp. calcarlongum *M. Qaiser in Fl. Libya,* 88: 56 (1982)—Libya.
tartessiana *(Vicioso) B. Valdés in Lagascalia,* 14(1): 157 (1986): L. heterophylla subsp. tartessiana.

LINDERA (Laurac.).
sect. Cephalodaphne *(Bl.) H.P. Tsui in Acta Phytotax. Sin.,* 25(3): 164 (1987): Daphnidium sect. Cephalodaphne.
subsect. Gemmiflorae *H.P. Tsui in Acta Phytotax. Sin.,* 25(3): 165 (1987).
subsect. Gracilipedes *H.P. Tsui in Acta Phytotax. Sin.,* 25(3): 165 (1987).
sect. Iteadaphne *H.P. Tsui in Acta Phytotax. Sin.,* 25(3): 167 (1987).
subsect. Pedunculares *H.P. Tsui in Acta Phytotax. Sin.,* 25(3): 167 (1987).
subsect. Sessiles *H.P. Tsui in Acta Phytotax. Sin.,* 25(3): 167 (1987).
sect. Trinerviae *H.P. Tsui in Acta Phytotax. Sin.,* 25(3): 167 (1987).
fragrans
var. linearifolia *Y.K. Li in Guihaia,* 5(4): 345 (1985)—China.
glabrata *(Bl.) H. Koyama in Acta Phytotax. Geobot.,* 38: 168 (1987): L. sericea var. glabrata.
henanensis *H.P. Tsui in Acta Phytotax. Sin.,* 25(5): 412 (1987)—China.
lancea *(Momiyama) H. Koyama in Acta Phytotax. Geobot.,* 38: 169 (1987): L. umbellata var. lancea.

LINDMANIA (Bromeliac.).
arachnoidea *(L.B. Smith, Steyermark & Robinson) L.B. Smith in Ann. Missouri Bot. Gard.,* 73(4): 697 (1986 publ. 1987): Cottendorfia arachnoidea.

LINDMANIA :–
atrorosea *(L.B. Smith, Steyermark & Robinson) L.B. Smith in Ann. Missouri Bot. Gard.*, 73(4): 695 (1986 publ. 1987): Cottendorfia atrorosea.
aurea *L.B. Smith, J.A. Steyermark & H. Robinson in Ann. Missouri Bot. Gard.*, 73(4): 697 (1986 publ. 1987)—Venezuela.
brachyphylla
var. angustior *(L.B. Smith, Steyermark & Robinson) R.W. Read exL.B. Smith in Ann. Missouri Bot. Gard.*, 73(4): 695 (1986 publ. 1987): Cottendorfia brachyphylla var. angustior.
dendritica *(L.B. Smith) L.B. Smith in Ann. Missouri Bot. Gard.*, 73(4): 695 (1986 publ. 1987): Cottendorfia dendritica.
dyckioides *(L.B. Smith) L.B. Smith in Ann. Missouri Bot. Gard.*, 73(4): 695 (1986 publ. 1987): Cottendorfia dyckioides.
geniculata
var. minor *(L.B. Smith, Steyermark & Robinson) L.B. Smith in Ann. Missouri Bot. Gard.*, 73(4): 695 (1986 publ. 1987): Cottendorfia geniculata var. minor.
gracillima *(L.B. Smith) L.B. Smith in Ann. Missouri Bot. Gard.*, 73(4): 693 (1986 publ. 1987): Cottendorfia gracillima.
guianensis
var. vestita *(L.B. Smith) L.B. Smith in Ann. Missouri Bot. Gard.*, 73(4): 695 (1986 publ. 1987): Cottendorfia guianensis var. vestita.
huberi *L.B. Smith, J.A. Steyermark & H. Robinson in Ann. Missouri Bot. Gard.*, 73(4): 699 (1986 publ. 1987)—Venezuela.
imitans *L.B. Smith, J.A. Steyermark & H. Robinson in Ann. Missouri Bot. Gard.*, 73(4): 697 (1986 publ. 1987)—Venezuela.
lateralis *(L.B. Smith & R.W. Read) L.B. Smith & H. Robinson in Ann. Missouri Bot. Gard.*, 73(4): 693 (1986 publ. 1987): Cottendorfia lateralis.
longipes *(L.B. Smith) L.B. Smith in Ann. Missouri Bot. Gard.*, 73(4): 695 (1986 publ. 1987): Cottendorfia longipes.
maguirei *(L.B. Smith) L.B. Smith in Ann. Missouri Bot. Gard.*, 73(4): 695 (1986 publ. 1987): Cottendorfia maguirei.
marahuacae *(L.B. Smith, Steyermark & Robinson) L.B. Smith in Ann. Missouri Bot. Gard.*, 73(4): 699 (1986 publ. 1987): Cottendorfia marahuacae.
nubigena *(L.B. Smith) L.B. Smith in Ann. Missouri Bot. Gard.*, 73(4): 695 (1986 publ. 1987): Cottendorfia nubigena.
piresii *L.B. Smith, J.A. Steyermark & H. Robinson in Ann. Missouri Bot. Gard.*, 73(4): 698 (1986 publ. 1987)—Brazil.
riparia *L.B. Smith, J.A. Steyermark & H. Robinson in Ann. Missouri Bot. Gard.*, 73(4): 695 (1986 publ. 1987)—Venezuela.
savannensis *(L.B. Smith) L.B. Smith in Ann. Missouri Bot. Gard.*, 73(4): 695 (1986 publ. 1987): Cottendorfia savannensis.

LINDMANIA :–
saxicola *L.B. Smith, J.A. Steyermark & H. Robinson in Ann. Missouri Bot. Gard.*, 73(4): 693 (1986 publ. 1987)—Venezuela.
sessilis *L.B. Smith, J.A. Steyermark & H. Robinson in Ann. Missouri Bot. Gard.*, 73(4): 693 (1986 publ. 1987)—Venezuela.
smithiana *(Steyermark & Luteyn) L.B. Smith in Ann. Missouri Bot. Gard.*, 73(4): 698 (1986 publ. 1987): Connellia smithiana.
terramarae *L.B. Smith, J.A. Steyermark & H. Robinson in Ann. Missouri Bot. Gard.*, 73(4): 698 (1986 publ. 1987)—Venezuela.

LINKAGROSTIS A.T. Romero García, G. Blanca & C. Morales Torres in Candollea, 42(1): 383 (1987). *GRAMINEAE.*
juressi *(Link) A.T. Romero García, G. Blanca & C. Morales Torres in Candollea,* 42(1): 383 (1987): Agrostis juressi.

LIPARIS (Orchidac.).
greenwoodiana *A. Espejo Serna in Orquídea (México),* 10(2): 370 (1987)—Mexico.
krameri
var. sasakii *(Hayata) T. Hashimoto in Bull. Nation. Sci. Mus., B (Tokyo),* 13(1): 39 (1987): L. sasakii.
truncata *F. Maekawa ex T. Hashimoto in Bull. Nation. Sci. Mus., B (Tokyo),* 13(1): 38 (1987)—Japan.

LIPOCARPHA (Cyperac.).
aristulata *(Cov.) G.C. Tucker in J. Arnold Arbor.*, 68(4): 410 (1987): Hemicarpha micrantha var. aristulata.
drummondii *(Nees) G.C. Tucker in J. Arnold Arbor.*, 68(4): 410 (1987): Hemicarpha drummondii.
micrantha *(Vahl) G.C. Tucker in J. Arnold Arbor.*, 68(4): 410 (1987): Scirpus micranthus.
occidentalis *(Gray) G.C. Tucker in J. Arnold Arbor.*, 68(4): 410 (1987): Hemicarpha occidentalis.
schomburgkii *(Friedl.) G.C. Tucker in J. Arnold Arbor.*, 68(4): 410 (1987): Hemicarpha schomburgkii.

LIPPIA (Verbenac.).
baumii
var. niykensis *R.B. Fernandes in Bol. Soc. Brot.*, 59: 252 (1986)—Malawi.
var. nyassensis *R.B. Fernandes in Bol. Soc. Brot.*, 59: 251 (1986)—Malawi.

LISSOCARPA (Lissocarpac.).
stenocarpa *J.A. Steyermark in Ann. Missouri Bot. Gard.*, 74(1): 104 (1987)—Venezuela.

LISTERA (Orchidac.).
cordata
f. tetraphylla *G. Lavoie in Provancheria,* 17: 92 (1984)—Canada (Quebec).

LITHOCARPUS (Fagac.).
arisanensis
var. globosa *C.F. Shen in C.F. Shen & T.S. Liu, Taxon. & Pollen Morph. Fagaceae Taiwan:* 26 (1984)—Taiwan.
kuarunensis *(Tomiya) C.F. Shen in C.F. Shen & T.S. Liu, Taxon. & Pollen Morph. Fagaceae Taiwan:* 36 (1984): Synaedrys kuarunensis.
var. lepidobalanus *C.F. Shen in C.F. Shen & T.S. Liu, Taxon. & Pollen Morph. Fagaceae Taiwan:* 36 (1984)—Taiwan.

LITSEA (Laurac.).
lakshmammaniana *K.R. Keshava Murthy & S.N. Yoganarasimhan in Current Sci.*, 56(8): 371 (1987)—India.

LOBELIA (Campanulac.).
gaudichaudii
var. albiflora *H. St. John & A.C. Medeiros in Phytologia,* 63(5): 366 (1987)—Hawaiian Is.
tortuosa
var. haupuensis *H. St. John in Phytologia,* 62(5): 435 (1987)—Hawaiian Is.

LOBULARIA (Crucif.).
canariensis *(DC.) L. Borgen in Op. Bot.*, 91: 66 (1987): Alyssum maritimum var. canariense.
subsp. fruticosa *(Webb) L. Borgen in Op. Bot.*, 91: 70 (1987): Koniga fruticosa.
subsp. intermedia *(Webb) L. Borgen in Op. Bot.*, 91: 73 (1987): L. intermedia.
subsp. marginata *(Webb) L. Borgen in Op. Bot.*, 91: 75 (1987): Koniga marginata.
subsp. microsperma *L. Borgen in Op. Bot.*, 91: 77 (1987)—Canary Is.
subsp. palmensis *(Christ) L. Borgen in Op. Bot.*, 91: 77 (1987): L. palmensis.
subsp. rosula-venti *(Svent.) L. Borgen in Op. Bot.,* 91: 80 (1987): L. maritima var. rosula-venti.
subsp. spathulata *(J.A. Schmidt) L. Borgen in Op. Bot.,* 91: 80 (1987): Koniga spathulata.
subsp. succulenta *L. Borgen in Op. Bot.*, 91: 82 (1987)—Salvage Is.

×LOCKOSTALIX hort. in Orchid Rev., 95(1127): centre page pull-out p. 9 (1987). *ORCHIDACEAE.* (Lockhartia × Sigmatostalix).

LOESENERIELLA (Celastrac.).
africana
var. obtusifolia *(Roxburgh) N. Hallé in Fl. Gabon,* 29: 255 (1986): Hippocratea obtusifolia.
var. togoensis *(Loesener) N. Hallé in Fl. Gabon,* 29: 255 (1986): Hippocratea cymosa var. togoensis.
apocynoides
var. guineensis *(Hutchinson & Moss) N. Hallé in Fl. Gabon,* 29: 248 (1986): Hippocratea guineensis.

LOESENERIELLA :–
camerunica *(Loesener) N. Hallé in Fl. Gabon,* 29: 242 (1986): Hippocratea camerunica.
crenata *(Klotzsch) Wilczek ex N. Hallé in Fl. Gabon,* 29: 238 (1986): Gymnema crenatum.
var. loandensis *(Exell) N. Hallé in Fl. Gabon,* 29: 239 (1986): Hippocratea loandensis.
var. rubiginosa *(Perrier) N. Hallé in Fl. Gabon,* 29: 240 (1986): Hippocratea rubiginosa.
iotricha
var. robida *N. Hallé in Fl. Gabon,* 29: 252 (1986)—Gabon.

LOLIUM (Gramin.).
gracile *(Dumort.) B.D. Jackson in Index Kewensis,* 2: 109 (1753–1885 publ. 1894): L. temulentum * (subsp.) gracile.

LOMATOCARPA (Umbellif.).
sect. Alposelinum *(M. Pimen.) M.G. Pimenov & T.V. Lavrova in Bot. Zhurn.*, 72(1): 35 (1987): gen. Alposelinum.
albomarginata *(Schrenk) M.G. Pimenov & T.V. Lavrova in Bot. Zhurn.*, 72(1): 35 (1987): Neogaya simplex β. albomarginata.

LOMATOGONIUM (Gentianac.).
sect. Pleurogynella *(Ikonn.) Sunita Garg, Gentianac. N.W. Himalayas (Internat. Biosci. Monog. 17):* 159 (1987): gen. Pleurogynella.

LOMATOPODIUM (Umbellif.).
staurophyllum *(Rech. f.) K.H. Rechinger in Fl. Iran.,* 162: 349 (1987): Seseli staurophyllum.

LONCHOCARPUS (Legumin.).
acuminatus *(Schldl.) M. Sousa S. in Ann. Missouri Bot. Gard.*, 73(4): 724 (1986 publ. 1987): Robinia acuminata.
ferrugineus *M. Sousa S. in Ann. Missouri Bot. Gard.*, 73(4): 725 (1986 publ. 1987)—Costa Rica, Nicaragua, Panama.
monticolus *M. Sousa S. in Ann. Missouri Bot. Gard.*, 73(4): 726 (1986 publ. 1987)—Nicaragua.
morenoi *M. Sousa S. in Ann. Missouri Bot. Gard.*, 73(4): 726 (1986 publ. 1987)—Nicaragua.
pilosus *M. Sousa S. in Ann. Missouri Bot. Gard.*, 73(4): 728 (1986 publ. 1987)—Nicaragua.
verrucosus *M. Sousa S. in Ann. Missouri Bot. Gard.*, 73(4): 730 (1986 publ. 1987)—Mexico, Guatemala, Nicaragua.

LONICERA (Caprifoliac.).
subsect. Uniflorae *V.A. Nedoluzhko in Sosud. Rast. Sovet. Dal'nego Vostoka,* 2: 284 (1987).
caerulea
subsp. stenantha *(Pojark.) Hultén ex A.K. Skvortsov in Symb. Bot. Upsal.*, 27(2): 100 (1986 publ. 1987): L. stenantha.

LONICERA :–
caucasica
subsp. govaniana *(Wall. ex DC.) H. Hara in Enum. Fl. Pl. Nepal,* 2: 195 (1979): L. govaniana.
glabrata
var. hirticaulis *C.Y. Wu & Y.F. Huang in Fl. Xizangica,* 4: 505 (1985)—Xizang.
pyrenaica
var. majoricensis *(Gand.) J.J. Pericás & J.A. Rosselló in Candollea,* 42(1): 397 (1987): L. majoricensis.
tomentella
var. conaensis *Hsu & Y.F. Huang in Fl. Xizangica,* 4: 495 (1985)—Xizang.

LOTONONIS (Legumin.).
brevicaulis *B-E. van Wyk in S. Afr. J. Bot.,* 53(2): 161 (1987)—South Africa (Cape Province).

LOTUS (Legumin.).
armeniacus *Kit Tan & F. Sorger in Pl. Syst. Evol.,* 154(1–2): 116 (1986)—Turkey.
digii *A. Chrtková in Pl. Syst. Evol.,* 155(1–4): 307 (1987)—Morocco, Algeria, Egypt.

LUCILIA (Compos.).
sect. Intermedieae *S.E. Freire in Cladistics, Int. J. Willi Hennig Soc.,* 3(3): 271 (1987).
ser. Lucilioides *S.E. Freire in Cladistics, Int. J. Willi Hennig Soc.,* 3(3): 272 (1987).
subsect. Lucilioides *S.E. Freire in Cladistics, Int. J. Willi Hennig Soc.,* 3(3): 272 (1987).
ser. Paralucilia *S.E. Freire in Darwiniana,* 27(1–4): 479 (1986).
subsect. Subspicata *S.E. Freire in Cladistics, Int. J. Willi Hennig Soc.,* 3(3): 271 (1987).
burkartii *(Cabr.) S.E. Freire in Darwiniana,* 27(1–4): 465 (1986): Gnaphalium burkartii.
catamarcensis *(Cabr.) S.E. Freire in Darwiniana,* 27(1–4): 465 (1986): Belloa catamarcensis.
lopezmirandae *(Cabr.) S.E. Freire in Darwiniana,* 27(1–4): 469 (1986): Belloa lopezmirandae.
santanica *(Cabr.) S.E. Freire in Darwiniana,* 27(1–4): 473 (1986): Gnaphalium santanicum.

LUDWIGIA (Onagrac.).
sect. Cinerascentes *T.P. Ramamoorthy & E.M. Zardini, Syst. & Evol. Ludwigia sect. Myrtocarpus s.l. (Monogr. Syst. Bot. Missouri Bot. Gard. 19):*101 (1987).
albiflora *T.P. Ramamoorthy, Syst. & Evol. Ludwigia sect. Myrtocarpus s.l. (Monogr. Syst. Bot. Missouri Bot. Gard. 19):*55 (1987)—Brazil.
filiformis *(Micheli) T.P. Ramamoorthy, Syst. & Evol. Ludwigia sect. Myrtocarpus s.l. (Monogr. Syst. Bot. Missouri Bot. Gard. 19):*86 (1987): Jussiaea filiformis.
grandiflora *(Michaux) W. Greuter & Burdet in Willdenowia,* 16(2): 448 (1987): Jussiaea grandiflora.

LUDWIGIA :–
hassleriana *(Chodat) T.P. Ramamoorthy, Syst. & Evol. Ludwigia sect. Myrtocarpus s.l. (Monogr. Syst. Bot. Missouri Bot. Gard. 19):*60 (1987): Jussiaea hassleriana.
irwinii *T.P. Ramamoorthy, Syst. & Evol. Ludwigia sect. Myrtocarpus s.l. (Monogr. Syst. Bot. Missouri Bot. Gard. 19):*66 (1987), nom. nov.: Jussiaea lanceolata *Camb.*
major *(Micheli) T.P. Ramamoorthy, Syst. & Evol. Ludwigia sect. Myrtocarpus s.l. (Monogr. Syst. Bot. Missouri Bot. Gard. 19):*84 (1987): Jussiaea longifolia var. major.
martii *(Micheli) T.P. Ramamoorthy, Syst. & Evol. Ludwigia sect. Myrtocarpus s.l. (Monogr. Syst. Bot. Missouri Bot. Gard. 19):*64 (1987): Jussiaea martii.
multinervia *(Hook. & Arn.) T.P. Ramamoorthy, Syst. & Evol. Ludwigia sect. Myrtocarpus s.l. (Monogr. Syst. Bot. Missouri Bot. Gard. 19):*40 (1987): Jussiaea multinervia.
pseudonarcissus *(Chodat & Hassler) T.P. Ramamoorthy, Syst. & Evol. Ludwigia sect. Myrtocarpus s.l. (Monogr. Syst. Bot. Missouri Bot. Gard. 19):*76 (1987): Jussiaea pseudonarcissus.

LUPINUS (Legumin.).
westianus
var. aridorum *(McFarlin ex Beckner) D. Isely in Brittonia,* 38(4): 356 (1986): L. aridorum.

LUZULA (Juncac.).
subsect. Alpinae *(Chrtek & Křísa) N.K. Kovtonyuk in Bot. Zhurn.,* 72(10): 1399 (1987): L. sect. Alpinae.
subsect. Confusae *N.K. Kovtonyuk in Bot. Zhurn.,* 72(10): 1399 (1987).
subsect. Nivales *N.K. Kovtonyuk in Bot. Zhurn.,* 72(10): 1399 (1987).
subsect. Parviflorae *N.K. Kovtonyuk in Bot. Zhurn.,* 72(10): 1398 (1987).
subsect. Tundricolae *N.K. Kovtonyuk in Bot. Zhurn.,* 72(10): 1399 (1987).

LYALLIA (Hectorellac.).
caespitosa *(J.D. Hooker) B.L. Nyananyo & V.H. Heywood in Taxon,* 36(3): 641 (1987): Hectorella caespitosa.

LYCHNIS (Caryophyllac.).
alpina
subsp. americana *(Fern.) J. Feilberg in Medd. Grønland, Biosci.,* 15: 12 (1984): Viscaria alpina var. americana.
var. serpentinicola *(Rune) P. Kallio & Y. Mäkinen in Ann. Univ. Turku, A, Biol. Geogr. Geol.,* 67: 63 (1982): Viscaria alpina var. serpentinicola.

LYCIANTHES (Solanac.).
acapulcensis *(Baill.) W.G. D'Arcy in Solanaceae Newsl.,* 2(4): 23 (1986), basionym page '133' in error for '338': Parascopolia acapulcensis.

LYCIANTHES :–
neglecta *(Dunal) A. Lourteig in Phytologia,* 62(6): 442 (1987): Solanum neglectum.
speciosa *(Dunal) A. Lourteig in Phytologia,* 62(6): 442 (1987): Solanum speciosum.

LYCIUM (Solanac.).
athium *L.M. Bernardello in Bol. Acad. Nac. Cienc. (Cordoba),* 57(3–4): 330 (1986)—Argentina.
chilense
var. comberi *(C. Hitchc.) L.M. Bernardello in Bol. Acad. Nac. Cienc. (Cordoba),* 57(3–4): 313 (1986): L. comberi.
var. filifolium *(Miers) L.M. Bernardello in Bol. Acad. Nac. Cienc. (Cordoba),* 57(3–4): 314 (1986): L. filifolium.
var. vergarae *(Phil.) L.M. Bernardello in Bol. Acad. Nac. Cienc. (Cordoba),* 57(3–4): 320 (1986): L. vergarae.
×ciliato-elongatum *L.M. Bernardello in Bol. Acad. Nac. Cienc. (Cordoba),* 57(3–4): 339 (1986), formula invalid acc. to ICBN(1983) Art. H.10.3—Argentina. (L. ciliatum × elongatum).

LYGODESMIA (Compos.).
grandiflora
var. arizonica *(Tomb) S.L. Welsh in Great Basin Nat.,* 43(2): 314 (1983): L. arizonica.
var. dianthopsis *(D.C. Eaton) S.L. Welsh in Great Basin Nat.,* 43(2): 314 (1983): L. juncea var. dianthopsis.

LYONIA (Ericac.).
ovalifolia
var. tomentosa *(Fang) C.Y. Wu in Fl. Xizangica,* 3: 689 (1986), without basionym ref.: Pieris ovalifolia var. tomentosa.

LYSILOMA (Legumin.).
latisiliquum
subsp. lataefoliolatum *A. Barreto Valdés & G.P. Yakovlev in Acta Bot. Cubana,* 41: 6 (1987)—Cuba.

LYSIMACHIA (Primulac.).
baoxingensis *(Chen & C.M. Hu) C.M. Hu in Acta Bot. Austro Sinica,* 2: 200 (1986): L. pterantha var. baoxingensis.
breviflora *C.M. Hu in Acta Bot. Austro Sinica,* 2: 201 (1986)—China.
jiangxiensis *C.M. Hu in Acta Bot. Austro Sinica,* 2: 200 (1986)—China.
patungensis
f. glabrifolia *C.M. Hu in Acta Bot. Austro Sinica,* 2: 200 (1986)—China.
sikokiana
subsp. petelotii *(Merr.) C.M. Hu in Acta Bot. Austro Sinica,* 2: 199 (1986): L. petelotii.

M

MAACKIA (Legumin.).
taiwanensis *H. Hoshi & H. Ohashi in J. Jap. Bot.,* 62(4): 98 (1987)—Taiwan.

MABEA (Euphorbiac.).
jefensis *M.J. Huft in Phytologia,* 62(4): 341 (1987)—Panama.

MACHAERANTHERA (Compos.).
bigelovii
var. mucronata *(Greene) B.L. Turner in Phytologia,* 62(3): 238 (1987): M. mucronata.
canescens
var. commixta *(Greene) S.L. Welsh in Great Basin Nat.,* 43(2): 316 (1983): M. commixta.
subsp. glabra *(A. Gray) B.L. Turner in Phytologia,* 62(3): 239 (1987): M. canescens var. glabra.
var. leucanthemifolia *(Greene) S.L. Welsh in Great Basin Nat.,* 43(2): 316 (1983): Aster leucanthemifolius.
var. vacans *(A. Nels.) S.L. Welsh in Great Basin Nat.,* 43(2): 316 (1983): M. pulverulenta var. vacans.
coulteri
var. arida *(Turner & Horne) B.L. Turner in Phytologia,* 61(3): 144 (1986): M. arida.
heterocarpa *R.L. Hartman & M.A. Lane in Brittonia,* 39(2): 253 (1987)—U.S.A. (Texas).

MACROBRIZA (Tzvel.) N.N. Tsvelev in Komarov. Chten., 37: 32 (1987): Briza sect. Macrobriza (Gramin.).

MACROPHARYNX (Apocynac.).
meyeri *(Ezcurra) C.C. Xifreda in Kurtziana,* 17: 164 (1984): Temnadenia meyeri.

MACROPTILIUM (Legumin.).
geophilum *(Burk.) D.G. Debouck, F. Juárez & S. Pérez in Kurtziana,* 18: 155 (1986): Phaseolus geophilus.

MAESA (Myrsinac.).
andamanica
var. longipedicellata *M.P. Nayar & G.S. Giri in Bull. Bot. Surv. India,* 27(1–4): 263 (1985 publ. 1987)—Andaman Is.
calcarea *H. Sleumer in Blumea,* 32(1): 48 (1987)—New Guinea (W. Irian).
corneri *H. Sleumer in Blumea,* 32(1): 55 (1987)—Bougainville Is.
hooglandii *H. Sleumer in Blumea,* 32(1): 45 (1987)—New Guinea (Papua New Guinea).
lineolata *H. Sleumer in Blumea,* 32(1): 54 (1987)—New Guinea (W. Irian).
loloruensis *H. Sleumer in Blumea,* 32(1): 46 (1987)—Bougainville Is.
regia *H. Sleumer in Blumea,* 32(1): 56 (1987)—British Solomon Is.

MAESA :–
rheophytica *H. Sleumer in Blumea,* 32(1): 47 (1987)—New Guinea (W. Irian).
sayersii *H. Sleumer in Blumea,* 32(1): 60 (1987)—New Britain, Admiralty Is., New Hanover, New Ireland.
spectabilis *H. Sleumer in Blumea,* 32(1): 58 (1987)—New Guinea (Papua New Guinea, W. Irian).

MAGNOLIA (Magnoliac.).
ashtonii *Dandy ex H.P. Nooteboom in Blumea,* 32(2): 363 (1987)—Borneo (Brunei, Sarawak, Sabah, Kalimantan), Sumatra.
bintuluensis *(Agostini) H.P. Nooteboom in Blumea,* 32(2): 362 (1987): Talauma bintuluensis.
borneensis *H.P. Nooteboom in Blumea,* 32(2): 366 (1987)—Borneo (Kalimantan, Sabah, Sarawak), Philippine Is.
candollei *(Blume) H.P. Nooteboom in Blumea,* 32(2): 369 (1987): Talauma candollii.
var. angatensis *(Blanco) H.P. Nooteboom in Blumea,* 32(2): 375 (1987): M. angatensis.
var. beccarii *(Ridley) H.P. Nooteboom in Blumea,* 32(2): 375 (1987): Talauma beccarii.
var. obovata *(Korth.) H.P. Nooteboom in Blumea,* 32(2): 374 (1987): Talauma obovata.
var. singapurensis *(Ridley) H.P. Nooteboom in Blumea,* 32(2): 376 (1987): Talauma singapurensis.
carsonii *Dandy ex H.P. Nooteboom in Blumea,* 32(2): 348 (1987)—Borneo (Sabah).
var. drymifolia *H.P. Nooteboom in Blumea,* 32(2): 351 (1987)—Borneo (Kalimantan, Sarawak, Sabah).
gigantifolia *(Miq.) H.P. Nooteboom in Blumea,* 32(2): 377 (1987): Talauma gigantifolia.
lasia *H.P. Nooteboom in Blumea,* 32(2): 377 (1987)—Borneo (Kalimantan, Sarawak, Sabah).
macklottii
var. beccariana *(Agostini) H.P. Nooteboom in Blumea,* 32(2): 348, 347 (1987): Michelia beccariana.
mariusjacobsia *H.P. Nooteboom in Blumea,* 32(2): 381 (1987)—Borneo (Sarawak).
pahangensis *H.P. Nooteboom in Blumea,* 32(2): 367 (1987)—Malaya.
persuaveolens
var. pubescens *H.P. Nooteboom in Blumea,* 32(2): 379 (1987)—Borneo (Sabah).
subsp. rigida *H.P. Nooteboom in Blumea,* 32(2): 379 (1987)—Borneo (Sabah).
var. rigida *H.P. Nooteboom in Blumea,* 32(2): 379 (1987)—Borneo (Sabah).
phaulanta *Dandy ex H.P. Nooteboom in Blumea,* 32(2): 359 (1987)—Celebes.
sarawakensis *(Agostini) H.P. Nooteboom in Blumea,* 32(2): 380 (1987): Talauma sarawakensis.
uvariifolia *Dandy ex H.P. Nooteboom in Blumea,* 32(2): 358 (1987)—Borneo (Sabah, Sarawak, Kalimantan).

MAHONIA (Berberidac.).
bracteolata
var. zhongdianensis *S.Y. Bao in Acta Phytotax. Sin.,* 25(2): 154 (1987)—China.
caelicolor *S.Y. Bao in Acta Phytotax. Sin.,* 25(2): 150 (1987)—China.
lomariifolia
var. estylis *C.Y. Wu ex S.Y. Bao in Acta Phytotax. Sin.,* 25(2): 154 (1987)—China.
paucijuga *C.Y. Wu ex S.Y. Bao in Acta Phytotax. Sin.,* 25(2): 151 (1987)—China.
reticulinervia *C.Y. Wu ex S.Y. Bao in Acta Phytotax. Sin.,* 25(2): 153 (1987)—China.

MAIANTHEMUM (Convallariac.).
amplexicaule *(Nuttall) W.A. Weber in Phytologia,* 62(5): 437 (1987): Smilacina amplexicaulis.

MALAXIS (Orchidac.).
apiculata *D.D. Dod in Moscosoa,* 4: 174 (1986)—Hispaniola (Dominican Republic).
megalantha *D.D. Dod in Moscosoa,* 4: 176 (1986)—Hispaniola (Haiti).
punctata *J.J. Wood in Kew Mag.,* 4(2): 76 (1987)—Borneo (Sabah).
shuicae *S.S. Ying in J. Jap. Bot.,* 62(3): 70 (1987), as 'shuicai'—Taiwan.

MALLOTUS (Euphorbiac.).
chuyenii *N.N. Thìn in J. Biol. (Vietnam),* 9(2): 38 (1987)—Vietnam.

MALOPE (Malvac.).
malacoides
subsp. asterotricha *(Pomel) Quézel & Santa ex W. Greuter & Burdet in Willdenowia,* 16(2): 447 (1987): M. asterotricha.
subsp. laevigata *(Pomel) Quézel & Santa ex W. Greuter & Burdet in Willdenowia,* 16(2): 448 (1987): M. laevigata.

MALPIGHIA (Malpighiac.).
emiliae *W.R. Anderson in Contrib. Univ. Mich. Herb.,* 16: 90 (1987)—Mexico.
novogaliciana *W.R. Anderson in Contrib. Univ. Mich. Herb.,* 16: 92 (1987)—Mexico.
rzedowskii *W.R. Anderson in Contrib. Univ. Mich. Herb.,* 16: 95 (1987)—Mexico.
stevensii *W.R. Anderson in Contrib. Univ. Mich. Herb.,* 16: 97 (1987)—Nicaragua, Costa Rica.
verruculosa *W.R. Anderson in Contrib. Univ. Mich. Herb.,* 16: 99 (1987)—Nicaragua, Honduras, Costa Rica.
wendtii *W.R. Anderson in Contrib. Univ. Mich. Herb.,* 16: 100 (1987)—Mexico, Guatemala.
wilburiorum *W.R. Anderson in Contrib. Univ. Mich. Herb.,* 16: 103 (1987)—Mexico.

MALUS (Rosac.).
angustifolia
var. spinosa *(Rehd.) C.F. Reed in Phytologia,* 63(5): 410 (1987): M. ioensis var. spinosa.

MALUS :–
coronaria
var. lancifolia *(Rehd.) C.F. Reed in Phytologia,* 63(5): 410 (1987): M. lancifolia.
dirangensis *M.C. Joshi & S.M. Singh in Pl. Sci. (Lucknow),* 12: 21 (1980), without latin descr. or type—India.
jinxianensis *J.Q. Deng & J.Y. Hong in Acta Phytotax. Sin.*, 25(4): 326 (1987)—China.

MALVAVISCUS (Malvac.).
urticifolius *(K. Presl) P.A. Fryxell in Syst. Bot.*, 12(2): 279 (1987): Pavonia urticifolia.

MAMMILLARIA (Cactac.).
dixanthocentron
var. rubrispina *R. Wolf in Kakt. And. Sukk.*, 38(2): 46 (1987)—Mexico.
freudenbergeri *W. Reppenhagen in Kakt. And. Sukk.*, 38(10): 255 (1987)—Mexico.
xaltianguensis *H. Sánchez-Mejorada R. in An. Inst. Biol. Univ. Nac. Autón. Méx., Bot.*, 44(1): 30 (1973)—Mexico.
zephyranthoides
var. heidiae *(Krainz) J. Lüthy in Kakt. And. Sukk.*, 38(1): 11 (1987): M. heidiae.

MANFREDA (Agavac.).
fusca *P. Ravenna in Herbertia,* 43(1): 17 (1987)—Guatemala.

MANGLIETIA (Magnoliac.).
glaucifolia *Y.W. Law & Y.F. Wu in Guihaia,* 6(4): 263 (1986)—China.

MARANTA (Marantac.).
amazonica *L. Andersson in Nordic J. Bot.*, 6(6): 752 (1986)—Ecuador, Colombia, Peru.
incrassata *L. Andersson in Nordic J. Bot.*, 6(6): 747 (1986)—Bolivia, Brazil.
lindmanii *L. Andersson in Nordic J. Bot.*, 6(6): 744 (1986)—Brazil.
linearis *L. Andersson in Nordic J. Bot.*, 6(6): 741 (1986)—Venezuela, Colombia.
rupicola *L. Andersson in Nordic J. Bot.*, 6(6): 751 (1986)—Surinam, Guyana, Venezuela, French Guiana, Brazil.
sobolifera *L. Andersson in Nordic J. Bot.*, 6(6): 742 (1986)—Brazil, Paraguay, Argentina.
tuberculata *L. Andersson in Nordic J. Bot.*, 6(6): 754 (1986)—Brazil.
zingiberina *L. Andersson in Nordic J. Bot.*, 6(6): 746 (1986)—Brazil.

MARICA (Iridac.).
coelestis *J.G.C. Lehmann in Sem. Hort. Bot. Hamb.:* 17 (1826)—Locality not stated. †
tigridia *J.G.C. Lehmann in Sem. Hort. Bot. Hamb.:* 17 (1826)—Locality not stated. †

MARISCUS (Cyperac.).
microbolbos *(C.B. Cl.) P. Vorster in Kew Bull.*, 42(1): 200 (1987): Cyperus microbolbos.

MARKEA (Solanac.).
lopezii *A.T. Hunziker in Lorentzia,* 5: 9 (1985)—Colombia.

MARSDENIA (Asclepiadac.).
carvalhoi *G. Morillo & G. Carnevali in Ernstia,* 45: 3 (1987)—Brazil.
fontellana *G. Morillo & G. Carnevali in Ernstia,* 45: 6 (1987)—Brazil.
hatschbachii *G. Morillo in Mem. Soc. Cienc. Nat. La Salle,* 37(107): 120 (1977)—Brazil.
neomanarae *G. Morillo in Ernstia,* 43: 21 (1987), as 'neo-manarae'—Brazil.
nubicola *Alain H. Liogier in Phytologia,* 63(1): 66 (1987)—Hispaniola (Dominican Republic).

MASCAGNIA (Malpighiac.).
mesoamericana *W.R. Anderson in Contrib. Univ. Mich. Herb.*, 16: 105 (1987)—Guatemala, Mexico, Panama.
paludicola *W.R. Anderson in Contrib. Univ. Mich. Herb.*, 16: 106 (1987)—Brazil.

MASDEVALLIA (Orchidac.).
manarana *G. Carnevali F.C. & I. Ramírez de Carnevali in Ernstia,* 44: 15 (1987)—Venezuela.
wubbenii *C.A. Luer in Amer. Orchid Soc. Bull.*, 56(11): 1141 (1987)—Venezuela.

MATAYBA (Sapindac.).
apetala
subsp. oppositifolia *(A. Rich.) Borhidi ex M.A. Vales & R. Carreras in Acta Bot. Hung.*, 32(1–4): 245 (1986): Cupania oppositifolia.

MATELEA (Asclepiadac.).
aliciae *G. Morillo in An. Jard. Bot. Madrid,* 43(2): 238 (1986 publ. 1987)—Peru.
amazonica *G. Morillo in Mem. Soc. Cienc. Nat. La Salle,* 37(107): 122 (1977)—Venezuela, Brazil.
asplundii *(Malme) G. Morillo in An. Jard. Bot. Madrid,* 43(2): 239 (1986 publ. 1987): Gonolobus asplundii.
bolivarensis *G. Morillo in Mem. Soc. Cienc. Nat. La Salle,* 37(107): 122 (1977), as 'bolivarense'—Venezuela.
borinquensis *Alain H. Liogier in Phytologia,* 61(6): 360 (1986)—Puerto Rico.
coriacea *G. Morillo in Mem. Soc. Cienc. Nat. La Salle,* 37(107): 124 (1977)—Venezuela.
costata *(Urb.) G. Morillo in An. Jard. Bot. Madrid,* 43(2): 239 (1986 publ. 1987): Poicilla ? costata.
dusenii *G. Morillo in An. Jard. Bot. Madrid,* 43(2): 239 (1986 publ. 1987)—Brazil.
harlingii *G. Morillo in An. Jard. Bot. Madrid,* 43(2): 241 (1986 publ. 1987)—Ecuador.
melinii *(Malme) G. Morillo in An. Jard. Bot. Madrid,* 43(2): 244 (1986 publ. 1987): Gonolobus melinii.
riparia *G. Morillo in Mem. Soc. Cienc. Nat. La Salle,* 37(107): 125 (1977)—Guyana.
rubra *(Karst.) Van der Aa & A.L. Stoffers in Proc. Kon. Nederl. Akad. Wetensch., C,* 84(3): 309 (1981): Omphalophthalmum ruber.

MATUCANA (Cactac.).
tuberculata *(Don.) R. Bregman, A. Meerstadt, P. Melis & A.B. Pullen in Succulenta,* 66(9): 175 (1987): Borzicactus tuberculatus.

MAYTENUS (Celastrac.).
drummondii *N.K.B. Robson & Sebsebe Demissew in Kew Bull.,* 42(2): 424 (1987)—Tanzania, Kenya, Madagascar.
keniensis *(Loes.) N.K.B. Robson & Sebsebe Demissew in Kew Bull.,* 42(2): 423 (1987): Gymnosporia serrata var. keniensis.
nguruensis *N.K.B. Robson & Sebsebe Demissew in Kew Bull.,* 42(2): 426 (1987)—Tanzania.
ocoensis *M.M. Mejía P. & T.A. Zanoni in Moscosoa,* 4: 112 (1986)—Hispaniola (Dominican Republic).
oleosa *A.E. van Wyk & R.H. Archer in S. Afr. J. Bot.,* 53(2): 155 (1987)—South Africa (Cape Province, Natal).
richardsiae *N.K.B. Robson & Sebsebe Demissew in Kew Bull.,* 42(2): 427 (1987)—Tanzania.

MECRANIUM (Melastomatac.).
neibense *J.D. Skean in Brittonia,* 39(2): 192 (1987)—Hispaniola (Dominican Republic).

MEDICAGO (Legumin.).
sect. Buceras *(Ser.) E. Small in C.H. Stirton (ed.), Advances in Legume System.,* 3: 180 (1987): Trigonella sect. Buceras.
sect. Geocarpa *E. Small in C.H. Stirton (ed.), Advances in Legume System.,* 3: 179 (1987).
sect. Lunatae *(Boiss.) E. Small in C.H. Stirton (ed.), Advances in Legume System.,* 3: 180 (1987): Trigonella Lunatae.
sect. Platycarpae *E. Small in C.H. Stirton (ed.), Advances in Legume System.,* 3: 180 (1987).
arenicola *(Huber-Mor.) E. Small in Willdenowia,* 16(2): 433 (1987): Trigonella arenicola.
biflora *(Griseb.) E. Small in Willdenowia,* 16(2): 433 (1987): Trigonella biflora.
carica *(Huber-Mor.) E. Small in Willdenowia,* 16(2): 433 (1987): Trigonella carica.
crassipes *(Boiss.) E. Small in Willdenowia,* 16(2): 433 (1987): Trigonella crassipes.
halophila *(Boiss.) E. Small in Willdenowia,* 16(2): 433 (1987): Trigonella halophila.
heldreichii *E. Small in Willdenowia,* 16(2): 434 (1987), nom. nov.: Trigonella polycarpa *Boiss. & Heldr.*
huberi *E. Small in Willdenowia,* 16(2): 434 (1987), nom. nov.: Trigonella sirjaevii *Huber-Mor.*
isthmocarpa *(Boiss. & Bal.) E. Small in Willdenowia,* 16(2): 433 (1987): Trigonella isthmocarpa.
medicaginoides *(Retz.) E. Small in Willdenowia,* 16(2): 434 (1987): Lotus medicaginoides.

MEDICAGO :–
pamphylica *(Huber-Mor. & Sirj.) E. Small in Willdenowia,* 16(2): 434 (1987): Trigonella pamphylica.
persica *(Boiss.) E. Small in Willdenowia,* 16(2): 434 (1987): Trigonella persica.
phrygia *(Boiss. & Bal.) E. Small in Willdenowia,* 16(2): 433 (1987): Trigonella phrygia.
polymorpha
subsp. polycarpa *(Willd.) C. Romero Zarco in Lagascalia,* 14(1): 146 (1986): M. polycarpa.
retrorsa *(Boiss.) E. Small in Willdenowia,* 16(2): 434 (1987): Trigonella retrorsa.
rhytidiocarpa *(Boiss. & Bal.) E. Small in Willdenowia,* 16(2): 434 (1987): Trigonella rhytidiocarpa.
rigida *(Boiss. & Bal.) E. Small in Willdenowia,* 16(2): 434 (1987): Trigonella rigida.
rostrata *(Boiss. & Bal.) E. Small in Willdenowia,* 16(2): 434 (1987): Pocockia rostrata.
tornata
subsp. striata *(Bast.) M. Kerguélen in Lejeunia,* 120: 127 (1987): M. striata.

MEDIUSELLA (Cavaco) L.J. Dorr in Ann. Missouri Bot. Gard., 73(4): 828 (1986 publ. 1987): Leptolaena subgen. Mediusella (Sarcolaenac.).
bernieri *(Baillon) L.J. Dorr in Ann. Missouri Bot. Gard.,* 73(4): 828 (1986 publ. 1987): Leptolaena bernieri.

MELALEUCA (Myrtac.).
acacioides
subsp. alsophila *(Cunn. ex Benth.) B.A. Barlow in Brunonia,* 9(2): 167 (1987): M. alsophila.
citrolens *B.A. Barlow in Brunonia,* 9(2): 168 (1987)—Australia (Queensland, N. Terr.).
minutifolia
subsp. monantha *B.A. Barlow in Brunonia,* 9(2): 176 (1987)—Australia (Queensland).
tamariscina
subsp. irbyana *(R. Baker) B.A. Barlow in Brunonia,* 9(2): 173 (1987): M. irbyana.
subsp. pallescens *(Byrnes) B.A. Barlow in Brunonia,* 9(2): 173 (1987): M. pallescens.

MELASTOMA (Melastomatac.).
pentapetalum *(Toyoda) T. Yamazaki & Toyoda in J. Jap. Bot.,* 62(4): 118 (1987): M. tetramerum var. pentapetalum.

MELICA (Gramin.).
atrovirens *(Thuill.) Le Turquier de Longchamp, Fl. Environs de Rouen, Suppl.:* 14 (1825): Aira atrovirens.
bocquetii *S. Talavera in Lagascalia,* 14(1): 118 (1986)—Spain.
caerulea *Lamarck, Tabl. Encycl. Méth. Bot.,* 1: 180 (1791)—Locality not stated.

MELICOPE (Rutac.).
capillacea *(Gillespie) A.C. Smith in Fl. Vitiensis Nova,* 3: 506 (1985): Euodia capillacea.
cucullata *(Gillespie) A.C. Smith in Fl. Vitiensis Nova,* 3: 502 (1985): Euodia cucullata.
var. robustior *(A.C. Sm.) A.C. Smith in Fl. Vitiensis Nova,* 3: 502 (1985): Euodia cucullata var. robustior.
evansensis *(A.C. Sm.) A.C. Smith in Fl. Vitiensis Nova,* 3: 505 (1985): Euodia evansensis.
seemannii *(Gillespie) A.C. Smith in Fl. Vitiensis Nova,* 3: 501 (1985): Euodia seemannii.
vitiensis *(A.C. Sm.) A.C. Smith in Fl. Vitiensis Nova,* 3: 503 (1985): Euodia vitiensis.
var. minor *(A.C. Sm.) A.C. Smith in Fl. Vitiensis Nova,* 3: 505 (1985): Euodia vitiensis var. minor.

MELILOTUS (Legumin.).
graeca *(Boiss. & Spruner) P. Lassen in Willdenowia,* 16(2): 445 (1987): Pocockia graeca.

MELINIS (Gramin.).
sect. Acutiglumae *G. Zizka, Revis. Melinideae Hitchcock:* 199 (1987).
sect. Drakensbergenses *G. Zizka, Revis. Melinideae Hitchcock:* 182 (1987).
sect. Gossweileri *G. Zizka, Revis. Melinideae Hitchcock:* 179 (1987).
sect. Reynaudiopsis *G. Zizka, Revis. Melinideae Hitchcock:* 186 (1987).
sect. Roseae *(Stapf & C.E. Hubbard) G. Zizka, Revis. Melinideae Hitchcock:* 84 (1987): Rhynchelytrum sect. Roseae.
sect. Setifoliae *(Stapf & C.E. Hubbard) G. Zizka, Revis. Melinideae Hitchcock:* 190 (1987): Rhynchelytrum sect. Setifoliae.
sect. Tanatrichae *(Stapf & C.E. Hubbard) G. Zizka, Revis. Melinideae Hitchcock:* 205 (1987): Rhynchelytrum sect. Tanatrichae.
sect. Trachytrophae *(Stapf & C.E. Hubbard) G. Zizka, Revis. Melinideae Hitchcock:* 121 (1987): Rhynchelytrum sect. Trachytrophae.
sect. Trichocraspedae *(Stapf & C.E. Hubbard) G. Zizka, Revis. Melinideae Hitchcock:* 214 (1987): Rhynchelytrum sect. Trichocraspedae.
ambigua
subsp. longicauda *(Mez) G. Zizka, Revis. Melinideae Hitchcock:* 152 (1987): Panicum longicaudum.
amethystea *(Franchet) G. Zizka, Revis. Melinideae Hitchcock:* 199 (1987): Tricholaena amethystea.
kallimorpha *(W.D. Clayton) G. Zizka, Revis. Melinideae Hitchcock:* 110 (1987): Rhynchelytrum kallimorphon.
longiseta *(A. Richard) G. Zizka, Revis. Melinideae Hitchcock:* 124 (1987): Tricholaena longiseta.

MELINIS :–
subsp. bellespicata *(Rendle) G. Zizka, Revis. Melinideae Hitchcock:* 130 (1987): Tricholaena bellespicata.
nerviglumis *(Franchet) G. Zizka, Revis. Melinideae Hitchcock:* 190 (1987): Tricholaena nerviglumis.
repens *(Willdenow) G. Zizka, Revis. Melinideae Hitchcock:* 86 (1987): Saccharum repens.
subsp. grandiflora *(Hochstetter) G. Zizka, Revis. Melinideae Hitchcock:* 101 (1987): Rhynchelytrum grandiflorum.
subsp. maroccana *(Maire & Samuelsson) G. Zizka, Revis. Melinideae Hitchcock:* 107 (1987): Tricholaena maroccana.
subsp. nigricans *(Mez) G. Zizka, Revis. Melinideae Hitchcock:* 109 (1987): M. nigricans.
reynaudioides *(C.E. Hubbard) G. Zizka, Revis. Melinideae Hitchcock:* 186 (1987): Rhynchelytrum reynaudioides.
rupicola *(Rendle) G. Zizka, Revis. Melinideae Hitchcock:* 210 (1987): Tricholaena rupicola.
tanatricha *(Rendle) G. Zizka, Revis. Melinideae Hitchcock:* 206 (1987): Tricholaena tanatricha.

MELIOSMA (Meliosmac.).
corymbosa *A.H. Gentry in Ann. Missouri Bot. Gard.,* 73(4): 821 (1986 publ. 1987)—Nicaragua.
hartshornii *A.H. Gentry in Ann. Missouri Bot. Gard.,* 73(4): 820 (1986 publ. 1987)—Costa Rica.
nanarum *A.H. Gentry in Ann. Missouri Bot. Gard.,* 73(4): 821 (1986 publ. 1987)—Nicaragua.
peytonii *A.H. Gentry in Ann. Missouri Bot. Gard.,* 73(4): 822 (1986 publ. 1987)—Peru.
solomonii *A.H. Gentry in Ann. Missouri Bot. Gard.,* 73(4): 823 (1986 publ. 1987)—Bolivia.
vasquezii *A.H. Gentry in Ann. Missouri Bot. Gard.,* 73(4): 823 (1986 publ. 1987)—Peru.

MELISSITUS (Legumin.).
ruthenicus
var. inschanicus *(H.C. Fu & Y.C. Tsiang) H.C. Fu & Y.C. Tsiang in Acta Sci. Nat. Univ. Intramongolicae,* 18(2): 391 (1987): Pocockia ruthenica var. inschanica.
var. liaosiensis *(P.Y. Fu & Y.A. Chen) H.C. Fu & Y.C. Tsiang in Acta Sci. Nat. Univ. Intramongolicae,* 18(2): 391 (1987): Pocockia liaosiensis.
var. oblongifolius *(Fr.) H.C. Fu & Y.C. Tsiang in Acta Sci. Nat. Univ. Intramongolicae,* 18(2): 391 (1987): Trigonella ruthenica f. oblongifolia.

MELOCACTUS (Cactac.).
ruestii
subsp. cintalapensis *J. Elizondo E. in Cact. Suc. Mex.,* 31(3): 54 (1986)—Mexico.

MELOCACTUS :–
subsp. maxonii *(Rose) J. Elizondo E. in Cact. Suc. Mex.,* 31(3): 53 (1986): Cactus maxonii.
subsp. oaxacensis *(Britton & Rose) J. Elizondo E. in Cact. Suc. Mex.,* 31(3): 52 (1986): Cactus oaxacensis.
subsp. sanctae-rosae *(L.D. Gómez) J. Elizondo E. in Cact. Suc. Mex.,* 31(3): 54 (1986): M. maxonii var. sanctae-rosae.

MELODINUS (Apocynac.).
hemsleyanus
var. bachthaiensis *Trân Đình Lý in Feddes Repert.,* 97(7–8): 413 (1986), without latin descr.—Vietnam.

MEMECYLON (Melastomatac.).
umbellatum
var. contractum *(Haines) D.M. Verma in D.M. Verma, P.C. Pant & M.I. Hanfi, Fl. Raipur, Durg & Rajnandgaon:* 148 (1985): M. edule var. contractum.

MENTZELIA (Loasac.).
cronquistii *H. Thompson & B. Prigge in Great Basin Nat.,* 46(3): 550 (1986)—U.S.A. (Utah, Colorado, Arizona, New Mexico).
marginata *(Osterhout) H. Thompson & B. Prigge in Great Basin Nat.,* 46(3): 549 (1986): Nuttallia marginata.
multicaulis
var. librina *K.H. Thorne & F.J. Smith in Great Basin Nat.,* 46(3): 556 (1986)—U.S.A. (Utah).
pumila
var. lagarosa *K.H. Thorne in Great Basin Nat.,* 46(3): 558 (1986)—U.S.A. (Utah, Nevada).

MERENDERA (Colchicac.).
mirzoevae *É.Ts. Gabriélyan in Biol. Zhurn. Arm.,* 39(7): 582 (1986)—U.S.S.R. (Caucasus).

MERIANIA (Melastomatac.).
brevipedunculata *W.S. Judd & J.D. Skean in Syst. Bot.,* 12(3): 374 (1987)—Hispaniola (Haiti).
crassiramis *(Naud.) J.J. Wurdack in Brittonia,* 39(2): 159 (1987): Davya crassiramis.
parvifolia *W.S. Judd & J.D. Skean in Syst. Bot.,* 12(3): 377 (1987)—Hispaniola (Haiti).

MERREMIA (Convolvulac.).
austinii *J.A. McDonald in Brittonia,* 39(1): 106 (1987)—Mexico.
sibirica
var. jiuhuaensis *B.A. Shen & X.L. Liu in Bull. Bot. Res. North-East. Forest. Inst.,* 7(3): 83 (1987)—China.
verruculosa *S.Y. Liu in Bull. Bot. Res. North-East. Forest. Inst.,* 7(2): 133 (1987)—China.

MESOSETUM (Gramin.).
alatum *T.S. Filgueiras in Brittonia,* 39(2): 306 (1987)—Brazil.

MESOSETUM :–
longiaristatum *T.S. Filgueiras in Brittonia,* 39(2): 308 (1987)—Brazil.

METROXYLON (Palmae).
sagu
f. longispinum *(Giseke) J.B. Rauwerdink in Principes,* 30(4): 175 (1986): Sagus genuina longispina.
f. micracanthum *(Blume) J.B. Rauwerdink in Principes,* 30(4): 175 (1986): Sagus micracantha.
f. tuberatum *J.B. Rauwerdink in Principes,* 30(4): 175 (1986)—New Guinea (Papua New Guinea).

METTERNICHIA (Solanac.).
princeps
var. macrocalyx *L. d'Avila Freire de Carvalho in W.G. D'Arcy, Solanaceae: biol. & syst.:* 12 (1986)—Brazil.

MEZILAURUS (Laurac.).
caatingae *H. van der Werff in Ann. Missouri Bot. Gard.,* 74(1): 159 (1987)—Brazil.
duckei *H. van der Werff in Ann. Missouri Bot. Gard.,* 74(1): 162 (1987)—Brazil.
glaucophylla *H. van der Werff in Ann. Missouri Bot. Gard.,* 74(1): 164 (1987)—Costa Rica.
mahuba *(Samp.) H. van der Werff in Ann. Missouri Bot. Gard.,* 74(1): 169 (1987): Acrodiclidium mahuba.
micrantha *H. van der Werff in Ann. Missouri Bot. Gard.,* 74(1): 170 (1987)—Brazil.
opaca *Kubitzki & H. van der Werff in Ann. Missouri Bot. Gard.,* 74(1): 171 (1987)—Peru.
palcazuensis *H. van der Werff in Ann. Missouri Bot. Gard.,* 74(1): 173 (1987)—Peru.
pyriflora *H. van der Werff in Ann. Missouri Bot. Gard.,* 74(1): 174 (1987)—Brazil.
quadrilocellata *H. van der Werff in Ann. Missouri Bot. Gard.,* 74(1): 176 (1987)—Colombia.
thoroflora *H. van der Werff in Ann. Missouri Bot. Gard.,* 74(1): 179 (1987), nom. nov.: Licaria maguireana *Allen.*

MICHELIA (Magnoliac.).
amoenna *Q.F. Zheng & M.M. Lin in Bull. Bot. Res. North-East. Forest. Inst.,* 7(1): 63 (1987)—China.
brevipes *Y.K. Li & X.M. Wang in Acta Phytotax. Sin.,* 25(5): 408 (1987)—China.
coriacea *Chang & B.L. Chen in Acta Sci. Nat. Univ. Sunyatseni,* 1987(3): 89 (1987)—China.
fulva *Chang & B.L. Chen in Acta Sci. Nat. Univ. Sunyatseni,* 1987(3): 87 (1987)—China.
oblongifolia *Chang & B.L. Chen in Acta Sci. Nat. Univ. Sunyatseni,* 1987(3): 86 (1987)—China.
opipara *Chang & B.L. Chen in Acta Sci. Nat. Univ. Sunyatseni,* 1987(3): 90 (1987)—China.

MICHELIA :–
pachycarpa *Y.W. Law & R.Z. Zhou in Bull. Bot. Res. North-East. Forest. Inst.*, 7(1): 85 (1987)—China.

MICONIA (Melastomatac.).
aliquantula *J.J. Wurdack in Brittonia*, 39(2): 161 (1987), nom. nov.: Tschudya pulverulenta *DC.*

MICRANDRA R Brown, Pterocymbium: 237 (4 June 1844), nom. rejic. *EUPHORBIACEAE*. †
ternata *R. Brown, Pterocymbium:* 238 (4 June 1844)—Brazil. †

MICROHOLMESIA P. Cribb in D.J. Mabberley, Plant-book: 371 (1987), nom. nov.: Holmesia Cribb (Orchidac.).
parva *(Cribb) P. Cribb in D.J. Mabberley, Plant-book:* 371 (1987): Holmesia parva.

MICROMERIA (Labiat.).
filiformis
subsp. minutifolia *(Chodat) M. Kerguélen in Lejeunia,* 120: 128 (1987): Satureja filiformis var. minutifolia.

MICROPHOLIS (Sapotac.).
polita
subsp. hotteana *W.S. Judd in Moscosoa,* 4: 223 (1986)—Hispaniola (Haiti).

MICROPYRUM (Gramin.).
tenellum
f. aristatum *(Tausch) J. Lambinon in Soc. Ech. Pl. Vasc. Eur. Bassin Médit. Bull.,* 21: 68 (1986): Triticum lolioides var. aristatum.

MIMOSA (Legumin.).
aureliana *R.C. Barneby & R.H. Fortunato in Brittonia,* 39(2): 165 (1987)—Paraguay.
castanoclada *R.C. Barneby & R.H. Fortunato in Brittonia,* 39(2): 167 (1987)—Paraguay, Bolivia.
chacoensis *R.C. Barneby & R.H. Fortunato in Brittonia,* 39(2): 169 (1987)—Paraguay.
cordobensis *L. Ariza Espinar in Lorentzia,* 6: 7 (1986)—Argentina.
nanchititlana *R. Grether & R.C. Barneby in Brittonia,* 39(3): 345 (1987)—Mexico.
pueblensis *R. Grether in J. Arnold Arbor.,* 68(3): 319 (1987), nom. nov.: Mimosopsis filipes *Britton & Rose.*
rhodocarpa *(Britton & Rose) R. Grether in J. Arnold Arbor.,* 68(3): 320 (1987): Mimosopsis rhodocarpa.
rubicaulis
subsp. himalayana *(Gamble) H. Ohashi in Enum. Fl. Pl. Nepal,* 2: 126 (1979): M. himalayana.
tobatiensis *R.C. Barneby & R.H. Fortunato in Brittonia,* 39(2): 171 (1987)—Paraguay.

MIMULUS (Scrophulariac.).
shevockii *L.R. Heckard & R. Bacigalupi in Madroño,* 33(4): 271 (1986)—U.S.A. (California).

MINUARTIA (Caryophyllac.).
biflora
f. versicolor *(Brügger) J. Cayouette in Nat. Canad.,* 113(4): 332 (1986): Alsine biflora var. versicolor.
rubella
f. epilis *(Fern.) J. Cayouette in Nat. Canad.,* 113(4): 332 (1986): Arenaria verna f. epilis.
verna
subsp. brevipetala *P. Hartvig & A. Strid in Bot. Jahrb.,* 108(2–3): 322 (1987)—Turkey.

MIRBELIA (Legumin.).
granitica *M.D. Crisp & J.M. Taylor in J. Adelaide Bot. Gard.,* 10(1): 135 (1987)—Australia (W. Australia).
rhagodioides *M.D. Crisp & J.M. Taylor in J. Adelaide Bot. Gard.,* 10(1): 139 (1987)—Australia (W. Australia).
stipitata *M.D. Crisp & J.M. Taylor in J. Adelaide Bot. Gard.,* 10(1): 141 (1987)—Australia (W. Australia).

MISOPATES (Scrophulariac.).
orontium
var. grandiflorum *(Chav.) B. Valdés in Lagascalia,* 14(1): 156 (1986): Antirrhinum orontium var. grandiflorum.

MITRIOSTIGMA (Rubiac.).
usambarense *B. Verdcourt in Kew Bull.,* 42(1): 245 (1987)—Tanzania.

MOLINIA (Gramin.).
ser. Caerulea *I. Milkovits in Symb. Bot. Upsal.,* 27(2): 140 (1986 publ. 1987).
ser. Hungaricae *I. Milkovits in Symb. Bot. Upsal.,* 27(2): 142 (1986 publ. 1987).
arundinacea
var. robusta *I. Milkovits in Symb. Bot. Upsal.,* 27(2): 141 (1986 publ. 1987)—Hungary.
horanszkyi *I. Milkovits in Symb. Bot. Upsal.,* 27(2): 143 (1986 publ. 1987)—Hungary.
hungarica *I. Milkovits in Symb. Bot. Upsal.,* 27(2): 143 (1986 publ. 1987)—Hungary.
pocsii *I. Milkovits in Symb. Bot. Upsal.,* 27(2): 141 (1986 publ. 1987)—Hungary.
simonii *I. Milkovits in Symb. Bot. Upsal.,* 27(2): 142 (1986 publ. 1987)—Hungary.
var. major *I. Milkovits in Symb. Bot. Upsal.,* 27(2): 142 (1986 publ. 1987)—Hungary.
ujhelyii *I. Milkovits in Symb. Bot. Upsal.,* 27(2): 142 (1986 publ. 1987)—Hungary.

MOLOPOSPERMUM (Umbellif.).
peloponnesiacum
subsp. bauhinii *I. Ullmann in Flora (Germany),* 179(4): 268 (1987)—Italy.

MONADENIUM (Euphorbiac.).
catenatum *S. Carter in Kew Bull.*, 42(4): 905 (1987)—Tanzania.
depauperatum *(Bally) S. Carter in Kew Bull.*, 42(4): 908 (1987): M. laeve f. depauperata.
elegans *S. Carter in Kew Bull.*, 42(4): 909 (1987)—Tanzania.
filiforme *(Bally) S. Carter in Kew Bull.*, 42(4): 905 (1987): M. chevalieri var. filiforme.
gladiatum *(Bally) S. Carter in Kew Bull.*, 42(4): 913 (1987): M. yattanum var. gladiatum.
globosum *Bally & S. Carter in Kew Bull.*, 42(4): 906 (1987)—Tanzania.
heteropodum
var. formosum *(Bally) S. Carter in Kew Bull.*, 42(4): 916 (1987): M. schubei var. formosum.
pedunculatum *S. Carter in Kew Bull.*, 42(4): 903 (1987)—Tanzania, Burundi, Zaïre, Zambia.
renneyi *S. Carter in Kew Bull.*, 42(4): 916 (1987)—Kenya.
ritchiei
subsp. marsabitense *S. Carter in Kew Bull.*, 42(4): 915 (1987)—Kenya.
subsp. nyambense *S. Carter in Kew Bull.*, 42(4): 915 (1987)—Kenya.
rubellum *(Bally) S. Carter in Kew Bull.*, 42(4): 913 (1987): M. montanum var. rubellum.
schaijesii *F. Malaisse in Bull. Jard. Bot. Nation. Belg.*, 56(3–4): 487 (1986)—Zaïre.
spectabile *S. Carter in Kew Bull.*, 42(4): 911 (1987)—Tanzania.
stapelioides
var. congestum *(Bally) S. Carter in Kew Bull.*, 42(4): 914 (1987): M. stapelioides f. congestum.
stoloniferum *(Bally) S. Carter in Kew Bull.*, 42(4): 913 (1987): M. rhizophorum var. stoloniferum.

MONECHMA (Acanthac.).
callothamnum *J. Munday in S. Afr. J. Bot.*, 53(2): 140 (1987)—Namibia.

MONOPSIS (Campanulac.).
alba *P.B. Phillipson in Bot. J. Linn. Soc.*, 93(4): 334 (1986)—South Africa (Cape Province).
debilis
var. depressa *(L.f.) P.B. Phillipson in Bot. J. Linn. Soc.*, 93(4): 338 (1986): Lobelia depressa.
var. gracilis *(C. Presl) P.B. Phillipson in Bot. J. Linn. Soc.*, 93(4): 340 (1986): M. conspicua var. gracilis.

MONOTROPASTRUM (Ericac.).
sciaphilum *(H. Andres) G.D. Wallace in Taxon*, 36(1): 129 (1987): Eremotropa sciaphila.

MORAEA (Iridac.).
brevistyla *(Goldblatt) P. Goldblatt, Moraeas of South. Africa (Ann. Kirstenbosch Bot. Gard., 14):* 162 (1986): M. pubiflora subsp. brevistyla.
exiliflora *P. Goldblatt, Moraeas of South. Africa (Ann. Kirstenbosch Bot. Gard., 14):* 128 (1986)—South Africa (Cape Province).
garipensis *P. Goldblatt, Moraeas of South. Africa (Ann. Kirstenbosch Bot. Gard., 14):* 48 (1986)—Namibia.
namaquamontana *P. Goldblatt, Moraeas of South. Africa (Ann. Kirstenbosch Bot. Gard., 14):* 50 (1986)—South Africa (Cape Province).
verecunda *P. Goldblatt, Moraeas of South. Africa (Ann. Kirstenbosch Bot. Gard., 14):* 116 (1986)—South Africa (Cape Province).

MORINA (Morinac.).
nepalensis
var. alba *(Hand.-Mazz.) Y.C. Tang ex C.H. Hsing in Fl. Reipubl. Popul. Sin.*, 73(1): 51 (1986): M. alba.
var. delavayi *(Franch.) C.H. Hsing in Fl. Reipubl. Popul. Sin.*, 73(1): 51 (1986): M. delavayi.

MORINDA (Rubiac.).
umbellata
subsp. boninensis *(Ohwi) T. Yamazaki in J. Jap. Bot.*, 61(11): 343 (1986): M. boninensis.
var. boninensis *(Ohwi) T. Yamazaki in J. Jap. Bot.*, 61(11): 343 (1986): M. boninensis.
var. hahazimensis *T. Yamazaki in J. Jap. Bot.*, 61(11): 343 (1986)—Bonin Is.

MORUS (Morac.).
chinlingensis *C.L. Min in Acta Bot. Bor.-Occid. Sin.*, 6(4): 277 (1986)—China.

MOSIERA (Myrtac.).
acunae *(Borhidi & Muñiz) J. Bisse in Rev. Jard. Bot. Nacion. Univ. Habana*, 6(3): 5 (1985 publ. 1986): Myrtus acunae.
araneosa *(Urb.) J. Bisse in Rev. Jard. Bot. Nacion. Univ. Habana*, 6(3): 5 (1985 publ. 1986): Psidium araneosum.
bullata *(Britt. & Wils.) J. Bisse in Rev. Jard. Bot. Nacion. Univ. Habana*, 6(3): 5 (1985 publ. 1986): Myrtus bullata.
subsp. leiophloea *(Urb.) J. Bisse in Rev. Jard. Bot. Nacion. Univ. Habana*, 6(3): 5 (1985 publ. 1986): Psidium leiophloeum.
buxoides *(Urb.) J. Bisse in Rev. Jard. Bot. Nacion. Univ. Habana*, 6(3): 4 (1985 publ. 1986): Eugenia buxoides.
ekmanii *(Urb.) J. Bisse in Rev. Jard. Bot. Nacion. Univ. Habana*, 6(3): 4 (1985 publ. 1986): Myrtus ekmanii.
elliptica *(Wr.) J. Bisse in Rev. Jard. Bot. Nacion. Univ. Habana*, 6(3): 5 (1985 publ. 1986): Myrtus elliptica.

MOSIERA :–
subsp. matanzasia *(Urb.) J. Bisse in Rev. Jard. Bot. Nacion. Univ. Habana,* 6(3): 5 (1985 publ. 1986): Myrtus matanzasia.
flavicans *(Urb. & Ekm.) J. Bisse in Rev. Jard. Bot. Nacion. Univ. Habana,* 6(3): 5 (1985 publ. 1986): Eugenia flavicans.
subsp. cabanasensis *(Britt. & Wils.) J. Bisse in Rev. Jard. Bot. Nacion. Univ. Habana,* 6(3): 5 (1985 publ. 1986): Eugenia cabanasensis.
subsp. pastelillensis *(Urb.) J. Bisse in Rev. Jard. Bot. Nacion. Univ. Habana,* 6(3): 5 (1985 publ. 1986): Eugenia pastelillensis.
guineensis *(Sw.) J. Bisse in Rev. Jard. Bot. Nacion. Univ. Habana,* 6(3): 4 (1985 publ. 1986): Psidium guineense.
havanensis *(Urb.) J. Bisse in Rev. Jard. Bot. Nacion. Univ. Habana,* 6(3): 4 (1985 publ. 1986): Psidium havanense.
jackii *(Urb.) J. Bisse in Rev. Jard. Bot. Nacion. Univ. Habana,* 6(3): 5 (1985 publ. 1986): Psidium jackii.
moaensis *(Britt. & Wils.) J. Bisse in Rev. Jard. Bot. Nacion. Univ. Habana,* 6(3): 4 (1985 publ. 1986): Eugenia moaensis.
munizii *(Borhidi) J. Bisse in Rev. Jard. Bot. Nacion. Univ. Habana,* 6(3): 5 (1985 publ. 1986): Myrtus munizii.
nummularioides *(Britt. & Wils.) J. Bisse in Rev. Jard. Bot. Nacion. Univ. Habana,* 6(3): 5 (1985 publ. 1986): Myrtus nummularioides.
oonophylla *(Urb.) J. Bisse in Rev. Jard. Bot. Nacion. Univ. Habana,* 6(3): 4 (1985 publ. 1986): Eugenia oonophylla.
ophiticola *(Britt. & Wils.) J. Bisse in Rev. Jard. Bot. Nacion. Univ. Habana,* 6(3): 5 (1985 publ. 1986): Myrtus ophiticola.
sagrae *(Berg) J. Bisse in Rev. Jard. Bot. Nacion. Univ. Habana,* 6(3): 4 (1985 publ. 1986): Myrtus sagrae.

MOTLEYIA J.T. Johansson in Blumea, 32(1): 149 (1987). *RUBIACEAE.*
borneensis *J.T. Johansson in Blumea,* 32(1): 150 (1987)—Borneo (Sarawak, Sabah).

MUCUNA (Legumin.).
pruriens
var. hirsuta *(Wight & Arn.) C.M. Wilmot-Dear in Kew Bull.,* 42(1): 44 (1987): M. hirsuta.

MUKIA G.A.W. Arnott in Madras J. Lit. Sci., 12: 50 (1840). *CUCURBITACEAE.*

MULTIDENTIA (Rubiac.).
castaneae *(Robyns) D.M. Bridson & B. Verdcourt in Kew Bull.,* 42(3): 651 (1987): Vangueriopsis castaneae.
concrescens *(Bullock) D.M. Bridson & B. Verdcourt in Kew Bull.,* 42(3): 653 (1987): Pygmaeothamnus concrescens.
crassa *(Hiern) D.M. Bridson & B. Verdcourt in Kew Bull.,* 42(3): 652 (1987): Canthium crassum.

MULTIDENTIA :–
var. ampla *(S. Moore) D.M. Bridson & B. Verdcourt in Kew Bull.,* 42(3): 653 (1987): Canthium amplum.
dichrophylla *(Mildbr.) D.M. Bridson in Kew Bull.,* 42(3): 646 (1987): Plectronia dichrophylla.
exserta *D.M. Bridson in Kew Bull.,* 42(3): 647 (1987)—Tanzania, Mozambique, Zimbabwe.
subsp. robsonii *D.M. Bridson in Kew Bull.,* 42(3): 649 (1987)—Zambia, Mozambique, Malawi.
fanshawei *(Tennant) D.M. Bridson in Kew Bull.,* 42(3): 647 (1987): Canthium fanshawei.
kingupirensis *D.M. Bridson in Kew Bull.,* 42(3): 649 (1987)—Tanzania.
pobeguinii *(Hutch. & Dalz.) D.M. Bridson in Kew Bull.,* 42(3): 646 (1987): Canthium pobeguinii.
sclerocarpa *(K. Schum.) D.M. Bridson in Kew Bull.,* 42(3): 650 (1987): Plectronia sclerocarpa.

MUNNOZIA (Compos.).
fournetii *H. Robinson in Phytologia,* 63(5): 407 (1987)—Bolivia.

MURCHISONIA (Anthericac.).
volubilis *N.H. Brittan in Fl. S. Australia,* 4: 1762 (1986)—Australia (W. Australia, N. Terr., S. Australia).

MURDANNIA (Commelinac.).
nimmoniana *(Graham) P.V. Bole & M.R. Almeida in J. Bombay Nat. Hist. Soc.,* 83(3): 593 (1986 publ. 1987): Commelina nimmoniana.

MUSA (Musac.).
dechangensis *J.L. Liu & M.G. Liu in Acta Bot. Yunnanica,* 9(2): 163 (1987)—China.

MUSCARI (Hyacinthac.).
botryoides
subsp. motelayi *(Fouc.) M. Kerguélen in Lejeunia,* 120: 130 (1987): M. motelayi.
matritensis *M. Ruíz Rejón, L. Pascual, C. Ruíz Rejón, B. Valdés & J.L. Oliver in Lagascalia,* 14(2): 295 (1986)—Spain. †
sandrasicum *T. Karlén in Willdenowia,* 16(2): 375 (1987)—Turkey.

MYOSURUS (Ranunculac.).
minimus
subsp. novae-zelandiae *(Oliver) P.J. Garnock-Jones in New Zealand J. Bot.,* 24(2): 352 (1986): M. novae-zelandiae.

MYRCIA (Myrtac.).
curassavica *(Amsh.) A.L. Stoffers in Proc. Kon. Nederl. Akad. Wetensch., C,* 84(3): 349 (1981): Aulomyrcia curassavica.

MYRIOCEPHALUS (Compos.).
pluriflorus *(J. Black) D.A. Cooke in Fl. S. Australia,* 3: 1563 (1986): M. rhizocephalus var. pluriflorus.

N

NABALUS (Compos.).
albus *(L.) C. Shih in Acta Phytotax. Sin.,* 25(3): 194 (1987): Prenanthes alba.

NAGEIA (Podocarpac.).
sect. Afrocarpus *(Buchholz & Gray) D.J. de Laubenfels in Blumea,* 32(1): 211 (1987): Podocarpus sect. Afrocarpus.
sect. Polypodiopsis *(Bertrand) D.J. de Laubenfels in Blumea,* 32(1): 210 (1987): Podocarpus sect. Polypodiopsis.
comptonii *(Buchholz) D.J. de Laubenfels in Blumea,* 32(1): 211 (1987): Podocarpus comptonii.
fleuryi *(Hickel) D.J. de Laubenfels in Blumea,* 32(1): 210 (1987): Podocarpus fleuryi.
maximus *(de Laub.) D.J. de Laubenfels in Blumea,* 32(1): 210 (1987): Decussocarpus maximus.
motleyi *(Parl.) D.J. de Laubenfels in Blumea,* 32(1): 210 (1987): Dammara motleyi.
piresii *(Silba) D.J. de Laubenfels in Blumea,* 32(1): 211 (1987): Decussocarpus piresii.
rospigliosii *(Pilger) D.J. de Laubenfels in Blumea,* 32(1): 211 (1987): Podocarpus rospigliosii.

NAJAS (Najadac.).
marina
subsp. arsenariensis *(Maire) L. Triest in Mém. Acad. Roy. Sci. Outre-Mer, Sci. Nat. Nouv. Sér.,* 21(4): 32 (1987): N. arsenariensis.
subsp. commersonii *L. Triest in Mém. Acad. Roy. Sci. Outre-Mer, Sci. Nat. Nouv. Sér.,* 21(4): 27 (1987)—Réunion, Mauritius, Madagascar.
subsp. ehrenbergii *(A. Br.) L. Triest in Mém. Acad. Roy. Sci. Outre-Mer, Sci. Nat. Nouv. Sér.,* 21(4): 24 (1987): N. major var. ehrenbergii.
subsp. microcarpa *(A. Br.) L. Triest in Mém. Acad. Roy. Sci. Outre-Mer, Sci. Nat. Nouv. Sér.,* 21(4): 25 (1987): N. major var. microcarpa.
poyangensis *S.F. Guan & Q. Lang in Bull. Bot. Res. North-East. Forest. Inst.,* 7(1): 78 (1987)—China.

NANDINA (Berberidac.).
domestica
var. linearifolia *C.Y. Wu ex S.Y. Bao in Acta Phytotax. Sin.,* 25(2): 154 (1987)—China.

NARAVELIA (Ranunculac.).
eichleri *M. Tamura in Acta Phytotax. Geobot.,* 37(4–6): 109 (1986)—Thailand.

NARCISSUS (Amaryllidac.).
bugei *(Fernández Casas) J. Fernández Casas in Lagascalia,* 14(1): 176 (1986): N. longispathus var. bugei.
×neocarpetanus
nothovar. romanensis *J.F. Ureña Plaza in An. Jard. Bot. Madrid,* 43(1): 185 (1986)—Spain. (N. bulbocodium var. bulbocodium × cantabricus).
×susannae
nothovar. toletanus *J. Fernández Casas & M. Luceño in An. Jard. Bot. Madrid,* 43(1): 187 (1986)—Spain. (N. cantabricus × pallidulus).

NASTURTIUM (Crucif.).
valdes-bermejoi *S. Castroviejo in An. Jard. Bot. Madrid,* 43(1): 17 (1986)—Spain.

NAUPLIUS (Compos.).
daltonii *(Webb) A. Wiklund, Some genera in Inula-group (Asterac.-Inuleae):* 78 (1986); *et in Nordic J. Bot.,* 7(1): 15 (1987): Odontospermum daltonii.
subsp. vogelii *(Webb) A. Wiklund, Some genera in Inula-group (Asterac.-Inuleae):* 79 (1986); *et in Nordic J. Bot.,* 7(1): 15 (1987): Odontospermum vogelii.
graveolens *(Forssk.) A. Wiklund, Some genera in Inula-group (Asterac.-Inuleae):* 80 (1986); *et in Nordic J. Bot.,* 7(1): 16 (1987): Buphthalmum graveolens.
subsp. odorus *(Schousb.) A. Wiklund, Some genera in Inula-group (Asterac.-Inuleae):* 92 (1986), as 'odorum'; *et in Nordic J. Bot.,* 7(1): 19 (1987): Buphthalmum odorum.
subsp. stenophyllus *(Link) A. Wiklund, Some genera in Inula-group (Asterac.-Inuleae):* 94 (1986); *et in Nordic J. Bot.,* 7(1): 19 (1987): Buphthalmum stenophyllum.
imbricatus *(Cav.) A. Wiklund, Some genera in Inula-group (Asterac.-Inuleae):* 60 (1986); *et in Nordic J. Bot.,* 7(1): 10 (1987): Buphthalmum imbricatum.
schultzii *(Bolle) A. Wiklund, Some genera in Inula-group (Asterac.-Inuleae):* 55 (1986); *et in Nordic J. Bot.,* 7(1): 8 (1987): Odontospermum schultzii.
smithii *(Webb) A. Wiklund, Some genera in Inula-group (Asterac.-Inuleae):* 77 (1986); *et in Nordic J. Bot.,* 7(1): 15 (1987): Odontospermum smithii.

NAVIA (Bromeliac.).
berryana *L.B. Smith, J.A. Steyermark & H. Robinson in Ann. Missouri Bot. Gard.,* 73(4): 712 (1986 publ. 1987)—Venezuela.
crassicaulis *L.B. Smith, J.A. Steyermark & H. Robinson in Ann. Missouri Bot. Gard.,* 73(4): 708 (1986 publ. 1987)—Venezuela.

NAVIA :–
culcitaria *L.B. Smith, J.A. Steyermark & H. Robinson in Ann. Missouri Bot. Gard.*, 73(4): 712 (1986 publ. 1987)—Venezuela.
delascionis *L.B. Smith, J.A. Steyermark & H. Robinson in Ann. Missouri Bot. Gard.*, 73(4): 709 (1986 publ. 1987)—Venezuela.
filifera *L.B. Smith, J.A. Steyermark & H. Robinson in Ann. Missouri Bot. Gard.*, 73(4): 713 (1986 publ. 1987)—Venezuela.
huberiana *L.B. Smith, J.A. Steyermark & H. Robinson in Ann. Missouri Bot. Gard.*, 73(4): 706 (1986 publ. 1987)—Venezuela.
igneosicola *L.B. Smith, J.A. Steyermark & H. Robinson in Ann. Missouri Bot. Gard.*, 73(4): 713 (1986 publ. 1987)—Venezuela.
liesneri *L.B. Smith, J.A. Steyermark & H. Robinson in Ann. Missouri Bot. Gard.*, 73(4): 709 (1986 publ. 1987)—Venezuela.
linearis *L.B. Smith, J.A. Steyermark & H. Robinson in Ann. Missouri Bot. Gard.*, 73(4): 709 (1986 publ. 1987)—Venezuela.
pedemontana *L.B. Smith, J.A. Steyermark & H. Robinson in Ann. Missouri Bot. Gard.*, 73(4): 708 (1986 publ. 1987)—Venezuela.
piresii *L.B. Smith, J.A. Steyermark & H. Robinson in Ann. Missouri Bot. Gard.*, 73(4): 712 (1986 publ. 1987)—Brazil.
plowmanii *L.B. Smith, J.A. Steyermark & H. Robinson in Ann. Missouri Bot. Gard.*, 73(4): 703 (1986 publ. 1987)—Venezuela.
polyglomerata *L.B. Smith, J.A. Steyermark & H. Robinson in Ann. Missouri Bot. Gard.*, 73(4): 704 (1986 publ. 1987)—Venezuela.
thomasii *L.B. Smith, J.A. Steyermark & H. Robinson in Ann. Missouri Bot. Gard.*, 73(4): 703 (1986 publ. 1987)—Venezuela.

NEANOTIS (Rubiac.).
oxyphylla *(Wall. ex G. Don) Ridsdale in Allertonia,* 4(5): 321 (1987), as '(Wall. ex Hook.f.)' but isonym, see *W.H. Lewis in Ann. Missouri Bot. Gard.*, 53: 39 (1966): Oldenlandia oxyphylla.

NEBLINARIA (Guttif.).
rubicunda *C. Sastre in Bull. Mus. Nation. Hist. Nat., B, Adansonia,* Ser. 4, 9(2): 129 (1987)—Guyana.

NEMCIA (Legumin.).
axillaris *(Meissner) M.D. Crisp in C.H. Stirton (ed.), Advances in Legume System.*, 3: 124 (1987): Gastrolobium axillare.
brownii *(Meissner) M.D. Crisp in C.H. Stirton (ed.), Advances in Legume System.*, 3: 124 (1987): Gastrolobium brownii.
carinata *M.D. Crisp in C.H. Stirton (ed.), Advances in Legume System.*, 3: 124 (1987), nom. nov.: Eutaxia reticulata *Meissner.*
crenulata *(Turcz.) M.D. Crisp in C.H. Stirton (ed.), Advances in Legume System.*, 3: 125 (1987): Gastrolobium crenulatum.
dilatata *(Benth.) M.D. Crisp in C.H. Stirton (ed.), Advances in Legume System.*, 3: 125 (1987): Oxylobium dilatatum.

NEMCIA :–
emarginata *(S. Moore) M.D. Crisp in C.H. Stirton (ed.), Advances in Legume System.*, 3: 126 (1987): Oxylobium emarginatum.
epacridoides *(Meissner) M.D. Crisp in C.H. Stirton (ed.), Advances in Legume System.*, 3: 126 (1987): Gastrolobium epacridoides.
hookeri *(Meissner) M.D. Crisp in C.H. Stirton (ed.), Advances in Legume System.*, 3: 126 (1987): Gastrolobium hookeri.
ilicifolia *(Meissner) M.D. Crisp in C.H. Stirton (ed.), Advances in Legume System.*, 3: 126 (1987): Gastrolobium ilicifolium.
leakeana *(Drumm.) M.D. Crisp in C.H. Stirton (ed.), Advances in Legume System.*, 3: 126 (1987): Gastrolobium leakeanum.
lehmannii *(Meissner) M.D. Crisp in C.H. Stirton (ed.), Advances in Legume System.*, 3: 127 (1987): Gastrolobium lehmanni.
obovata *(Benth.) M.D. Crisp in C.H. Stirton (ed.), Advances in Legume System.*, 3: 127 (1987): Gastrolobium obovatum.
pauciflora *(C. Gardner) M.D. Crisp in C.H. Stirton (ed.), Advances in Legume System.*, 3: 127 (1987): Gastrolobium pauciflorum.
plicata *(Turcz.) M.D. Crisp in C.H. Stirton (ed.), Advances in Legume System.*, 3: 127 (1987): Gastrolobium plicatum.
pulchella *(Turcz.) M.D. Crisp in C.H. Stirton (ed.), Advances in Legume System.*, 3: 127 (1987): Gastrolobium pulchellum.
punctata *(Turcz.) M.D. Crisp in C.H. Stirton (ed.), Advances in Legume System.*, 3: 127 (1987): Eutaxia punctata.
pyramidalis *(T. Moore) M.D. Crisp in C.H. Stirton (ed.), Advances in Legume System.*, 3: 127 (1987): Gastrolobium pyramidale.
rubra *M.D. Crisp in C.H. Stirton (ed.), Advances in Legume System.*, 3: 128 (1987), nom. nov.: N. atropurpurea var. minorifolia *Domin.*
spathulata *(Benth.) M.D. Crisp in C.H. Stirton (ed.), Advances in Legume System.*, 3: 128 (1987): Gastrolobium spathulatum.
stipularis *(Meissner) M.D. Crisp in C.H. Stirton (ed.), Advances in Legume System.*, 3: 128 (1987): Gastrolobium stipulare.
tricuspidata *(Meissner) M.D. Crisp in C.H. Stirton (ed.), Advances in Legume System.*, 3: 128 (1987): Gastrolobium tricuspidatum.
truncata *(Benth.) M.D. Crisp in C.H. Stirton (ed.), Advances in Legume System.*, 3: 128 (1987): Gastrolobium truncatum.

NEOASTELIA J.B. Williams in Fl. Australia, 45: 493, 173 (1987). *ASTELIACEAE.*
spectabilis *J.B. Williams in Fl. Australia,* 45: 493, 175 (1987)—Australia (New South Wales).

NEOCONOPODIUM (K.-Pol.) M.G. Pimenov & E.V. Kljuykov in Feddes Repert., 98(7–8): 377 (1987): Conopodium III. Neoconopodium (Umbellif.).

NEOCONOPODIUM :–
capnoides *(Decne.) M.G. Pimenov & E.V. Kljuykov in Feddes Repert.*, 98(7–8): 377 (1987): Butinia capnoides.
laseroides *(Hedge & Lamond) M.G. Pimenov & E.V. Kljuykov in Feddes Repert.*, 98(7–8): 378 (1987): Chaerophyllum laseroides.

NEOCRYPTODISCUS I.C. Hedge & J.M. Lamond in Fl. Iran., 162: 207 (1987), nom. nov.: Cryptodiscus Schrenk ex C.A. Mey. (Umbellif.).
ammophilus *(Bunge) I.C. Hedge & J.M. Lamond in Fl. Iran.*, 162: 208 (1987): Cryptodiscus ammophilus.
didymus *(Regel) I.C. Hedge & J.M. Lamond in Fl. Iran.*, 162: 208 (1987): Cachrys didyma.
persicus *(Boiss.) I.C. Hedge & J.M. Lamond in Fl. Iran.*, 162: 207 (1987): Cryptodiscus persicus.

NEOGOEZIA (Umbellif.).
breedlovei *L. Constance in Op. Bot.*, 92: 71 (1987)—Mexico.
macvaughii *L. Constance in Op. Bot.*, 92: 68 (1987)—Mexico.

NEOHOUZEAUA (Gramin.).
helferi *(Munro) J.S. Gamble in Bull. Misc. Inf. Kew*, 1923: 91 (1923): Teinostachyum helferi.

NEOKOEHLERIA (Orchidac.).
corydaloides *(Krzl.) K. Senghas in Orchidee*, 38(3): 114 (1987): Rodriguezia corydaloides.

NEOPORTERIA (Cactac.).
centeteria *(Lehm.) G.D. Rowley in Excelsa*, 12: 39 (1986): Echinocactus centeterius.
horrida *(Remy ex Gay) D. Hunt in Bradleya*, 5: 93 (1987): Echinocactus horridus.

NEOREGELIA (Bromeliac.).
angustibracteolata *E. Pereira & E.M.C. Leme in Bradea*, 4(34): 269 (1986)—Brazil.

NEOTORULARIA Hedge & J. Léonard in Bull. Jard. Bot. Nation. Belg., 56(3–4): 393 (1986), nom. nov.: Sisymbrium sect. Torularia Cosson (Crucif.).
aculeolata *(Boiss.) Hedge & J. Léonard in Bull. Jard. Bot. Nation. Belg.*, 56(3–4): 393 (1986): Sisymbrium aculeolatum.
afghanica *(Gilli) Hedge & J. Léonard in Bull. Jard. Bot. Nation. Belg.*, 56(3–4): 393 (1986): Sisymbrium afghanicum.
brachycarpa *(Vassilcz.) Hedge & J. Léonard in Bull. Jard. Bot. Nation. Belg.*, 56(3–4): 393 (1986): Torularia brachycarpa.
brevipes *(Karelin & Kir.) Hedge & J. Léonard in Bull. Jard. Bot. Nation. Belg.*, 56(3–4): 393 (1986): Sisymbrium brevipes.
contortuplicata *(Stephan ex Willd.) Hedge & J. Léonard in Bull. Jard. Bot. Nation. Belg.*, 56(3–4): 393 (1986): Cheiranthus contortuplicatus.

NEOTORULARIA :–
dentata *(Freyn & Sint.) Hedge & J. Léonard in Bull. Jard. Bot. Nation. Belg.*, 56(3–4): 394 (1986): Cryptospora dentata.
humilis *(C. Meyer) Hedge & J. Léonard in Bull. Jard. Bot. Nation. Belg.*, 56(3–4): 394 (1986): Sisymbrium humile.
korolkowii *(Regel & Schmalh.) Hedge & J. Léonard in Bull. Jard. Bot. Nation. Belg.*, 56(3–4): 394 (1986): Sisymbrium korolkowii.
rossica *(O. Schulz) Hedge & J. Léonard in Bull. Jard. Bot. Nation. Belg.*, 56(3–4): 394 (1986): Torularia rossica.
tetracmoides *(Boiss. & Hausskn.) Hedge & J. Léonard in Bull. Jard. Bot. Nation. Belg.*, 56(3–4): 394 (1986): Sisymbrium tetracmoides.
torulosa *(Desf.) Hedge & J. Léonard in Bull. Jard. Bot. Nation. Belg.*, 56(3–4): 395 (1986): Sisymbrium torulosum.
var. scorpiuroides *(Boiss.) Hedge & J. Léonard in Bull. Jard. Bot. Nation. Belg.*, 56(3–4): 395 (1986): Sisymbrium scorpiuroides.

NEOTTIA (Orchidac.).
mackinnonii *S. Deva & H.B. Naithani, Orchid Fl. N.W. Himalaya:* 75 (1986)—India.

NEPETA (Labiat.).
jomdaensis *H.W. Li in Fl. Xizangica*, 4: 132 (1985)—Xizang.
kunlunshanica *C.Y. Yang & B. Wang in Bull. Bot. Res. North-East. Forest. Inst.*, 7(1): 97 (1987)—Xinjiang.
orphanidea
var. parnidea *C. Baden in Op. Bot.*, 93: 16 (1987)—Greece.
sibthorpii
subsp. dirphya *(Boiss.) C. Baden in Op. Bot.*, 93: 22 (1987): N. sibthorpii var. (*β*) dirphya.
subsp. malacotrichos *C. Baden in Op. Bot.*, 93: 19 (1987)—Greece.
subsp. vourinensis *C. Baden in Op. Bot.*, 93: 23 (1987)—Greece.
tarbagataica *C.Y. Yang & B. Wang in Bull. Bot. Res. North-East. Forest. Inst.*, 7(1): 99 (1987)—Xinjiang.
troitzkyi *Sosn. ex Sh.I. Kutateladze in Fl. Gruzii, ed. 2*, 11: 237 (1987)—U.S.S.R. (Caucasus).
zandaensis *H.W. Li in Fl. Xizangica*, 4: 126 (1985)—Xizang.

NERIACANTHUS (Acanthac.).
harlingii *D.C. Wasshausen in Op. Bot.*, 92: 271 (1987)—Ecuador.
var. longifolius *D.C. Wasshausen in Op. Bot.*, 92: 271 (1987)—Ecuador.

NERVILIA (Orchidac.).
gammieana *(Hook. f.) E. Pfitzer in Engler & Prantl, Nat. Pflanzenfam.*, 2(6): 56, 106 (1888): Pogonia gammieana.

NESOGORDONIA (Sterculiac.).
abrahamii *L.C. Barnett in Bull. Mus. Nation. Hist. Nat., B, Adansonia,* Ser. 4, 9(1): 95 (1987)—Madagascar.
pachyneura *Capuron ex L.C. Barnett in Bull. Mus. Nation. Hist. Nat., B, Adansonia,* Ser. 4, 9(1): 98 (1987)—Madagascar.

NEWTONIA (Legumin.).
warmingii *(Benth.) G.P. Lewis, Legumes of Bahia:* 111 (1987): Mimosa warmingii.

NIDULARIUM (Bromeliac.).
angustibracteatum *E.M.C. Leme in Bradea,* 4(34): 271 (1986)—Brazil.
edmundoi *E.M.C. Leme in Bradea,* 4(34): 270 (1986)—Brazil.

NIGRITELLA (Orchidac.).
corneliana *(Beauverd) P. Gölz & H.R. Reinhard in Jahresber. Naturwiss. Ver. Wuppertal,* 39: 39 (1986): N. nigra subsp. corneliana.

NOGRA (Legumin.).
guangxiensis *C.F. Wei in Guihaia,* 5(4): 351 (1985)—China.

NONEA (Boraginac.).
ovczinnikovii *A.P. Chukavina in Izv. Akad. Nauk Tadzh. SSR, Biol. Nauk,* 3(92): 91 (1983)—U.S.S.R. (Middle Asia).

NORONHIA (Oleac.).
emarginata *(Lam.) Poiret in Dict. Sci. Nat.,* 35: 151 (1825): Olea emarginata. †

NOTELAEA (Oleac.).
lloydii *G.P. Guymer in Austrobaileya,* 2(4): 341 (1987)—Australia (Queensland).
pungens *G.P. Guymer in Austrobaileya,* 2(4): 339 (1987)—Australia (Queensland).

NOTHOCESTRUM (Solanac.).
inconcinnum *H. St. John in Phytologia,* 61(5): 343 (1986)—Hawaiian Is.

NOTOCACTUS (Cactac.).
catarinensis *(Ritter) L. Scheinvar in Fl. Illustr. Catar.,* 1 (Cactác.): 50 (1985): Brasilparodia catarinensis.
ottonis
var. nigrispinus *H.D. Lück in Internoto,* 8(4): 124 (1987), without latin descr. or type—Locality not stated.
pseudopulvinatus *N. Gerloff in Internoto,* 8(4): 127 (1987), without latin descr. or type—Locality not stated.
scopa
var. machadoensis *R. Menges in Internoto,* 8(4): 109 (1987), without latin descr. or type—Brazil.

NOTONIA (Compos.).
hebdingii *W. Rauh & Buchloh in Kakt. And. Sukk.,* 38(11): 274 (1987)—Madagascar.

NOTOSERIS C. Shih in Acta Phytotax. Sin., 25(3): 196 (1987). *COMPOSITAE.*
formosana *(Kitam.) C. Shih in Acta Phytotax. Sin.,* 25(3): 201 (1987): Prenanthes formosana.
glandulosa *(Dunn) C. Shih in Acta Phytotax. Sin.,* 25(3): 198 (1987): Prenanthes glandulosa.
gracilipes *C. Shih in Acta Phytotax. Sin.,* 25(3): 198 (1987)—China.
guizhouensis *C. Shih in Acta Phytotax. Sin.,* 25(3): 196 (1987)—China.
henryi *(Dunn) C. Shih in Acta Phytotax. Sin.,* 25(3): 202 (1987): Prenanthes henryi.
melanantha *(Franch.) C. Shih in Acta Phytotax. Sin.,* 25(3): 198 (1987): Lactuca melanantha.
porphyrolepis *C. Shih in Acta Phytotax. Sin.,* 25(3): 201 (1987)—China.
psilolepis *C. Shih in Acta Phytotax. Sin.,* 25(3): 197 (1987)—China.
rhombiformis *C. Shih in Acta Phytotax. Sin.,* 25(3): 200 (1987)—China.
triflora *(Hemsl.) C. Shih in Acta Phytotax. Sin.,* 25(3): 202 (1987): Lactuca triflora.
wilsonii *(Chang) C. Shih in Acta Phytotax. Sin.,* 25(3): 202 (1987): Prenanthes wilsoni.
yunnanensis *C. Shih in Acta Phytotax. Sin.,* 25(3): 200 (1987)—China.

×NOTYLOPSIS hort. in Orchid Rev., 95(1127): centre page pull-out p. 9 (1987). *ORCHIDACEAE.* (Ionopsis × Notylia).

NOVOSIEVERSIA (Rosac.).
reptans *(L.) E.I. Golubkova in Bot. Zhurn.,* 72(6): 775 (1987): Geum reptans.

NUTTALLIA (Loasac.).
cronquistii *(Thompson & Prigge) W.A. Weber in Phytologia,* 62(5): 437 (1987): Mentzelia cronquistii.

NYMPHAEA (Nymphaeac.).
amazonum
subsp. pedersenii *J.H. Wiersema, Monogr. Nymphaea subgen. Hydrocallis (Nymphaeac.) (Syst. Bot. Monog., 16):* 86 (1987)—Argentina, Brazil.

NYMPHOIDES (Menyanthac.).
beaglensis *H.I. Aston in Muelleria,* 6(5): 359 (1987)—Australia (W. Australia).

O

OBERNA (Caryophyllac.).
behen
subsp. angustifolia *(Guss.) J. Soják in Čas. Nár. Muz. (Prague),* 153(4): 216 (1984 publ. 1986): Silene angustifolia.
subsp. antelopum *(Vest) J. Soják in Čas. Nár. Muz. (Prague),* 153(4): 216 (1984 publ. 1986): Cucubalus antelopum.
subsp. commutata *(Guss.) J. Soják in Čas. Nár. Muz. (Prague),* 153(4): 216 (1984 publ. 1986): Silene commutata.
subsp. glareosa *(Jord.) J. Soják in Čas. Nár. Muz. (Prague),* 153(4): 216 (1984 publ. 1986): Silene glareosa.
subsp. humilis *(Schubert) J. Soják in Čas. Nár. Muz. (Prague),* 153(4): 216 (1984 publ. 1986): Silene vulgaris var. humilis.
subsp. macrocarpa *(Turrill) J. Soják in Čas. Nár. Muz. (Prague),* 153(4): 216 (1984 publ. 1986): Silene vulgaris subsp. macrocarpa.
subsp. marginata *(Kit.) J. Soják in Čas. Nár. Muz. (Prague),* 153(4): 216 (1984 publ. 1986): Cucubalus marginatus.
subsp. maritima *(With.) J. Soják in Čas. Nár. Muz. (Prague),* 153(4): 216 (1984 publ. 1986): Silene maritima.
subsp. prostrata *(Gaud.) J. Soják in Čas. Nár. Muz. (Prague),* 153(4): 216 (1984 publ. 1986): Silene inflata prostrata.
subsp. suffrutescens *(Greuter, Matthäs & Risse) J. Soják in Čas. Nár. Muz. (Prague),* 153(4): 216 (1984 publ. 1986): Silene vulgaris subsp. suffrutescens.
subsp. thorei *(Duf.) J. Soják in Čas. Nár. Muz. (Prague),* 153(4): 216 (1984 publ. 1986): Silene thorei.

OBERONIA (Orchidac.).
rangannana *K.R. Keshava Murthy, S.N. Yoganarasimhan & K. Vasudevan Nair in Current Sci.,* 56(12): 621 (1987), as 'rangannaiana'—India.

OCTOMERIA (Orchidac.).
heleneana *G. Carnevali F.C. & F. Delascio Chitty in Ernstia,* 45: 12 (1987)—Venezuela.

ODONTOGLOSSUM (Orchidac.).
nobile
var. leucoxanthum *(Rchb. f.) L. Bockemühl in Orchidee,* 38(4): 177 (1987): O. pescatorei var. leucoxanthum.
var. veitchianum *(Rchb. f.) L. Bockemühl in Orchidee,* 38(4): 177 (1987): O. pescatorei var. veitchianum.
sceptrum
var. masereelianum *(Rchb. f.) L. Bockemühl in Orchidee,* 38(5): 229 (1987): O. masereelianum.

×ODONTOPILIA hort. in Orchid Rev., 95(1119): centre page pull-out p. 8 (1987), as 'Odontopolia'. *ORCHIDACEAE.* (Odontoglossum × Trichopilia).

OEDIBASIS (Umbellif.).
pachyphylla *I.C. Hedge & J.M. Lamond in Fl. Iran.,* 162: 437 (1987)—Afghanistan.

OENOTHERA (Onagrac.).
ser. Candela *W. Dietrich & W.L. Wagner in Ann. Missouri Bot. Gard.,* 74(1): 147 (1987).
subsect. Nutantigemma *W. Dietrich & W.L. Wagner in Ann. Missouri Bot. Gard.,* 74(1): 145 (1987).
ser. Raimannia *(Rose ex Britton & A. Brown) W. Dietrich & W.L. Wagner in Ann. Missouri Bot. Gard.,* 74(1): 148 (1987): gen. Raimannia.
breedlovei *W. Dietrich & W.L. Wagner in Ann. Missouri Bot. Gard.,* 74(1): 146 (1987)—Mexico.
drummondii
subsp. thalassaphila *(Brandegee) W. Dietrich & W.L. Wagner in Ann. Missouri Bot. Gard.,* 74(1): 150 (1987): O. thalassaphila.
elata
subsp. hookeri *(Torrey & A. Gray) W. Dietrich & W.L. Wagner in Ann. Missouri Bot. Gard.,* 74(1): 152 (1987): O. hookeri.
subsp. texensis *W. Dietrich & W.L. Wagner in Ann. Missouri Bot. Gard.,* 74(1): 152 (1987)—U.S.A. (Texas).
falfurriae *W. Dietrich & W.L. Wagner in Ann. Missouri Bot. Gard.,* 74(1): 149 (1987)—U.S.A. (Texas).
nissensis
var. fiedleri *P. Gutte & K. Rostański in Ber. Arbeitsgemeinsch. Sächs. Bot.,* 11(2): 185 (1981)—Germany.
×rigirubata *Renner ex P. Gutte & K. Rostański in Ber. Arbeitsgemeinsch. Sächs. Bot.,* 11(2): 188 (1981)—Germany. (O. ammophila var. germanica × biennis).
rubricaulis
var. longistylis *P. Gutte & K. Rostański in Ber. Arbeitsgemeinsch. Sächs. Bot.,* 11(2): 187 (1981)—Germany.
tamrae *W. Dietrich & W.L. Wagner in Ann. Missouri Bot. Gard.,* 74(1): 146 (1987)—Mexico.

OLEARIA (Compos.).
pannosa
subsp. cardiophylla *(F. Muell.) D.A. Cooke in Fl. S. Australia,* 3: 1483 (1986): Eurybia cardiophylla.

OMPHALODES (Boraginac.).
davisiana *Kit Tan & F. Sorger in Pl. Syst. Evol.,* 154(1–2): 125 (1986)—Turkey.

OMPHALOTRIGONOTIS (Boraginac.).
vaginata *Y.Y. Fang in Bull. Bot. Res. North-East. Forest. Inst.,* 7(1): 89 (1987)—China.

ONCOSTEMA (Hyacinthac.).
subgen. Petranthe *(Salisb.) F. Speta in Phyton (Austria),* 26(2): 272 (1987): gen. Petranthe.
africana *(Borzi & Mattei) F. Speta in Phyton (Austria),* 26(2): 264 (1987): Scilla africana.
barba-caprae *(Aschers. & Barbey ex Durand & Barratte) F. Speta in Phyton (Austria),* 26(2): 264 (1987): Ornithogalum barba-caprae.
beirana *(Samp.) F. Speta in Phyton (Austria),* 26(2): 264 (1987): Scilla beirana.
cerulea *(Rafin.) F. Speta in Phyton (Austria),* 26(2): 265 (1987): Ornithogalum ceruleum.
elongata *(Parlatore) F. Speta in Phyton (Austria),* 26(2): 265 (1987): Scilla elongata.
hughii *(Tineo ex Guss.) F. Speta in Phyton (Austria),* 26(2): 265 (1987): Scilla hughii.
liliohyacinthus *(L.) F. Speta in Phyton (Austria),* 26(2): 274 (1987): Scilla liliohyacinthus.
monophyllos *(Link) F. Speta in Phyton (Austria),* 26(2): 274 (1987): Scilla monophyllos.
odorata *(Link) F. Speta in Phyton (Austria),* 26(2): 274 (1987): Scilla odorata.
peruviana *(L.) F. Speta in Phyton (Austria),* 26(2): 264 (1987): Scilla peruviana.
ramburei *(Boiss.) F. Speta in Phyton (Austria),* 26(2): 274 (1987): Scilla ramburei.
sicula *(Tineo ex Guss.) F. Speta in Phyton (Austria),* 26(2): 265 (1987): Scilla sicula.
tingitana *(Schousb.) F. Speta in Phyton (Austria),* 26(2): 274 (1987): Scilla tingitana.
umbellata *(Ramond) F. Speta in Phyton (Austria),* 26(2): 274 (1987): Scilla umbellata.
verna *(Hudson) F. Speta in Phyton (Austria),* 26(2): 272 (1987): Scilla verna.

ONOBRYCHIS (Legumin.).
alba
subsp. tenoreana *(Lacaita) Pignatti ex W. Greuter & Burdet in Willdenowia,* 16(2): 446 (1987): O. tenoreana.
cilicica *Kit Tan & F. Sorger in Pl. Syst. Evol.,* 154(1–2): 119 (1986)—Turkey.
mutensis *Kit Tan & F. Sorger in Pl. Syst. Evol.,* 154(1–2): 117 (1986)—Turkey.
sivasica *Kit Tan & F. Sorger in Pl. Syst. Evol.,* 154(1–2): 119 (1986)—Turkey.

ONONIS (Legumin.).
azcaratei *J.A. Devesa in Lagascalia,* 14(1): 143 (1986)—Spain.
baetica
var. donanensis *J.A. Devesa in Lagascalia,* 14(1): 83 (1986)—Spain.
pendula
subsp. boissieri *(Širj.) J.A. Devesa in Lagascalia,* 14(1): 144 (1986): O. pendula f. boissieri.
subsp. munbyi *(Širj.) W. Greuter & Burdet in Willdenowia,* 16(2): 446 (1987): O. pendula var. munbyi.

ONONIS :-
repens
subsp. australis *(Širj.) J.A. Devesa in Lagascalia,* 14(1): 145 (1986): O. repens var. australis.
subsp. austriaca *(G. Beck) W. Greuter & Burdet in Willdenowia,* 16(2): 446 (1987): O. austriaca.
var. hispanica *(Lange) J.A. Devesa in Lagascalia,* 14(1): 145 (1986): O. antiquorum var. hispanica.
varelae *J.A. Devesa in Lagascalia,* 14(1): 84 (1986)—Spain.

ONOPORDUM (Compos.).
sarrafii *C.C. Townsend in Kew Bull.,* 42(2): 439 (1987)—Iraq.

OPHIOPOGON (Convallariac.).
compressus *Y. Wan & C.C. Huang in Acta Phytotax. Sin.,* 25(5): 400 (1987)—China.
longipedicellatus *Y. Wan & C.C. Huang in Acta Phytotax. Sin.,* 25(5): 399 (1987)—China.
sinensis *Y. Wan & C.C. Huang in Acta Phytotax. Sin.,* 25(5): 398 (1987)—China.

OPHIORRHIZA (Rubiac.).
exigua *(Li) H.S. Lo in Acta Bot. Yunnanica,* 9(3): 301 (1987): Geophila exigua.

OPHRYS (Orchidac.).
aegaea *M. Kalteisen & H.R. Reinhard in Mitt. Arbeitskr. Heim. Orch. Baden-Württemberg,* 19(4): 918 (1987)—Karpathos.
subsp. lucis *M. Kalteisen & H.R. Reinhard in Mitt. Arbeitskr. Heim. Orch. Baden-Württemberg,* 19(4): 922 (1987)—Rhodes.
apifera
var. fulvofusca *M.P. Grasso & A. Scrugli in Orchidee,* 38(1): 47 (1987)—Sardinia.
archipelagi *P. Gölz & H.R. Reinhard in Mitt. Arbeitskr. Heim. Orch. Baden-Württemberg,* 18(4): 731 (1986)—Jugoslavia, Italy.
×bernardii *H. Van Looken in Orchidophile,* 75: 1211 (1987)—France. (O. aveyronensis × scolopax subsp. scolopax).
×duvigneaudiana *P. & C. Delforge in Orchidophile,* 78: 1364 (1987)—France. (O. araneola × scolopax).
×japygia *R. Lorenz & C. Gembardt in Mitt. Arbeitskr. Heim. Orch. Baden-Württemberg,* 19(3): 515 (1987)—Italy. (O. garganica × sipontensis).
mammosa
subsp. transhyrcana *(Czernjakowska) H.F. Paulus & C. Gack in Jahresber. Naturwiss. Ver. Wuppertal,* 39: 68 (1986), without basionym page: O. transhyrcana.
maxima *(Fleischmann) H.F. Paulus & C. Gack in Jahresber. Naturwiss. Ver. Wuppertal,* 39: 79 (1986): O. fuciflora var. maxima.
×messeniensis
nothosubsp. baselicei *R. Lorenz & C. Gembardt in Mitt. Arbeitskr. Heim. Orch. Baden-Württemberg,* 19(3): 516 (1987)—Italy. (O. oestrifera subsp. oestrifera × tenthredinifera subsp. tenthredinifera).

OPHRYS :–
minor
subsp. galilaea *(Fleischm. & Bornm.) H.F. Paulus & C. Gack in Jahresber. Naturwiss. Ver. Wuppertal,* 39: 66 (1986): O. galilaea.
sipontensis *R. Lorenz & C. Gembardt in Mitt. Arbeitskr. Heim. Orch. Baden-Württemberg,* 19(3): 472 (1987)—Italy.
×sommieri
nothosubsp. chivotii *D. Tyteca in Orchidophile,* 75: 1219 (1987)—Portugal. (O. bombyliflora × tenthredinifera subsp. praecox).
tetraloniae *W. Teschner in Orchidee,* 38(5): 223 (1987)—Jugoslavia.
×turrica *R. Lorenz & C. Gembardt in Mitt. Arbeitskr. Heim. Orch. Baden-Württemberg,* 19(3): 514 (1987)—Italy. (O. apulica × oestrifera).
vernixia
subsp. ciliata *(Biv.) M. Kerguélen in Lejeunia,* 120: 135 (1987): O. ciliata.

OPITHANDRA (Gesneriac.).
sect. Briggsioides *W.T. Wang in Bull. Bot. Res. North-East. Forest. Inst.,* 7(2): 10 (1987).
sect. Microstigma *W.T. Wang in Bull. Bot. Res. North-East. Forest. Inst.,* 7(2): 9 (1987).
sect. Stenosiphon *W.T. Wang in Bull. Bot. Res. North-East. Forest. Inst.,* 7(2): 9 (1987).
dinghushanensis *W.T. Wang in Bull. Bot. Res. North-East. Forest. Inst.,* 7(2): 10 (1987)—China.

OPLISMENUS (Gramin.).
undulatifolius *(Ard.) Roemer & Schultes, Syst. Veg.,* 2: 482 (1817): Panicum undulatifolium.

OPUNTIA (Cactac.).
articulata *(Pfeiffer) D. Hunt in Bradleya,* 5: 93 (1987): Cereus articulatus.
basilaris
var. heilii *S.L. Welsh & E. Neese in Great Basin Nat.,* 43(4): 700 (1983)—U.S.A. (Utah).
lindheimeri
var. subarmata *(Griff.) J.L. Elizondo & J.A. Wehbe in Cact. Suc. Mex.,* 31(4): 100 (1986), without basionym ref.: Basionym not stated.

ORBEXILUM (Legumin.).
lupinellum *(Michx.) D. Isely in Sida,* 11(4): 432 (1986): Psoralea lupinella.
pedunculatum
var. eglandulosum *(Elliott) D. Isely in Sida,* 11(4): 432 (1986): Psoralea eglandulosa.

ORCHIS (Orchidac.).
×menosii *C. Bernard & G. Fabre in Orchidophile,* 78: 1345 (1987)—France. (O. coriophora subsp. fragrans × papilionacea).

ORCHIS :–
papilionacea
subsp. grandiflora *(Boiss.) H. Baumann in Jahresber. Naturwiss. Ver. Wuppertal,* 39: 94 (1986): O. papilionacea var. grandiflora.
subsp. heroica *(E.D. Clarke) H. Baumann in Jahresber. Naturwiss. Ver. Wuppertal,* 39: 95 (1986): O. heroica.
pugeensis *K.Y. Lang in Acta Phytotax. Sin.,* 25(5): 403 (1987)—China.
×sarcidani *A. Scrugli & M.P. Grasso in Webbia,* 40(2): 290 (1986)—Sardinia. (O. laxiflora × longicornu).
sichuanica *K.Y. Lang in Acta Phytotax. Sin.,* 25(5): 401 (1987)—China.

OREOCEREUS (Cactac.).
hempelianus *(Guerke) D. Hunt in Bradleya,* 5: 93 (1987): Echinopsis hempelianus.

OREOCHARIS (Gesneriac.).
sect. Orthoanthera *K.Y. Pan in Acta Phytotax. Sin.,* 25(4): 281 (1987).
sect. Platyanthera *K.Y. Pan in Acta Phytotax. Sin.,* 25(4): 289 (1987).
argyreia *Chun ex K.Y. Pan in Acta Phytotax. Sin.,* 25(4): 283 (1987)—China.
var. angustifolia *K.Y. Pan in Acta Phytotax. Sin.,* 25(4): 285 (1987)—China.
auricula
var. denticulata *K.Y. Pan in Acta Phytotax. Sin.,* 25(4): 276 (1987)—China.
dasyantha
var. ferruginosa *K.Y. Pan in Acta Phytotax. Sin.,* 25(4): 283 (1987)—Hainan.
elliptica
var. parvifolia *W.T. Wang & K.Y. Pan in Acta Phytotax. Sin.,* 25(4): 282 (1987)—China.
magnidens *Chun ex K.Y. Pan in Acta Phytotax. Sin.,* 25(4): 276 (1987)—China.
rotundifolia *K.Y. Pan in Acta Phytotax. Sin.,* 25(4): 280 (1987)—China.
tubiflora *K.Y. Pan in Acta Phytotax. Sin.,* 25(4): 287 (1987)—China.
xiangguiensis *W.T. Wang & K.Y. Pan in Acta Phytotax. Sin.,* 25(4): 285 (1987)—China.

OREOGEUM (Ser.) E.I. Golubkova in Bot. Zhurn., 72(6): 772 (1987): Geum sect. Oreogeum (Rosac.).
bulgaricum *(Panč.) E.I. Golubkova in Bot. Zhurn.,* 72(6): 774 (1987): Geum bulgaricum.
montanum *(L.) E.I. Golubkova in Bot. Zhurn.,* 72(6): 774 (1987): Geum montanum.

ORNITHOGALUM (Hyacinthac.).
kochii
subsp. gorenflotii *J. Moret in Bull. Mus. Nation. Hist. Nat., B, Adansonia,* Ser. 4, 9(2): 191 (1987)—Morocco.

ORNITHOPUS (Legumin.).
sativus
var. macrorrhynchus *(Willk.) M. Alsina Aser in Taxon,* 36(3): 646 (1987): O. roseus var. macrorrhynchus.
subsp. roseus *(Duf.) M. Alsina Aser in Taxon,* 36(3): 645 (1987): O. roseus.

OROBANCHE (Scrophulariac.).
baumanniorum *W. Greuter in Willdenowia,* 16(2): 448 (1987)—Greece.
crenata
var. callieri *N.N. Tsvelev in Novosti Sist. Vyssh. Rast.,* 23: 262 (1986)—U.S.S.R. (Crimea).

ORTHAEA (Ericac.).
subgen. Lysiclesia *(A.C. Smith) J.L. Luteyn in Nordic J. Bot.,* 7(1): 33 (1987): gen. Lysiclesia.
brachysiphon *(Sleumer) J.L. Luteyn in Nordic J. Bot.,* 7(1): 33 (1987): Empedoclesia brachysiphon.
caudata *(A.C. Smith) J.L. Luteyn in Nordic J. Bot.,* 7(1): 33 (1987): Lysiclesia caudata.
coriacea *J.L. Luteyn in Nordic J. Bot.,* 7(1): 36 (1987)—Ecuador.
glandulifera *J.L. Luteyn in Nordic J. Bot.,* 7(1): 34 (1987)—Colombia.
minor *(A.C. Smith) J.L. Luteyn in Nordic J. Bot.,* 7(1): 33 (1987): Lysiclesia minor.
oedipus *J.L. Luteyn in Nordic J. Bot.,* 7(1): 36 (1987)—Colombia.
panamensis *(Luteyn & Wilbur) J.L. Luteyn in Nordic J. Bot.,* 7(1): 33 (1987): Lysiclesia panamensis.
paniculata *J.L. Luteyn in Nordic J. Bot.,* 7(1): 33 (1987)—Colombia.
stipitata *(Luteyn) J.L. Luteyn in Nordic J. Bot.,* 7(1): 33 (1987): Empedoclesia stipitata.

ORTHOPHYTUM (Bromeliac.).
disjunctum
var. angustobracteatum *W. Rauh in Abh. Akad. Wiss. Lit. Mainz, Math.-Naturwiss. Klasse, Trop. Subtrop. Pflanzenwelt,* 58: 56 (1986)—Brazil.
var. variegatum *W. Rauh in Abh. Akad. Wiss. Lit. Mainz, Math.-Naturwiss. Klasse, Trop. Subtrop. Pflanzenwelt,* 58: 58 (1986)—Brazil.
var. viridiflorum *W. Rauh in Abh. Akad. Wiss. Lit. Mainz, Math.-Naturwiss. Klasse, Trop. Subtrop. Pflanzenwelt,* 58: 60 (1986)—Brazil.

ORYZA (Gramin.).
indandamanica *J.L. Ellis in Bull. Bot. Surv. India,* 27(1–4): 225 (1985 publ. 1987)—Andaman Is.
meyeriana
var. granulata *(Watt) H. Duistermaat in Blumea,* 32(1): 185 (1987): O. granulata.

OSTERICUM (Umbellif.).
citriodorum *(Hance) R.H. Shan & C.Q. Yuan in Bull. Nanjing Bot. Gard. Mem. Sun Yat Sen,* 1984–1985: 3 (1986): Angelica citriodora.
maximowiczii
var. alpinum *R.H. Shan & C.Q. Yuan in Bull. Nanjing Bot. Gard. Mem. Sun Yat Sen,* 1984–1985: 3 (1986)—China.
var. filisectum *(Chu) R.H. Shan & C.Q. Yuan in Bull. Nanjing Bot. Gard. Mem. Sun Yat Sen,* 1984–1985: 3 (1986): O. filisectum.
scaberulum *(Franch.) R.H. Shan & C.Q. Yuan in Bull. Nanjing Bot. Gard. Mem. Sun Yat Sen,* 1984–1985: 3 (1986): Angelica scaberula.

OSTODES (Euphorbiac.).
paniculata
var. katherinae *(Pax) T. Chakrabarty & N.P. Balakrishnan in Bull. Bot. Surv. India,* 27(1–4): 260 (1985 publ. 1987): O. katherinae.
var. thyrsantha *(Pax) T. Chakrabarty & N.P. Balakrishnan in Bull. Bot. Surv. India,* 27(1–4): 260 (1985 publ. 1987): O. thyrsanthus.

OTHONNA (Compos.).
armiana *E.J. van Jaarsveld in S. Afr. J. Bot.,* 52(6): 569 (1986)—South Africa (Cape Province).
schaeferi *Muschler ex K. Dinter in Fedde Repert.,* 20: 316 (1924), without descr.—Namibia.
surculosa *Muschler ex K. Dinter in Fedde Repert.,* 20: 316 (1924), without descr.—Namibia.
torulosa *Muschler ex K. Dinter in Fedde Repert.,* 20: 316 (1924), without descr.—Namibia.

OTOCHILUS (Orchidac.).
lancilabius *G. Seidenfaden in Op. Bot.,* 89: 94 (1986)—Sikkim, Bhutan, India, Nepal, Laos. †

OURATEA (Ochnac.).
bipartita *C. Sastre in Bull. Mus. Nation. Hist. Nat., B, Adansonia,* Ser. 4, 8(4): 420 (1986 publ. 1987)—Guyana.
brevipes *(van Tieghem) C. Sastre in Bull. Mus. Nation. Hist. Nat., B, Adansonia,* Ser. 4, 8(4): 420 (1986 publ. 1987): Cercouratea brevipes.
cerebroidea *C. Sastre in Bull. Mus. Nation. Hist. Nat., B, Adansonia,* Ser. 4, 8(4): 413 (1986 publ. 1987)—French Guiana, Guyana, Brazil.
francineae *C. Sastre in Bull. Mus. Nation. Hist. Nat., B, Adansonia,* Ser. 4, 8(4): 417 (1986 publ. 1987), as 'francinae'—French Guiana.
maasiorum *C. Sastre in Bull. Mus. Nation. Hist. Nat., B, Adansonia,* Ser. 4, 8(4): 419 (1986 publ. 1987), as 'maasorum'—Guyana.

OURATEA :–
macrocarpa *C. Sastre in Bull. Mus. Nation. Hist. Nat., B, Adansonia,* Ser. 4, 8(4): 414 (1986 publ. 1987)—French Guiana, Guyana, Surinam.

OXALIS (Oxalidac.).
dillenii
subsp. recurva *(Ell.) C.F. Reed in Phytologia,* 63(5): 411 (1987): O. recurva.

OXYANTHUS (Rubiac.).
lepidus
subsp. kigogoensis *D.M. Bridson in Kew Bull.,* 42(1): 251 (1987)—Tanzania.

OXYCOCCUS (Ericac.).
oxycoccos *(L.) MacMillan in Bull. Torrey Bot. Club,* 19: 15 (1892), as 'Oxycoccus oxycoccus'; *see also K. Adolphi in Taxon,* 36(1): 127 (1987): Vaccinium oxycoccos. †
palustris
var. macrophyllus *W. Gugnacka-Fiedor in Stud. Soc. Sci. Torun, D,* 11(4): 345 (1986 publ. 1987)—Poland.

OXYGONUM (Polygonac.).
alatum
var. longisquamatum *G. Germishuizen in Bothalia,* 17(2): 186 (1987)—Namibia, Botswana.
robustum *G. Germishuizen in Bothalia,* 17(1): 40 (1987)—Natal.

OXYTROPIS (Legumin.).
sect. Auricula *C.W. Chang in Acta Bot. Bor.-Occid. Sin.,* 7(3): 197 (1987).
auriculata *C.W. Chang in Acta Bot. Bor.-Occid. Sin.,* 7(3): 197 (1987)—China.
bicolor
var. luteola *C.W. Chang in Acta Bot. Bor.-Occid. Sin.,* 6(3): 188 (1986)—China.
falcata
var. maquensis *C.W. Chang in Acta Bot. Bor.-Occid. Sin.,* 7(3): 202 (1987)—China.
glabra
var. drakeana *(Franch.) C.W. Chang in Acta Bot. Bor.-Occid. Sin.,* 6(3): 185 (1986): O. drakeana.
xinglongshanica *C.W. Chang in Acta Bot. Bor.-Occid. Sin.,* 7(3): 199 (1987), with two types cited—China.

P

PACHYCEREUS (Cactac.).
militaris *(Audot) D. Hunt in Bradleya,* 5: 93 (1987): Cereus militaris.

PACHYCEREUS :–
schottii *(Engelmann) D. Hunt in Bradleya,* 5: 93 (1987): Cereus schottii.

PACHYSTACHYS (Acanthac.).
badiospica *D.C. Wasshausen in Proc. Biol. Soc. Wash.,* 99(1): 168 (1986)—Peru.
fosteri *D.C. Wasshausen in Proc. Biol. Soc. Wash.,* 99(1): 168 (1986)—Peru.
incarnata *D.C. Wasshausen in Proc. Biol. Soc. Wash.,* 99(1): 171 (1986)—Peru.
killipii *D.C. Wasshausen in Proc. Biol. Soc. Wash.,* 99(1): 179 (1986)—Peru.
longibracteata *D.C. Wasshausen in Proc. Biol. Soc. Wash.,* 99(1): 163 (1986)—Peru.
ossolae *D.C. Wasshausen in Proc. Biol. Soc. Wash.,* 99(1): 180 (1986)—Peru.
puberula *D.C. Wasshausen in Proc. Biol. Soc. Wash.,* 99(1): 173 (1986)—Peru.
rosea *D.C. Wasshausen in Proc. Biol. Soc. Wash.,* 99(1): 178 (1986)—Peru.
schunkei *D.C. Wasshausen in Proc. Biol. Soc. Wash.,* 99(1): 166 (1986)—Peru.
spicata *(R. & P.) D.C. Wasshausen in Proc. Biol. Soc. Wash.,* 99(1): 175 (1986): Justicia spicata.

PACHYSTIGMA (Rubiac.).
albosetulosum *B. Verdcourt in Kew Bull.,* 42(1): 139 (1987)—Zambia.
gillettii *(Tennant) B. Verdcourt in Kew Bull.,* 42(1): 140 (1987): Rytigynia gillettii.
loranthifolium *(K. Schum.) B. Verdcourt in Kew Bull.,* 42(1): 140 (1987): Vangueria loranthifolia.
subsp. salaense *B. Verdcourt in Kew Bull.,* 42(1): 140 (1987)—Kenya.
micropyren *B. Verdcourt in Kew Bull.,* 42(1): 139 (1987)—Zambia, Angola.

PACHYSTOMA (Orchidac.).
hirsuta *(Joseph & Vajravelu) C. Sathish Kumar & K.S. Manilal in Kew Bull.,* 42(4): 942 (1987): Eulophia hirsuta.

PADELLUS (Rosac.).
maximowiczii *(Rupr.) G.V. Eremin & A.A. Yushev in Byull. Vses. Ord. Lenina Inst. Rast. N.I. Vavilova,* 166: 42 (1986): Prunus maximowiczii.
pennsylvanica *(L.f.) G.V. Eremin & A.A. Yushev in Byull. Vses. Ord. Lenina Inst. Rast. N.I. Vavilova,* 166: 43 (1986): Prunus pennsylvanica.

PAEONIA (Paeoniac.).
ser. Anomalae *Kem.-Nath. ex M.S. Uspenskaya in Byull. Mosk. Obshch. Ispŷt. Prir., Biol.,* 92(3): 84 (1987).
ser. Arietinae *Stern ex M.S. Uspenskaya in Byull. Mosk. Obshch. Ispŷt. Prir., Biol.,* 92(3): 84 (1987).
subsect. Flavonia *(Kem.-Nath.) M.S. Uspenskaya in Byull. Mosk. Obshch. Ispŷt. Prir., Biol.,* 92(3): 83 (1987): P. sect. Flavonia.

PAEONIA :-
ser. Masculae *F.C. Stern ex M.S. Uspenskaya in Byull. Mosk. Obshch. Ispȳt. Prir., Biol.,* 92(3): 83 (1987).
subsect. Obovatae *Komarov ex M.S. Uspenskaya in Byull. Mosk. Obshch. Ispȳt. Prir., Biol.,* 92(3): 84 (1987).
ser. Officinalis *Kem.-Nath. ex M.S. Uspenskaya in Byull. Mosk. Obshch. Ispȳt. Prir., Biol.,* 92(3): 84 (1987).
ser. Tenuifoliae *Stern ex M.S. Uspenskaya in Byull. Mosk. Obshch. Ispȳt. Prir., Biol.,* 92(3): 84 (1987).
emodi
f. glabrata *(Hook. f. & Thoms.) H. Hara in Enum. Fl. Pl. Nepal,* 2: 23 (1979): P. emodi var. glabrata.

PAEPALANTHUS (Eriocaulac.).
amoenus
var. bolivianus *H.N. Moldenke in Phytologia,* 61(7): 444 (1987)—Bolivia.
bosseri *(P. Morat) T. Stützel in Pl. Syst. Evol.,* 156(3–4): 138 (1987): Moldenkeanthus bosseri.
itremensis *(P. Morat) T. Stützel in Pl. Syst. Evol.,* 156(3–4): 135 (1987): Moldenkeanthus itremensis.

PAGAMEOPSIS (Rubiac.).
maguirei
var. glabrus *J.A. Steyermark in Ann. Missouri Bot. Gard.,* 74(1): 108 (1987)—Venezuela.
var. pirapucuensis *J.A. Steyermark in Ann. Missouri Bot. Gard.,* 74(1): 108 (1987)—Brazil.
subsp. pusillus *J.A. Steyermark in Ann. Missouri Bot. Gard.,* 74(1): 106 (1987)—Venezuela.

PANDANUS (Pandanac.).
sect. Hastatistigma *B.C. Stone in Blumea,* 32(2): 434 (1987).
sect. Macrokurzia *B.C. Stone in Blumea,* 32(2): 439 (1987).
abbiwii *K.L. Huynh in Bot. Helvetica,* 97(1): 81 (1987)—Ghana.
benignus *H. St. John, Account of Fl. Pitcairn Is. with new Pandanus spp. (Pacific Pl. Stud. 46):* 9 (1987)—Pitcairn Is.
calvus *B.C. Stone in Blumea,* 32(2): 427 (1987)—Malaya.
columellatus *K.L. Huynh in Candollea,* 42(1): 140 (1987)—Senegal.
crassicollis *K.-L. Huynh in Bull. Soc. Neuchâtel. Sci. Nat.,* 110: 9 (1987)—Togo.
denudatus *K.-L. Huynh in Bull. Soc. Neuchâtel. Sci. Nat.,* 110: 11 (1987)—Togo.
djalonensis *K.L. Huynh in Webbia,* 41(1): 155 (1987)—Guinea.
gabonensis *K.-L. Huynh in Fl. Gabon,* 28: 12 (1986)—Gabon.
guineabissauensis *K.L. Huynh in Candollea,* 42(1): 140 (1987)—Guinea Bissau.

PANDANUS :-
insolitus *K.L. Huynh in Webbia,* 41(1): 160 (1987)—Congo.
kosteri *B.C. Stone in Blumea,* 32(2): 435 (1987)—New Guinea (W. Irian).
parvicentralis *K.-L. Huynh in Fl. Gabon,* 28: 10 (1986)—Gabon.
pitcairnensis *H. St. John, Account of Fl. Pitcairn Is. with new Pandanus spp. (Pacific Pl. Stud. 46):* 12 (1987)—Pitcairn Is.
senegalensis *St. John ex K.L. Huynh in Candollea,* 42(1): 135 (1987)—Senegal.

PANICUM (Gramin.).
ambiguum *(DC.) Le Turquier de Longchamp, Fl. Environs de Rouen:* 28 (1816): Paspalum ambiguum.
annuale *H. St. John in Phytologia,* 63(5): 368 (1987)—Hawaiian Is.
assurgens *H. St. John in Phytologia,* 63(5): 368 (1987)—Hawaiian Is.
baltodes *H. St. John in Phytologia,* 63(5): 368 (1987)—Hawaiian Is.
bifurcatum *H. St. John in Phytologia,* 63(5): 368 (1987)—Hawaiian Is.
cookei *H. St. John in Phytologia,* 63(5): 368 (1987)—Hawaiian Is.
ekeanum *H. St. John in Phytologia,* 63(5): 369 (1987)—Hawaiian Is.
furtivum *H. St. John in Phytologia,* 63(5): 369 (1987)—Hawaiian Is.
hobdyi *H. St. John in Phytologia,* 63(5): 369 (1987)—Hawaiian Is.
honokowaiense *H. St. John in Phytologia,* 63(5): 369 (1987)—Hawaiian Is.
infraventale *H. St. John in Phytologia,* 63(5): 369 (1987)—Hawaiian Is.
kahiliense *H. St. John in Phytologia,* 63(5): 369 (1987), as 'kahiliensis'—Hawaiian Is.
kahoolawense *H. St. John in Phytologia,* 63(5): 369 (1987)—Hawaiian Is.
kanaioense *H. St. John in Phytologia,* 63(5): 370 (1987)—Hawaiian Is.
kaonohuaense *H. St. John in Phytologia,* 63(5): 370 (1987)—Hawaiian Is.
knudsenii *H. St. John in Phytologia,* 63(5): 370 (1987)—Hawaiian Is.
kokeeense *H. St. John in Phytologia,* 63(5): 370 (1987)—Hawaiian Is.
kukaiwaaense *H. St. John in Phytologia,* 63(5): 370 (1987)—Hawaiian Is.
latzii *R.D. Webster, Australian Paniceae (Poac.):* 132 (1987)—Australia (N. Terr., W. Australia, Queensland).
lihauense *H. St. John in Phytologia,* 63(5): 370 (1987)—Hawaiian Is.
lineale *H. St. John in Phytologia,* 63(5): 370 (1987)—Hawaiian Is.
longivaginatum *H. St. John in Phytologia,* 63(5): 370 (1987)—Hawaiian Is.
malikoense *H. St. John in Phytologia,* 63(5): 371 (1987)—Hawaiian Is.
mokuleiaense *H. St. John in Phytologia,* 63(5): 371 (1987)—Hawaiian Is.

PANICUM :–
nephelophilum
var. levius *H. St. John in Phytologia,* 63(5): 371 (1987)—Hawaiian Is.
ninoleense *H. St. John in Phytologia,* 63(5): 371 (1987)—Hawaiian Is.
ooense *H. St. John in Phytologia,* 63(5): 371 (1987)—Hawaiian Is.
pepeopaeense *H. St. John in Phytologia,* 63(5): 371 (1987), nom. nov.: P. imbricatum var. molokaiense *Skottsb.*
pilgerianum *(Schweick.) W.D. Clayton in Kew Bull.,* 42(2): 402 (1987): Acroceras pilgerianum.
simplex *H. St. John in Phytologia,* 63(5): 371 (1987)—Hawaiian Is.
subglabrum *H. St. John in Phytologia,* 63(5): 372 (1987)—Hawaiian Is.
subtiliracemosum *S.A. Renvoize in Kew Bull.,* 42(4): 922 (1987)—Brazil.
sylvanum *H. St. John in Phytologia,* 63(5): 372 (1987)—Hawaiian Is.
waikoloaense *H. St. John in Phytologia,* 63(5): 372 (1987)—Hawaiian Is.
waimeaense *H. St. John in Phytologia,* 63(5): 372 (1987)—Hawaiian Is.
wilburii *H. St. John in Phytologia,* 63(5): 372 (1987), as 'wilburi'—Hawaiian Is.
wiliwilinuiense *H. St. John in Phytologia,* 63(5): 372 (1987)—Hawaiian Is.

PAPAVER (Papaverac.).
montenegrinum *(J. Rohlena) A.W. Hill in Index Kewensis,* Supp. 6: 145 (1916–1920 publ. 1926): P. rhoeas var. montenegrinum.
persicum
var. brachycarpum *(O. Kuntze) Kit Tan in Pl. Syst. Evol.,* 154(1–2): 113 (1986): P. pilosum var. brachycarpum.
subsp. fulvum *Kit Tan & F. Sorger in Pl. Syst. Evol.,* 154(1–2): 111 (1986)—Turkey.

PAPHIOPEDILUM (Orchidac.).
sect. Parvisepalum *(Karasawa & Saito) P. Cribb, Genus Paphiopedilum:* 86 (1987): P. subgen. Parvisepalum.
bullenianum
var. celebesense *(Fowlie & Birk) P. Cribb, Genus Paphiopedilum:* 167 (1987): P. celebesense.
callosum
var. sublaeve *(Reichb. f.) P. Cribb, Genus Paphiopedilum:* 188 (1987): Cypripedium callosum var. sublaeve.
dollii *E. Lückel in Orchidee,* 38(5): 266 (1987)—China.
glanduliferum
var. wilhelminae *(L.O. Williams) P. Cribb, Genus Paphiopedilum:* 111 (1987): P. wilhelminae.
hainanense *J.A. Fowlie in Orchid Dig.,* 51(2): 69 (1987), as 'hainanensis'—Hainan.
hennisianum
var. fowliei *(Birk) P. Cribb, Genus Paphiopedilum:* 190 (1987): P. fowliei.

PAPHIOPEDILUM :–
henryanum *G.J. Braem in Schlechteriana,* 1(1): 4 (1987)—China, Vietnam.
hirsutissimum
var. chiwuanum *(Tang & Wang) P. Cribb, Genus Paphiopedilum:* 140 (1987): P. chiwuanum.
var. esquirolei *(Schltr.) P. Cribb, Genus Paphiopedilum:* 140 (1987): P. esquirolei.
philippinense
var. roebelenii *(Veitch) P. Cribb, Genus Paphiopedilum:* 100 (1987): Cypripedium philippinense var. roebelenii.
primulinum
var. purpurascens *(M. Wood) P. Cribb, Genus Paphiopedilum:* 131 (1987): P. victoria-regina f. purpurascens.
sangii *G.J. Braem in Orchidee,* 38(4): 169 (1987)—Celebes.

PARACTAENUM (Gramin.).
novae-hollandiae
subsp. reversum *(F. Muell.) R.D. Webster, Australian Paniceae (Poac.):* 147 (1987): Panicum reversum.
refractum *(F. Muell.) R.D. Webster, Australian Paniceae (Poac.):* 148 (1987): Setaria refracta.

PARAEREMOSTACHYS T.A. Adylov, R.V. Kamelin & A.M. Makhmedov in Novosti Sist. Vyssh. Rast., 23: 112 (1986). *LABIATAE.*
sect. Thyrsiflorae *(Rech. f.) T.A. Adylov, R.V. Kamelin & A.M. Makhmedov in Novosti Sist. Vyssh. Rast.,* 23: 113 (1986): Eremostachys sect. Thyrsiflorae.
ammophila *(Rech. f.) T.A. Adylov, R.V. Kamelin & A.M. Makhmedov in Novosti Sist. Vyssh. Rast.,* 23: 114 (1986): Eremostachys ammophila.
anisochila *(Pazij & Vved.) T.A. Adylov, R.V. Kamelin & A.M. Makhmedov in Novosti Sist. Vyssh. Rast.,* 23: 113 (1986): Eremostachys anisochila.
aralensis *(Bunge) T.A. Adylov, R.V. Kamelin & A.M. Makhmedov in Novosti Sist. Vyssh. Rast.,* 23: 113 (1986): Eremostachys aralensis.
badakhschanica *(Hedge) T.A. Adylov, R.V. Kamelin & A.M. Makhmedov in Novosti Sist. Vyssh. Rast.,* 23: 113 (1986): Eremostachys badakhschanica.
desertorum *(Regel) T.A. Adylov, R.V. Kamelin & A.M. Makhmedov in Novosti Sist. Vyssh. Rast.,* 23: 113 (1986): Eremostachys desertorum.
dshungarica *(M. Pop.) T.A. Adylov, R.V. Kamelin & A.M. Makhmedov in Novosti Sist. Vyssh. Rast.,* 23: 113 (1986): Eremostachys phlomoides f. dshungarica.
ghorana *(Rech. f.) T.A. Adylov, R.V. Kamelin & A.M. Makhmedov in Novosti Sist. Vyssh. Rast.,* 23: 114 (1986): Eremostachys ghorana.

PARAEREMOSTACHYS :-
karatavica *(Pavl.) T.A. Adylov, R.V. Kamelin & A.M. Makhmedov in Novosti Sist. Vyssh. Rast.*, 23: 113 (1986): Eremostachys karatavica.
lindbergii *(Rech. f.) T.A. Adylov, R.V. Kamelin & A.M. Makhmedov in Novosti Sist. Vyssh. Rast.*, 23: 114 (1986): Eremostachys lindbergii.
paniculata *(Regel) T.A. Adylov, R.V. Kamelin & A.M. Makhmedov in Novosti Sist. Vyssh. Rast.*, 23: 113 (1986): Eremostachys paniculata.
persimilis *(Aitch. & Hemsl.) T.A. Adylov, R.V. Kamelin & A.M. Makhmedov in Novosti Sist. Vyssh. Rast.*, 23: 114 (1986): Eremostachys persimilis.
phlomoides *(Bunge) T.A. Adylov, R.V. Kamelin & A.M. Makhmedov in Novosti Sist. Vyssh. Rast.*, 23: 113 (1986): Eremostachys phlomoides.
sogdiana *(Pazij & Vved.) T.A. Adylov, R.V. Kamelin & A.M. Makhmedov in Novosti Sist. Vyssh. Rast.*, 23: 113 (1986): Eremostachys sogdiana.
thyrsiflora *(Benth.) T.A. Adylov, R.V. Kamelin & A.M. Makhmedov in Novosti Sist. Vyssh. Rast.*, 23: 113 (1986): Eremostachys thyrsiflora.
transoxana *(Bunge) T.A. Adylov, R.V. Kamelin & A.M. Makhmedov in Novosti Sist. Vyssh. Rast.*, 23: 113 (1986): Eremostachys transoxana.

PARAHEBE (Scrophulariac.).
decorosa *(F. Muell.) B.G. Briggs & F. Ehrendorfer in Fl. S. Australia*, 3: 1301 (1986): Veronica decorosa.

PARALEPISTEMON (Convolvulac.).
curtoi *(Rendle) J. Lejoly & S. Lisowski in Bull. Jard. Bot. Nation. Belg.*, 57(1–2): 272 (1987): Ipomoea curtoi.

PARAPHOLIS (Gramin.).
longiflora *(Grossh.) S.G. Musaev in Bot. Zhurn.*, 72(1): 94 (1987): Lepturus incurvatus var. longiflorus.

PARASTEMON (Chrysobalanac.).
grandifructus *G.T. Prance in Brittonia*, 39(3): 366 (1987)—Borneo (Sarawak, Sabah).

PARATHESIS (Myrsinac.).
longipetiolata *C.L. Lundell in Phytologia*, 63(2): 78 (1987)—Costa Rica.
obovalifolia *C.L. Lundell in Phytologia*, 63(2): 78 (1987)—Costa Rica.

PARINARI (Chrysobalanac.).
costata
subsp. polyneura *(Miq.) G.T. Prance in Brittonia*, 39(3): 368 (1987): P. polyneurum.
subsp. rubiginosa *(Ridley) G.T. Prance in Brittonia*, 39(3): 368 (1987): P. costatum var. rubiginosum.

PARINARI :-
papuana
subsp. salomonensis *(C.T. White) G.T. Prance in Brittonia*, 39(3): 369 (1987): P. salomonensis.
subsp. whitei *G.T. Prance in Brittonia*, 39(3): 369 (1987)—New Guinea (Papua New Guinea, W. Irian).

PARKIA (Legumin.).
balslevii *H.C.F. Hopkins in Fl. Neotrop.*, 43: 103 (1986)—Ecuador.

PARNASSIA (Parnassiac.).
sect. Allolobos *T.C. Ku in Bull. Bot. Res. North-East. Forest. Inst.*, 7(1): 46 (1987).
ser. Amoenae *T.C. Ku in Bull. Bot. Res. North-East. Forest. Inst.*, 7(1): 51 (1987).
ser. Fimbripetalum *T.C. Ku in Bull. Bot. Res. North-East. Forest. Inst.*, 7(1): 51 (1987).
ser. Gansuenses *T.C. Ku in Bull. Bot. Res. North-East. Forest. Inst.*, 7(1): 52 (1987).
ser. Lijiangense *T.C. Ku in Bull. Bot. Res. North-East. Forest. Inst.*, 7(1): 43 (1987).
ser. Luteae *T.C. Ku in Bull. Bot. Res. North-East. Forest. Inst.*, 7(1): 31 (1987).
sect. Nectarobilobos *T.C. Ku in Bull. Bot. Res. North-East. Forest. Inst.*, 7(1): 46 (1987).
sect. Nectaroquinquelobos *T.C. Ku in Bull. Bot. Res. North-East. Forest. Inst.*, 7(1): 52 (1987).
ser. Nectaroquinquelobos *T.C. Ku in Bull. Bot. Res. North-East. Forest. Inst.*, 7(1): 55 (1987).
subsect. Nectarotrilobos *T.C. Ku in Bull. Bot. Res. North-East. Forest. Inst.*, 7(1): 31 (1987).
ser. Nectarotrilobos *T.C. Ku in Bull. Bot. Res. North-East. Forest. Inst.*, 7(1): 32 (1987).
sect. Odontohymen *T.C. Ku in Bull. Bot. Res. North-East. Forest. Inst.*, 7(1): 29 (1987).
ser. Rosulatae *T.C. Ku in Bull. Bot. Res. North-East. Forest. Inst.*, 7(1): 28 (1987).
ser. Saxifragastrum *T.C. Ku in Bull. Bot. Res. North-East. Forest. Inst.*, 7(1): 22 (1987).
subsect. Xiphosandra *(Franch.) T.C. Ku in Bull. Bot. Res. North-East. Forest. Inst.*, 7(1): 44 (1987): P. sect. Xiphosandra.
angustipetala *T.C. Ku in Bull. Bot. Res. North-East. Forest. Inst.*, 7(1): 23 (1987)—China.
aphylla *T.C. Ku in Bull. Bot. Res. North-East. Forest. Inst.*, 7(1): 31 (1987)—Xizang, China.
cacuminum
f. yushuensis *T.C. Ku in Bull. Bot. Res. North-East. Forest. Inst.*, 7(1): 37 (1987)—China.
chengkouensis *T.C. Ku in Bull. Bot. Res. North-East. Forest. Inst.*, 7(1): 49 (1987)—China.
cordata *(Drude) Jien ex T.C. Ku in Bull. Bot. Res. North-East. Forest. Inst.*, 7(1): 37 (1987): P. nubicola var. cordata.
degeensis *T.C. Ku in Bull. Bot. Res. North-East. Forest. Inst.*, 7(1): 30 (1987)—China.
deqenensis *T.C. Ku in Bull. Bot. Res. North-East. Forest. Inst.*, 7(1): 41 (1987)—China.

PARNASSIA :–
gansuensis *T.C. Ku in Bull. Bot. Res. North-East. Forest. Inst.*, 7(1): 52 (1987)—China.
humilis *T.C. Ku in Bull. Bot. Res. North-East. Forest. Inst.*, 7(1): 27 (1987)—Xizang.
kangdingensis *T.C. Ku in Bull. Bot. Res. North-East. Forest. Inst.*, 7(1): 35 (1987)—China.
lanceolata *T.C. Ku in Bull. Bot. Res. North-East. Forest. Inst.*, 7(1): 34 (1987)—China.
lijiangensis *T.C. Ku in Bull. Bot. Res. North-East. Forest. Inst.*, 7(1): 43 (1987)—China.
longipetala
var. brevipetala *Jien ex T.C. Ku in Bull. Bot. Res. North-East. Forest. Inst.*, 7(1): 23 (1987)—China.
longshengensis *T.C. Ku in Bull. Bot. Res. North-East. Forest. Inst.*, 7(1): 54 (1987)—China.
omeiensis *T.C. Ku in Bull. Bot. Res. North-East. Forest. Inst.*, 7(1): 47 (1987)—China.
palustris
var. izuinsularis *H. Ohba in J. Jap. Bot.*, 62(5): 139 (1987)—Japan.
f. nana *T.C. Ku in Bull. Bot. Res. North-East. Forest. Inst.*, 7(1): 56 (1987)—Xinjiang.
f. rhodanthera *H. Ohba & Y. Umezu in J. Jap. Bot.*, 62(7): 208 (1987)—Japan.
var. yakusimensis *(Masamune) H. Ohba in J. Fac. Sci. Univ. Tokyo, Bot.*, 14(2): 88 (1987): P. yakusimensis.
subscaposa *C.Y. Wu ex T.C. Ku in Bull. Bot. Res. North-East. Forest. Inst.*, 7(1): 40 (1987)—China.
tibetana *Jien ex T.C. Ku in Bull. Bot. Res. North-East. Forest. Inst.*, 7(1): 38 (1987)—Xizang.
yanyuanensis *T.C. Ku in Bull. Bot. Res. North-East. Forest. Inst.*, 7(1): 26 (1987)—China.
yiliangensis *T.C. Ku in Bull. Bot. Res. North-East. Forest. Inst.*, 7(1): 48 (1987)—China.
yulongshanensis *T.C. Ku in Bull. Bot. Res. North-East. Forest. Inst.*, 7(1): 25 (1987)—China.

PARODIA (Cactac.).
allosiphon *(Marchesi) N.P. Taylor in Bradleya*, 5: 93 (1987): Notocactus allosiphon.
buiningii *(F. Buxbaum) N.P. Taylor in Bradleya*, 5: 93 (1987): Notocactus buiningii.
caespitosa *(Spegazzini) N.P. Taylor in Bradleya*, 5: 93 (1987): Echinocactus caespitosus.
concinna *(Monville) N.P. Taylor in Bradleya*, 5: 93 (1987): Echinocactus concinnus.
crassigibba *(F. Ritter) N.P. Taylor in Bradleya*, 5: 93 (1987): Notocactus crassigibbus.
erinacea *(Haworth) N.P. Taylor in Bradleya*, 5: 93 (1987): Cactus erinaceus.
herteri *(Werdermann) N.P. Taylor in Bradleya*, 5: 93 (1987): Echinocactus herteri.
horstii *(F. Ritter) N.P. Taylor in Bradleya*, 5: 93 (1987): Notocactus horstii.

PARODIA :–
liliputana *(Werdermann) N.P. Taylor in Bradleya*, 5: 93 (1987): Blossfeldia liliputana.
mammulosa *(Lemaire) N.P. Taylor in Bradleya*, 5: 93 (1987): Echinocactus mammulosus.
mueller-melchersii *(Backeberg) N.P. Taylor in Bradleya*, 5: 93 (1987): Notocactus mueller-melchersii.
neohorstii *(Theunissen) N.P. Taylor in Bradleya*, 5: 93 (1987): Notocactus neohorstii.
ottonis *(Lehmann) N.P. Taylor in Bradleya*, 5: 93 (1987): Cactus ottonis.
var. tortuosa *(Link & Otto) N.P. Taylor in Bradleya*, 5: 93 (1987): Echinocactus tortuosus.
rutilans *(Daeniker & Krainz) N.P. Taylor in Bradleya*, 5: 93 (1987): Notocactus rutilans.
scopa *(Sprengel) N.P. Taylor in Bradleya*, 5: 93 (1987): Cactus scopa.
succinea *(F. Ritter) N.P. Taylor in Bradleya*, 5: 93 (1987): Notocactus succineus.
werdermanniana *(Herter) N.P. Taylor in Bradleya*, 5: 93 (1987): Notocactus werdermannianus.

PASACCARDOA (Compos.).
jeffreyi
subsp. kasaiensis *S. Lisowski in Bull. Jard. Bot. Nation. Belg.*, 57(1–2): 257 (1987)—Zaïre, Angola.
subsp. procumbens *S. Lisowski in Bull. Jard. Bot. Nation. Belg.*, 57(1–2): 258 (1987)—Zaïre.

PASPALIDIUM (Gramin.).
reflexum *R.D. Webster, Australian Paniceae (Poac.)*: 166 (1987)—Australia (N. Terr., W. Australia, S. Australia).

PASPALUM (Gramin.).
concavum *S.A. Renvoize in Kew Bull.*, 42(4): 921 (1987)—Brazil.
perspicinervium *S.A. Renvoize in Kew Bull.*, 42(4): 922 (1987)—Brazil.
potamophilum *S.A. Renvoize in Kew Bull.*, 42(4): 922 (1987)—Brazil.

PASSERINA (Thymelaeac.).
drakensbergensis *O.M. Hilliard & B.L. Burtt, Bot. South. Natal Drakensberg*: 233 (1987)—South Africa (Natal).

PASSIFLORA (Passiflorac.).
sect. Discophora *P.M. Jørgensen & J.E. Lawesson in Nordic J. Bot.*, 7(2): 129 (1987).
sect. Leptopoda *Killip ex C. Feuillet & G. Cremers in Proc. Kon. Nederl. Akad. Wetensch., C*, 87(4): 378 (1984), without type.
crenata *C. Feuillet & G. Cremers in Proc. Kon. Nederl. Akad. Wetensch., C*, 87(4): 378 (1984)—French Guiana.

PASSIFLORA :–
cumbalensis
var. caucana *L.K. Escobar in Syst. Bot.*, 12(2): 242 (1987)—Colombia.
var. goudotiana *(Triana & Planchon) L.K. Escobar in Syst. Bot.*, 12(2): 242 (1987): P. goudotiana.
var. macrochlamys *(Harms) L.K. Escobar in Syst. Bot.*, 12(2): 246 (1987): P. macrochlamys.
var. mesadenia *(Killip) L.K. Escobar in Syst. Bot.*, 12(2): 246 (1987): P. mesadenia.
var. peruana *L.K. Escobar in Syst. Bot.*, 12(2): 249 (1987)—Peru.
var. pilosa *L.K. Escobar in Syst. Bot.*, 12(2): 249 (1987)—Ecuador.
discophora *P.M. Jørgensen & J.E. Lawesson in Nordic J. Bot.*, 7(2): 127 (1987)—Ecuador.
hirtiflora *P.M. Jørgensen & L.B. Holm-Nielsen in Nordic J. Bot.*, 7(2): 132 (1987)—Ecuador.
monadelpha *P.M. Jørgensen & L.B. Holm-Nielsen in Nordic J. Bot.*, 7(2): 129 (1987)—Ecuador.
plumosa *C. Feuillet & G. Cremers in Proc. Kon. Nederl. Akad. Wetensch., C*, 87(4): 381 (1984)—French Guiana.
sanctae-barbarae *L.B. Holm-Nielsen & P.M. Jørgensen in Nordic J. Bot.*, 7(2): 132 (1987)—Ecuador.
subpurpurea *P.M. Jørgensen & L.B. Holm-Nielsen in Nordic J. Bot.*, 7(2): 131 (1987)—Ecuador.

PAULLINIA (Sapindac.).
anomophylla *L. Radlkofer, Monogr. Paullinia:* 106 (1895–96); *et in Abh. Königl. Bayer. Akad. Wiss., Math.-Phys. Kl.*, 19(1): 172 (1895–96)— Brazil.
apoda *L. Radlkofer, Monogr. Paullinia:* 192 (1895–96); *et in Abh. Königl. Bayer. Akad. Wiss., Math.-Phys. Kl.*, 19(1): 258 (1895–96)— Colombia.
bidentata *L. Radlkofer, Monogr. Paullinia:* 129 (1895–96); *et in Abh. Königl. Bayer. Akad. Wiss., Math.-Phys. Kl.*, 19(1): 195 (1895–96)— Peru.
caloptera *L. Radlkofer, Monogr. Paullinia:* 208 (1895–96); *et in Abh. Königl. Bayer. Akad. Wiss., Math.-Phys. Kl.*, 19(1): 274 (1895–96)— Brazil, Venezuela.
castaneifolia *L. Radlkofer, Monogr. Paullinia:* 116 (1895–96); *et in Abh. Königl. Bayer. Akad. Wiss., Math.-Phys. Kl.*, 19(1): 182 (1895–96)— Brazil.
clathrata *L. Radlkofer, Monogr. Paullinia:* 172 (1895–96); *et in Abh. Königl. Bayer. Akad. Wiss., Math.-Phys. Kl.*, 19(1): 238 (1895–96)— Brazil.
conduplicata *L. Radlkofer, Ergänz. Monogr. Serjania:* 100 (1886–87); *et in Abh. Königl. Bayer. Akad. Wiss., Math.-Phys. Kl.*, 16(1): 100 (1886)— Guyana.

PAULLINIA :–
costaricensis *L. Radlkofer, Ergänz. Monogr. Serjania:* 157 (1886–87); *et in Abh. Königl. Bayer. Akad. Wiss., Math.-Phys. Kl.*, 16(1): 157 (1886)— Nicaragua.
cristata *L. Radlkofer, Monogr. Paullinia:* 256 (1895–96); *et in Abh. Königl. Bayer. Akad. Wiss., Math.-Phys. Kl.*, 19(1): 322 (1895–96)— Brazil.
curvicuspis *L. Radlkofer, Monogr. Paullinia:* 171 (1895–96); *et in Abh. Königl. Bayer. Akad. Wiss., Math-Phys. Kl.*, 19(1): 237 (1895–96)— Peru.
dasygonia *L. Radlkofer, Monogr. Paullinia:* 257 (1895–96); *et in Abh. Königl. Bayer. Akad. Wiss., Math.-Phys. Kl.*, 19(1): 323 (1895–96)— French Guiana, Guyana, Trinidad.
dasystachya *L. Radlkofer, Monogr. Paullinia:* 204 (1895–96); *et in Abh. Königl. Bayer. Akad. Wiss., Math.-Phys. Kl.*, 19(1): 270 (1895–96)— Peru, Bolivia.
elongata *L. Radlkofer, Monogr. Paullinia:* 172 (1895–96); *et in Abh. Königl. Bayer. Akad. Wiss., Math.-Phys. Kl.*, 19(1): 238 (1895–96)— Peru.
excisa *L. Radlkofer, Monogr. Paullinia:* 207 (1895–96); *et in Abh. Königl. Bayer. Akad. Wiss., Math.-Phys. Kl.*, 19(1): 273 (1895–96)— Tobago.
fasciculata *L. Radlkofer, Monogr. Paullinia:* 58 (1895–96); *et in Abh. Königl. Bayer. Akad. Wiss., Math.-Phys. Kl.*, 19(1): 124 (1895–96)— French Guiana, Tobago.
firma *L. Radlkofer, Monogr. Paullinia:* 168 (1895–96); *et in Abh. Königl. Bayer. Akad. Wiss., Math.-Phys. Kl.*, 19(1): 234 (1895–96)— Brazil.
fistulosa *L. Radlkofer, Monogr. Paullinia:* 193 (1895–96); *et in Abh. Königl. Bayer. Akad. Wiss., Math.-Phys. Kl.*, 19(1): 259 (1895–96)— Peru.
fusiformis *L. Radlkofer, Monogr. Paullinia:* 155 (1895–96); *et in Abh. Königl. Bayer. Akad. Wiss., Math.-Phys. Kl.*, 19(1): 221 (1895–96)— Brazil.
glomerulosa *L. Radlkofer, Monogr. Paullinia:* 191 (1895–96); *et in Abh. Königl. Bayer. Akad. Wiss., Math.-Phys. Kl.*, 19(1): 257 (1895–96)— Mexico, Panama, Venezuela.
imberbis *L. Radlkofer, Monogr. Paullinia:* 111 (1895–96); *et in Abh. Königl. Bayer. Akad. Wiss., Math.-Phys. Kl.*, 19(1): 177 (1895–96)— Brazil, French Guiana.
laeta *L. Radlkofer, Monogr. Paullinia:* 129 (1895–96); *et in Abh. Königl. Bayer. Akad. Wiss., Math.-Phys. Kl.*, 19(1): 195 (1895–96)— Peru.
linearis *L. Radlkofer, Monogr. Paullinia:* 157 (1895–96); *et in Abh. Königl. Bayer. Akad. Wiss., Math.-Phys. Kl.*, 19(1): 223 (1895–96)— Peru.
livescens *L. Radlkofer, Monogr. Paullinia:* 182 (1895–96); *et in Abh. Königl. Bayer. Akad. Wiss., Math.-Phys. Kl.*, 19(1): 248 (1895–96)— Brazil.

PAULLINIA :–
macrocarpa *L. Radlkofer, Monogr. Paullinia:* 260 (1895–96); *et in Abh. Königl. Bayer. Akad. Wiss., Math.-Phys. Kl.,* 19(1): 326 (1895–96)— Costa Rica.
mallophylla *L. Radlkofer, Monogr. Paullinia:* 188 (1895–96); *et in Abh. Königl. Bayer. Akad. Wiss., Math.-Phys. Kl.,* 19(1): 254 (1895–96)— Costa Rica.
microsepala *L. Radlkofer, Monogr. Paullinia:* 187 (1895–96); *et in Abh. Königl. Bayer. Akad. Wiss., Math.-Phys. Kl.,* 19(1): 253 (1895–96)— Windward Is., Leeward Is.
neuroptera *L. Radlkofer, Monogr. Paullinia:* 178 (1895–96); *et in Abh. Königl. Bayer. Akad. Wiss., Math.-Phys. Kl.,* 19(1): 244 (1895–96)— Windward Is.
nobilis *L. Radlkofer, Monogr. Paullinia:* 205 (1895–96); *et in Abh. Königl. Bayer. Akad. Wiss., Math.-Phys. Kl.,* 19(1): 271 (1895–96)— Peru, Brazil.
parvibractea *L. Radlkofer, Monogr. Paullinia:* 152 (1895–96); *et in Abh. Königl. Bayer. Akad. Wiss., Math.-Phys. Kl.,* 19(1): 218 (1895–96)— Brazil.
platymisca *L. Radlkofer, Monogr. Paullinia:* 163 (1895–96); *et in Abh. Königl. Bayer. Akad. Wiss., Math.-Phys. Kl.,* 19(1): 229 (1895–96)— Brazil.
quitensis *L. Radlkofer, Monogr. Paullinia:* 203 (1895–96); *et in Abh. Königl. Bayer. Akad. Wiss., Math.-Phys. Kl.,* 19(1): 269 (1895–96)— Ecuador.
revoluta *L. Radlkofer, Monogr. Paullinia:* 251 (1895–96); *et in Abh. Königl. Bayer. Akad. Wiss., Math.-Phys. Kl.,* 19(1): 317 (1895–96)— Brazil.
rhomboidea *L. Radlkofer, Ergänz. Monogr. Serjania:* 23 (1886–87); *et in Abh. Königl. Bayer. Akad. Wiss., Math.-Phys. Kl.,* 16(1): 23 (1886)— Brazil.
selenoptera *L. Radlkofer, Monogr. Paullinia:* 237 (1895–96); *et in Abh. Königl. Bayer. Akad. Wiss., Math.-Phys. Kl.,* 19(1): 303 (1895–96)— Peru.
stellata *L. Radlkofer, Monogr. Paullinia:* 153 (1895–96); *et in Abh. Königl. Bayer. Akad. Wiss., Math.-Phys. Kl.,* 19(1): 219 (1895–96)— French Guiana.
subauriculata *L. Radlkofer, Monogr. Paullinia:* 130 (1895–96); *et in Abh. Königl. Bayer. Akad. Wiss., Math.-Phys. Kl.,* 19(1): 196 (1895–96)— Peru.
subnuda *L. Radlkofer, Monogr. Paullinia:* 208 (1895–96); *et in Abh. Königl. Bayer. Akad. Wiss., Math.-Phys. Kl.,* 19(1): 274 (1895–96)— French Guiana, Brazil, Panama, Costa Rica.
ternata *L. Radlkofer, Monogr. Paullinia:* 189 (1895–96); *et in Abh. Königl. Bayer. Akad. Wiss., Math.-Phys. Kl.,* 19(1): 255 (1895–96)— Brazil.
trilatera *L. Radlkofer, Monogr. Paullinia:* 188 (1895–96); *et in Abh. Königl. Bayer. Akad. Wiss., Math.-Phys. Kl.,* 19(1): 254 (1895–96)— Brazil.

PAULLINIA :–
uloptera *L. Radlkofer, Monogr. Paullinia:* 255 (1895–96); *et in Abh. Königl. Bayer. Akad. Wiss., Math.-Phys. Kl.,* 19(1): 321 (1895–96)— Brazil.
urvilleoides *L. Radlkofer, Monogr. Paullinia:* 125 (1895–96); *et in Abh. Königl. Bayer. Akad. Wiss., Math.-Phys. Kl.,* 19(1): 191 (1895–96)— Brazil.
venezuelana *L. Radlkofer, Monogr. Paullinia:* 176 (1895–96); *et in Abh. Königl. Bayer. Akad. Wiss., Math.-Phys. Kl.,* 19(1): 242 (1895–96)— Venezuela.
venosa *L. Radlkofer, Monogr. Paullinia:* 164 (1895–96); *et in Abh. Königl. Bayer. Akad. Wiss., Math.-Phys. Kl.,* 19(1): 230 (1895–96)— Surinam, Brazil.
verrucosa *L. Radlkofer, Monogr. Paullinia:* 181 (1895–96); *et in Abh. Königl. Bayer. Akad. Wiss., Math.-Phys. Kl.,* 19(1): 247 (1895–96)— Guyana.
xestophylla *L. Radlkofer, Monogr. Paullinia:* 163 (1895–96); *et in Abh. Königl. Bayer. Akad. Wiss., Math.-Phys. Kl.,* 19(1): 229 (1895–96)— Brazil.

PAVETTA (Rubiac.).
diversicalyx *D.M. Bridson in Kew Bull.,* 42(1): 251 (1987)—Tanzania.
johnstonii
subsp. breviloba *D.M. Bridson in Kew Bull.,* 42(1): 253 (1987)—Tanzania, Zambia.
lynesii *D.M. Bridson in Kew Bull.,* 42(1): 253 (1987)—Tanzania.
nitidissima *D.M. Bridson in Kew Bull.,* 42(1): 255 (1987)—Tanzania.
subumbellata
var. subcoriacea *D.M. Bridson in Kew Bull.,* 42(1): 257 (1987)—Tanzania, Malawi.

PAVONIA (Malvac.).
uniflora *(Sessé & Mociño) P.A. Fryxell in Syst. Bot.,* 12(2): 279 (1987): Urena uniflora.
venusta *P.A. Fryxell in Syst. Bot.,* 12(2): 279 (1987), nom. nov.: P. discolor *Fryxell.*

PECTIS (Compos.).
×floridana *D.J. Keil in Sida,* 11(4): 389 (1986)— U.S.A. (Florida). (P. glaucescens × prostrata).
glaucescens *(Cassini) D.J. Keil in Sida,* 11(4): 386 (1986): Chthonia glaucescens.

PEDICULARIS (Scrophulariac.).
chamissonis
var. rebunensis *T. Yamazaki in J. Jap. Bot.,* 62(2): 54 (1987)—Japan.

PEDIOMELUM (Legumin.).
digitatum *(Nutt. ex T. & G.) D. Isely in Sida,* 11(4): 430 (1986): Psoralea digitata.
hypogaeum
var. scaposum *(Gray) W.F. Mahler in Sida,* 12(1): 250 (1987): Psoralea hypogaea var. scaposa.
latestipulatum *(Shinners) W.F. Mahler in Sida,* 12(1): 250 (1987): Psoralea latestipulata.

PELARGONIUM (Geraniac.).
caroli-henrici *B. Nordenstam in Pl. Syst. Evol.*, 155(1–4): 333 (1987)—South Africa (Cape Province).
divisifolium *P. Vorster in S. Afr. J. Bot.*, 53(1): 71 (1987)—South Africa (Cape Province).

PENIOCEREUS (Cactac.).
serpentinus *(Lagasca & Rodrigues) N.P. Taylor in Bradleya*, 5: 93 (1987): Cactus serpentinus.

PENSTEMON (Scrophulariac.).
acaulis
var. yampaensis *(Penl.) E.C. Neese in Great Basin Nat.*, 46(3): 459 (1986): P. yampaensis.
angustifolius
var. dulcis *E.C. Neese in Great Basin Nat.*, 46(3): 459 (1986)—U.S.A. (Utah).
duchesnensis *(N. Holmgren) E.C. Neese in Great Basin Nat.*, 46(3): 459 (1986): P. dolius var. duchesnensis.
flowersii *E. Neese & S.L. Welsh in Great Basin Nat.*, 43(3): 429 (1983)—U.S.A. (Utah).
leonardii
var. higginsii *E.C. Neese in Great Basin Nat.*, 46(3): 459 (1986)—U.S.A. (Utah).
var. patricus *(N. Holmgren) E.C. Neese in Great Basin Nat.*, 46(3): 460 (1986): P. patricus.
pachyphyllus
var. mucronatus *(N. Holmgren) E.C. Neese in Great Basin Nat.*, 46(3): 460 (1986): P. mucronatus.
scariosus
var. cyanomontanus *E.C. Neese in Great Basin Nat.*, 46(3): 460 (1986)—U.S.A. (Utah, Colorado).
thompsoniae
var. desperatus *E.C. Neese in Great Basin Nat.*, 46(3): 460 (1986)—U.S.A. (Utah, Nevada).

PENTARHOPALOPILIA (Engler) P. Hiepko in Bot. Jahrb., 108(2–3): 280 (1987): Rhopalopilia sect. Pentarhopalopilia (Opiliac.).
madagascariensis *(Cavaco & Keraudren) P. Hiepko in Bot. Jahrb.*, 108(2–3): 285 (1987): Rhopalopilia madagascariensis.
marquesii *(Engler) P. Hiepko in Bot. Jahrb.*, 108(2–3): 284 (1987): Rhopalopilia marquesii.
perrieri *(Cavaco & Keraudren) P. Hiepko in Bot. Jahrb.*, 108(2–3): 287 (1987): Rhopalopilia perrieri.
umbellulata *(Baillon) P. Hiepko in Bot. Jahrb.*, 108(2–3): 281 (1987): Opilia umbellulata.

PENTZIA (Compos.).
peduncularis *B. Nordenstam in Bot. Jahrb.*, 108(2–3): 196 (1987)—South Africa (Cape Province).

PEPEROMIA (Piperac.).
calderoniae *M.A. Barrios, J.H. Cota & J.M. Medina-Cota in Phytologia*, 62(1): 54 (1987)—Mexico.
dolabella *W. Rauh & M. Kimnach in Cact. Succ. J. (U.S.A.)*, 59(1): 5 (1987)—Peru.
gentryi *J.A. Steyermark in Ann. Missouri Bot. Gard.*, 74(1): 87 (1987)—Venezuela.
marahuacensis *J.A. Steyermark in Ann. Missouri Bot. Gard.*, 74(1): 87 (1987)—Venezuela.

PERESKIA (Cactac.).
grandifolia
var. violacea *B.E. Leuenberger in Mem. New York Bot. Gard.*, 41: 116 (1986)—Brazil.
humboldtii
var. rauhii *(Backeberg) B.E. Leuenberger in Mem. New York Bot. Gard.*, 41: 69 (1986): P. vargasii var. rauhii.

PERIPTERA (Malvac.).
lobelioides *P.A. Fryxell & S.D. Koch in Aliso*, 11(4): 552 (1987)—Mexico.

PERISTYLUS (Orchidac.).
duthiei *(Hook. f.) S. Deva & H.B. Naithani, Orchid Fl. N.W. Himalaya:* 181 (1986): Herminium duthiei.
var. inayatii *S. Deva & H.B. Naithani, Orchid Fl. N.W. Himalaya:* 185 (1986)—India.
fallax
var. dwarikae *S. Deva & H.B. Naithani, Orchid Fl. N.W. Himalaya:* 187 (1986), as 'dwarikaii'—India.

PERITYLE (Compos.).
specuicola *S.L. Welsh & E. Neese in Great Basin Nat.*, 43(2): 373 (1983)—U.S.A. (Utah).

PEROVSKIA (Labiat.).
pamirica *C.Y. Yang & B. Wang in Bull. Bot. Res. North-East. Forest. Inst.*, 7(1): 95 (1987)—Xinjiang.

PERSEA (Laurac.).
parviflora *(Meissn.) K. Haridasan & R.R. Rao, Forest Fl. Meghalaya*, 2: 739 (1987): Machilus parviflora.

PERSICARIA (Polygonac.).
amphibia
var. natans *(Leyss.) A.H. Munshi & G.N. Javeid, Syst. Stud. Polygonac. Kashmir Himalaya (J. Econ. Taxon. Bot. Addit. Ser. 2):* 63 (1986): Polygonum amphibium var. natans.
var. terrestris *(Leyss.) A.H. Munshi & G.N. Javeid, Syst. Stud. Polygonac. Kashmir Himalaya (J. Econ. Taxon. Bot. Addit. Ser. 2):* 63 (1986): Polygonum amphibium var. terrestre.

PERSICARIA :–
caespitosa
var. longiseta *(DeBruyn) C.F. Reed in Phytologia,* 63(5): 410 (1987): Polygonum longisetum.
glacialis *(Hook. f.) A.H. Munshi & G.N. Javeid, Syst. Stud. Polygonac. Kashmir Himalaya (J. Econ. Taxon. Bot. Addit. Ser. 2):* 76 (1986): Polygonum glaciale.
hydropiper
subsp. flaccida *(Meisn.) A.H. Munshi & G.N. Javeid, Syst. Stud. Polygonac. Kashmir Himalaya (J. Econ. Taxon. Bot. Addit. Ser. 2):* 76 (1986): Polygonum flaccidum.
subsp. mite *(Schrank) A.H. Munshi & G.N. Javeid, Syst. Stud. Polygonac. Kashmir Himalaya (J. Econ. Taxon. Bot. Addit. Ser. 2):* 75 (1986): Polygonum mite.
salicifolia *(Brouss. ex Willd.) A.H. Munshi & G.N. Javeid, Syst. Stud. Polygonac. Kashmir Himalaya (J. Econ. Taxon. Bot. Addit. Ser. 2):* 73 (1986): Polygonum salicifolium.
snobarii *A.H. Munshi & G.N. Javeid, Syst. Stud. Polygonac. Kashmir Himalaya (J. Econ. Taxon. Bot. Addit. Ser. 2):* 73 (1986)—India.

PERTYA (Compos.).
ser. Paniculatea *Y.C. Tseng in Guihaia,* 5(4): 327 (1985).
angustifolia *Y.C. Tseng in Guihaia,* 5(4): 328 (1985)—China.
berberidoides *(Hand.-Mazz.) Y.C. Tseng in Guihaia,* 5(4): 328 (1985): P. bodinieri var. berberidoides.
corymbosa *Y.C. Tseng in Guihaia,* 5(4): 332 (1985)—China.
henanensis *Y.C. Tseng in Guihaia,* 5(4): 330 (1985)—China.
pungens *Y.C. Tseng in Guihaia,* 5(4): 334 (1985)—China.

PERYMENIUM (Compos.).
tamaulipense *B.L. Turner in Phytologia,* 63(5): 396 (1987)—Mexico.

PETRADORIA (Compos.).
pumila
var. graminea *(Woot. & Standl.) S.L. Welsh in Great Basin Nat.,* 43(2): 324 (1983): P. graminea.

PETROEDMONDIA S.G. Tamamschian in Fl. Iran., 162: 167 (1987). *UMBELLIFERAE.*
syriaca *(Boiss.) S.G. Tamamschian in Fl. Iran.,* 162: 167 (1987): Colladonia syriaca.

PETUNIA (Solanac.).
exserta *J.R. Stehmann in Napaea, Rev. Bot.,* 2: 19 (1987)—Brazil.

PEUCEDANUM (Umbellif.).
articulatum *C.C. Townsend in Kew Bull.,* 42(3): 602 (1987)—Malawi.
dispersum *C.C. Townsend in Kew Bull.,* 42(3): 596 (1987)—Tanzania, Uganda.

PEUCEDANUM :–
friesiorum
var. bipinnatum *C.C. Townsend in Kew Bull.,* 42(3): 591 (1987)—Kenya.
harmsianum
subsp. australe *C.C. Townsend in Kew Bull.,* 42(3): 593 (1987)—Tanzania.
oblongisectum *C.C. Townsend in Kew Bull.,* 42(3): 598 (1987)—Tanzania.
stenospermum *C.C. Townsend in Kew Bull.,* 42(3): 600 (1987)—Zambia.
tenue *C.C. Townsend in Kew Bull.,* 42(3): 602 (1987)—Zambia, Tanzania.

PFEIFFERA (Cactac.).
miyagawae *W. Barthlott & W. Rauh in Cact. Succ. J. (U.S.A.),* 59(2): 63 (1987)—Bolivia.

PHACELLARIA (Santalac.).
glomerata *D.D. Tao in Acta Phytotax. Sin.,* 25(5): 407 (1987)—China.

PHAEDRANASSA (Amaryllidac.).
brevifolia *A.W. Meerow in Sida,* 12(1): 30 (1987)—Ecuador.

PHAGNALON (Compos.).
rechingeri *H.W. Lack & M. Qaiser in Pl. Syst. Evol.,* 155(1–4): 45 (1987)—Iran.

PHALAENOPSIS (Orchidac.).
sect. Deliciosae *E.A. Christenson in Indian Orchid J.,* 1(4): 152 (1985); *et in Selbyana,* 9(1): 167 (1986).
braceana *(J.D. Hook.) E.A. Christenson in Selbyana,* 9(1): 169 (1986): Doritis braceana.
philippinensis *Golamco ex J.A. Fowlie & C.Z. Tang in Orchid Dig.,* 51(2): 92 (1987), as 'philippinense'—Philippine Is.
taenialis *(Lindl.) E.A. Christenson & U.C. Pradhan in Indian Orchid J.,* 1(4): 154 (1985); *et in Selbyana,* 9(1): 168 (1986): Aerides taeniale.

PHEBALIUM (Rutac.).
equestre *D.A. Cooke in J. Adelaide Bot. Gard.,* 10(2): 241 (1987)—Australia (S. Australia).

PHILODENDRON (Arac.).
azulitense *T.B. Croat in Aroideana,* 9(1–4): 79 (1986)—Venezuela.
benitezii *T.B. Croat in Aroideana,* 9(1–4): 84 (1986)—Venezuela.
brevispathum
var. wurdackii *T.B. Croat in Aroideana,* 9(1–4): 88 (1986)—Venezuela.
buntingianum *T.B. Croat in Aroideana,* 9(1–4): 89 (1986)—Venezuela.
effusilobum *T.B. Croat in Aroideana,* 9(1–4): 99 (1986)—Venezuela.
glanduliferum
subsp. camiloanum *T.B. Croat in Aroideana,* 9(1–4): 109 (1986)—Venezuela.
kautskyi *G.S. Bunting in Phytologia,* 61(7): 441 (1987)—Brazil.

PHILODENDRON :-
milleri *T.B. Croat in Aroideana,* 9(1–4): 132 (1986)—Venezuela.
panduriforme
var. reichenbachianum *(Schott) T.B. Croat in Aroideana,* 9(1–4): 140 (1986): P. reichenbachianum.
spiritus-sancti *G.S. Bunting in Phytologia,* 61(7): 441 (1987)—Brazil.
venosum *(Willd.) T.B. Croat in Aroideana,* 9(1–4): 157 (1986): Pothos venosa.
yaracuyense *T.B. Croat in Aroideana,* 9(1–4): 164 (1986)—Venezuela.

PHILYDRELLA (Philydrac.).
drummondii *L.G. Adams in Fl. Australia,* 45: 454, 42 (1987)—Australia (W. Australia).
pygmaea
subsp. minima *L.G. Adams in Fl. Australia,* 45: 454, 43 (1987)—Australia (W. Australia).

PHLEBOCARYA (Haemodorac.).
pilosissima
subsp. teretifolia *T.D. Macfarlane in Fl. Australia,* 45: 465, 131 (1987)—Australia (W. Australia).

PHLEUM (Gramin.).
parviceps *(Briq.) A.W. Hill in Index Kewensis,* Suppl. 6: 152 (1916–20 publ. 1926): P. alpinum var. parviceps.
vulgare *(Čelak.) A. Chase & C.D. Niles in Index Grass Sp.,* 3: 81 (1962): P. pratense vulgare.

PHLOMIS (Labiat.).
brevidentata *H.W. Li in Fl. Xizangica,* 4: 157 (1985)—Xizang.
×margaritae *A. Aparicio & S. Silvestre in Lagascalia,* 14(1): 100 (1986)—Spain. (P. composita × purpurea).
nana *C.Y. Wu in Fl. Xizangica,* 4: 154 (1985)—Xizang.
nyalamensis *H.W. Li in Fl. Xizangica,* 4: 157 (1985)—Xizang.

PHLOMOIDES (Labiat.).
sect. Filipendula *(M. Pop.) T.A. Adylov, R.V. Kamelin & A.M. Makhmedov in Novosti Sist. Vyssh. Rast.,* 23: 111 (1986): Eremostachys sect. Filipendula.
cordifolia *(Regel) T.A. Adylov, R.V. Kamelin & A.M. Makhmedov in Novosti Sist. Vyssh. Rast.,* 23: 115 (1986): Eremostachys cordifolia.
lanatifolia *A.M. Makhmedov in Novosti Sist. Vyssh. Rast.,* 23: 115 (1986)—U.S.S.R. (Middle Asia).

PHOLIDOTA (Orchidac.).
aidiolepis *G. Seidenfaden & E.F. de Vogel in Op. Bot.,* 89: 106 (1986)—Thailand.
bracteata *(D. Don) G. Seidenfaden in Op. Bot.,* 89: 100 (1986): Ptilocnema bracteatum.

PHORADENDRON (Viscac.).
balslevii *J. Kuijt in Fl. Ecuador,* 24(32B): 31 (1986)—Ecuador.
burkartii *C.T. Rizzini & E.A. Ulibarri in Darwiniana,* 27(1–4): 499 (1986)—Argentina.
macrocarpum *J. Kuijt in Fl. Ecuador,* 24(32B): 37 (1986)—Ecuador.
madisonii *J. Kuijt in Fl. Ecuador,* 24(32B): 46 (1986)—Ecuador.
wiensii *J. Kuijt in Fl. Ecuador,* 24(32B): 44 (1986)—Ecuador.

PHOTINIA (Rosac.).
dabeishanensis *M.B. Deng & G. Yao in Bull. Nanjing Bot. Gard. Mem. Sun Yat Sen,* 1984–1985: 126 (1986)—China.

PHRAGMITES (Gramin.).
australis *(Cav.) Steudel, Nomencl. Bot.,* ed. 2, 1: 143 (1840): Arundo australis. †
var. berlanderi *(Fourn.) C.F. Reed in Phytologia,* 63(5): 410 (1987): P. berlanderi.
subsp. chrysanthus *(Mabille) M. Kerguélen in Lejeunia,* 120: 139 (1987): P. chrysanthus.
egmontianus *(Roemer & Schultes) Steudel, Nomencl. Bot.,* ed. 2, 1: 144 (1840), as 'egmontiana': Arundo egmontiana. †
pungens *Chevallier ex A. Chevalier in Bull. Mus. Hist. Nat. (Paris),* Sér. 2, 5: 84 (1933), nom. nov.: P. communis var. stenophylla *Boiss.*

PHYLLAGATHIS (Melastomatac.).
magnifica *A. Weber in Pl. Syst. Evol.,* 157(3–4): 189 (1987)—Malaya.
stonei *A. Weber in Pl. Syst. Evol.,* 157(3–4): 192 (1987)—Malaya.

PHYLLANTHODENDRON (Euphorbiac.).
sect. Tetrandrum *P.T. Li in Bull. Bot. Res. North-East. Forest. Inst.,* 7(3): 8 (1987).
breynioides *P.T. Li in Bull. Bot. Res. North-East. Forest. Inst.,* 7(3): 6 (1987)—China.
caudatifolium *P.T. Li in Bull. Bot. Res. North-East. Forest. Inst.,* 7(3): 7 (1987)—China.
moi *(P.T. Li) P.T. Li in Bull. Bot. Res. North-East. Forest. Inst.,* 7(3): 9 (1987): Phyllanthus moi.
orbicularifolium *P.T. Li in Bull. Bot. Res. North-East. Forest. Inst.,* 7(3): 5 (1987)—China.
petraeum *P.T. Li in Bull. Bot. Res. North-East. Forest. Inst.,* 7(3): 4 (1987)—China.

PHYLLANTHUS (Euphorbiac.).
subgen. Metaphyllanthus *P.T. Li in Acta Phytotax. Sin.,* 25(5): 382 (1987).
borenensis *M.G. Gilbert in Kew Bull.,* 42(2): 354 (1987)—Ethiopia.
dewildeorum *M.G. Gilbert in Kew Bull.,* 42(2): 356 (1987), as 'dewildiorum'—Ethiopia.
dongfangensis *P.T. Li in Acta Phytotax. Sin.,* 25(5): 382 (1987)—Hainan.

PHYLLANTHUS :-
fangchengensis *P.T. Li in Acta Phytotax. Sin.*, 25(5): 377 (1987)—China.
fimbricalyx *P.T. Li in Acta Phytotax. Sin.*, 25(5): 380 (1987)—China.
guangdongensis *P.T. Li in Acta Phytotax. Sin.*, 25(5): 376 (1987)—China.
mooneyi *M.G. Gilbert in Kew Bull.*, 42(2): 357 (1987)—Ethiopia.
nanellus *P.T. Li in Acta Phytotax. Sin.*, 25(5): 376 (1987)—Hainan.
oblongiglans *M.G. Gilbert in Kew Bull.*, 42(2): 359 (1987)—Ethiopia.
tsiangii *P.T. Li in Acta Phytotax. Sin.*, 25(5): 375 (1987), nom. nov.: Nymphanthus ruber *Lour.*

PHYLLOSTEGIA (Labiat.).
adenophora *H. St. John in Phytologia*, 63(3): 172 (1987)—Hawaiian Is.
alba *H. St. John in Phytologia*, 63(3): 172 (1987)—Hawaiian Is.
arenicola *H. St. John in Phytologia*, 63(3): 172 (1987)—Hawaiian Is.
atomifera *H. St. John in Phytologia*, 63(3): 172 (1987)—Hawaiian Is.
axillaris *H. St. John in Phytologia*, 63(3): 172 (1987)—Hawaiian Is.
brevicalycis *H. St. John in Phytologia*, 63(3): 172 (1987)—Hawaiian Is.
brevilobata *H. St. John in Phytologia*, 63(3): 173 (1987)—Hawaiian Is.
brevis *H. St. John in Phytologia*, 63(3): 173 (1987)—Hawaiian Is.
brighamii *H. St. John in Phytologia*, 63(3): 173 (1987)—Hawaiian Is.
bryanii *(Sherff) H. St. John in Phytologia*, 63(3): 173 (1987): P. racemosa var. bryanii.
capitata *H. St. John in Phytologia*, 63(3): 173 (1987)—Hawaiian Is.
catenulata *H. St. John in Phytologia*, 63(3): 173 (1987)—Hawaiian Is.
cernua *H. St. John in Phytologia*, 63(3): 173 (1987)—Hawaiian Is.
chartacea *H. St. John in Phytologia*, 63(3): 173 (1987)—Hawaiian Is.
cordata *H. St. John in Phytologia*, 63(3): 173 (1987)—Hawaiian Is.
cowanii *H. St. John in Phytologia*, 63(3): 174 (1987)—Hawaiian Is.
curta *H. St. John in Phytologia*, 63(3): 174 (1987)—Hawaiian Is.
decemiflorifer *H. St. John in Phytologia*, 63(3): 174 (1987)—Hawaiian Is.
decumbens *H. St. John in Phytologia*, 63(3): 174 (1987)—Hawaiian Is.
deltoidea *H. St. John in Phytologia*, 63(3): 174 (1987)—Hawaiian Is.
elliptica *H. St. John in Phytologia*, 63(3): 174 (1987)—Hawaiian Is.
fagerlindii *(Sherff) H. St. John in Phytologia*, 63(3): 174 (1987): P. mollis var. fagerlindii.
fayi *H. St. John in Phytologia*, 63(3): 174 (1987)—Hawaiian Is.
foliosa *H. St. John in Phytologia*, 63(3): 175 (1987)—Hawaiian Is.

PHYLLOSTEGIA :-
glabriuscula *(A. Gray) H. St. John in Phytologia*, 63(3): 175 (1987): P. parviflora var. glabriuscula.
glandulosa *H. St. John in Phytologia*, 63(3): 175 (1987)—Hawaiian Is.
grandiflora
var. majoricalycis *H. St. John in Phytologia*, 63(3): 175 (1987)—Hawaiian Is.
heterodoxa *(Sherff) H. St. John in Phytologia*, 63(3): 175 (1987): P. brevidens var. heterodoxa.
hiati *H. St. John in Phytologia*, 63(3): 175 (1987)—Hawaiian Is.
hitchcockii *(Sherff) H. St. John in Phytologia*, 63(3): 175 (1987): P. stachyoides var. hitchcockii.
hobdyi *H. St. John in Phytologia*, 63(3): 175 (1987)—Hawaiian Is.
hualalaiensis *H. St. John in Phytologia*, 63(3): 175 (1987)—Hawaiian Is.
imminuta *(Sherff) H. St. John in Phytologia*, 63(3): 175 (1987): P. helleri var. imminuta.
insignis *H. St. John in Phytologia*, 63(3): 175 (1987)—Hawaiian Is.
interrupta *H. St. John in Phytologia*, 63(3): 176 (1987)—Hawaiian Is.
kaalaensis *H. St. John in Phytologia*, 63(3): 176 (1987)—Hawaiian Is.
kahiliensis *H. St. John in Phytologia*, 63(3): 176 (1987)—Hawaiian Is.
kamokuensis *H. St. John in Phytologia*, 63(3): 176 (1987)—Hawaiian Is.
kauaiensis *H. St. John in Phytologia*, 63(3): 176 (1987)—Hawaiian Is.
kauensis *H. St. John in Phytologia*, 63(3): 176 (1987)—Hawaiian Is.
kilaueaensis *H. St. John in Phytologia*, 63(3): 176 (1987)—Hawaiian Is.
kohalaensis *H. St. John in Phytologia*, 63(3): 176 (1987)—Hawaiian Is.
konaensis *H. St. John in Phytologia*, 63(3): 176 (1987)—Hawaiian Is.
lanaiensis *(Sherff) H. St. John in Phytologia*, 63(3): 177 (1987): P. glabra var. lanaiensis.
lantanoides
var. konahuanuiana *H. St. John in Phytologia*, 63(3): 177 (1987)—Hawaiian Is.
laxior *(Deg. & Sherff) H. St. John in Phytologia*, 63(3): 177 (1987): P. hirsuta var. laxior.
lebishopii *H. St. John in Phytologia*, 63(3): 177 (1987)—Hawaiian Is.
lehuaensis *H. St. John in Phytologia*, 63(3): 177 (1987)—Hawaiian Is.
leukantha *H. St. John in Phytologia*, 63(3): 177 (1987)—Hawaiian Is.
longitubata *H. St. John in Phytologia*, 63(3): 177 (1987)—Hawaiian Is.
lydgatei *(Sherff) H. St. John in Phytologia*, 63(3): 177 (1987): P. mollis var. lydgatei.
makawaoensis *H. St. John in Phytologia*, 63(3): 177 (1987)—Hawaiian Is.
manoana *H. St. John in Phytologia*, 63(3): 177 (1987)—Hawaiian Is.

PHYLLOSTEGIA :-
micrantha *H. St. John in Phytologia,* 63(3): 178 (1987)—Hawaiian Is.
molokaiensis *H. St. John in Phytologia,* 63(3): 178 (1987)—Hawaiian Is.
moniliformis *H. St. John in Phytologia,* 63(3): 178 (1987)—Hawaiian Is.
montana *H. St. John in Phytologia,* 63(3): 178 (1987)—Hawaiian Is.
multiflora *H. St. John in Phytologia,* 63(3): 178 (1987)—Hawaiian Is.
oahuensis *H. St. John in Phytologia,* 63(3): 178 (1987)—Hawaiian Is.
obatae *H. St. John in Phytologia,* 63(3): 178 (1987)—Hawaiian Is.
occidentalis *H. St. John in Phytologia,* 63(3): 178 (1987)—Hawaiian Is.
odorata *H. St. John in Phytologia,* 63(3): 179 (1987)—Hawaiian Is.
olokeleensis *H. St. John in Phytologia,* 63(3): 179 (1987)—Hawaiian Is.
olokuiensis *H. St. John in Phytologia,* 63(3): 179 (1987)—Hawaiian Is.
orientalis *H. St. John in Phytologia,* 63(3): 179 (1987)—Hawaiian Is.
ovata *H. St. John in Phytologia,* 63(3): 179 (1987)—Hawaiian Is.
phytolaccoides *(Sherff) H. St. John in Phytologia,* 63(3): 179 (1987): P. macrophylla var. phytolaccoides.
var. salebrosa *H. St. John in Phytologia,* 63(3): 179 (1987)—Hawaiian Is.
pilosa *H. St. John in Phytologia,* 63(3): 179 (1987)—Hawaiian Is.
pilosicincta *H. St. John in Phytologia,* 63(3): 179 (1987)—Hawaiian Is.
pilosula *H. St. John in Phytologia,* 63(3): 180 (1987)—Hawaiian Is.
pluriflora *H. St. John in Phytologia,* 63(3): 180 (1987)—Hawaiian Is.
plurinodosa *H. St. John in Phytologia,* 63(3): 180 (1987)—Hawaiian Is.
polyantha *H. St. John in Phytologia,* 63(3): 180 (1987)—Hawaiian Is.
pubens *H. St. John in Phytologia,* 63(3): 180 (1987)—Hawaiian Is.
reflexa *H. St. John in Phytologia,* 63(3): 180 (1987)—Hawaiian Is.
remyi *(Sherff) H. St. John in Phytologia,* 63(3): 180 (1987): P. macrophylla var. remyi.
repanda *H. St. John in Phytologia* 63(3): 180 (1987)—Hawaiian Is.
retrorsa *H. St. John in Phytologia,* 63(3): 180 (1987)—Hawaiian Is.
rhuakos *H. St. John in Phytologia,* 63(3): 181 (1987)—Hawaiian Is.
rubescens *H. St. John in Phytologia,* 63(3): 181 (1987)—Hawaiian Is.
rubritincta *H. St. John in Phytologia,* 63(3): 181 (1987)—Hawaiian Is.
secunda *H. St. John in Phytologia,* 63(3): 181 (1987)—Hawaiian Is.
serrata *H. St. John in Phytologia,* 63(3): 181 (1987)—Hawaiian Is.
sexiflora *H. St. John in Phytologia,* 63(3): 181 (1987)—Hawaiian Is.

PHYLLOSTEGIA :-
suaveolens *H. St. John in Phytologia,* 63(3): 181 (1987)—Hawaiian Is.
swezeyi *H. St. John in Phytologia,* 63(3): 181 (1987)—Hawaiian Is.
ternaria *H. St. John in Phytologia,* 63(3): 182 (1987)—Hawaiian Is.
tongaensis *H. St. John in Phytologia,* 63(3): 182 (1987)—Tonga.
triangularis *H. St. John in Phytologia,* 63(3): 182 (1987)—Hawaiian Is.
triquetra *H. St. John in Phytologia,* 63(3): 182 (1987)—Hawaiian Is.
velutina *(Sherff) H. St. John in Phytologia,* 63(3): 182 (1987): P. macrophylla var. velutina.
waianaeensis *H. St. John in Phytologia,* 63(3): 182 (1987)—Hawaiian Is.
warshaueri *H. St. John in Phytologia,* 63(3): 182 (1987)—Hawaiian Is.

PHYLOHYDRAX C. Puff in Pl. Syst. Evol., 154(3–4): 362 (1986). *RUBIACEAE.*
carnosa *(Hochst.) C. Puff in Pl. Syst. Evol.,* 154(3–4): 362 (1986): Diodia carnosa.
madagascariensis *(Roem. & Schult.) C. Puff in Pl. Syst. Evol.,* 154(3–4): 363 (1986): Hydrophylax madagascariensis.

PHYSOTRICHIA (Umbellif.).
muriculata *(Welw. ex Hiern) S. Droop & C.C. Townsend in Kew Bull.,* 42(3): 587 (1987): Peucedanum muriculatum.

PICRIS (Compos.).
wagenitzii *H.W. Lack in Bot. Jahrb.,* 108(2–3): 189 (1987)—Australia (W. Australia).

PILEA (Urticac.).
perexigua *(Griseb.) I.A. Grudzinskaya & P. Herrera in Novosti Sist. Vyssh. Rast.,* 23: 52 (1986): P. microphylla var. perexigua.

PILOSELLA (Compos.).
sandrasica *P. Hartvig & A. Strid in Bot. Jahrb.,* 108(2–3): 309 (1987)—Turkey.

PILOSOCEREUS (Cactac.).
albisummus *P. Braun & E. Esteves Pereira in Kakt. And. Sukk.,* 38(5): 130 (1987)—Brazil.
braunii *E. Esteves Pereira in Kakt. And. Sukk.,* 38(5): 132 (1987)—Brazil.
cristalinensis *P. Braun & E. Esteves Pereira in Kakt. And. Sukk.,* 38(5): 132 (1987); *et in Kakt. And. Sukk.,* 38(7): 166 (1987)—Brazil.
kanukuensis *(Alexander) B.E. Leuenberger in Willdenowia,* 16(2): 506 (9th March 1987): Cephalocereus kanukuensis.
kanukuensis *(Alexander) P.J. Braun in Succulenta,* 66(5): 106 (May 1987), received at Kew 21st May 1987: Cephalocereus kanukuensis.
lindaianus *P. Braun & E. Esteves Pereira in Kakt. And. Sukk.,* 38(5): 132 (1987)—Brazil.

PILOSOCEREUS :–
pusillibaccatus *P.J. Braun & E. Esteves Pereira in Cact. Succ. J. (U.S.A.),* 58(6): 240 (1986)—Brazil.
schoebelii *P.J. Braun in Cact. Succ. J. (U.S.A.),* 59(4): 150 (1987)—Brazil.

PIMELEA (Thymelaeac.).
curviflora
var. gracilis *(R. Br.) S. Threlfall in Fl. S. Australia,* 4: 2147 (1986): P. gracilis.

PIMENTA (Myrtac.).
podocarpoides *(Areces) L.R. Landrum in Fl. Neotrop.,* 45: 99 (1986): Myrtekmania podocarpoides.
pseudocaryophyllus
var. fulvescens *(A.P. de Candolle) L.R. Landrum in Fl. Neotrop.,* 45: 104 (1986): Eugenia fulvescens.
var. hoehnei *(Burret) L.R. Landrum in Fl. Neotrop.,* 45: 104 (1986): Pseudocaryophyllus hoehnei.

PIMPINELLA (Umbellif.).
hadacii *L. Engstrand in Fl. Iran.,* 162: 327 (1987)—Iraq.
khorasanica *L. Engstrand in Fl. Iran.,* 162: 328 (1987)—Iran.
zhongdianensis *C.Y. Wu in Fl. Xizangica,* 3: 475 (1986)—China, Xizang.

PINUS (Pinac.).
californiarum *D.K. Bailey in Notes Roy. Bot. Gard. Edinburgh,* 44(2): 278 (1987)—U.S.A. (California), Mexico.
subsp. fallax *(Little) D.K. Bailey in Notes Roy. Bot. Gard. Edinburgh,* 44(2): 279 (1987): Pinus edulis var. fallax.
cembroides
subsp. lagunae *(Robert-Passini) D.K. Bailey in Phytologia,* 54(2): 98 (1983): P. cembroides var. lagunae.
lagunae *(M.-F. Robert-Passini) M.-F. Passini in Phytologia,* 63(5): 337 (1987): P. cembroides var. lagunae.
mugo
f. applanata *(Willk.) K.I. Christensen in Nordic J. Bot.,* 7(4): 392 (1987): P. pumilio var. applanata.
f. mughoides *(Willk.) K.I. Christensen in Nordic J. Bot.,* 7(4): 393 (1987): P. uncinata f. mughoides.
nubicola *J.P. Perry in J. Arnold Arbor.,* 68(4): 447 (1987)—Guatemala, Mexico, Salvador, Honduras.

PIPER (Piperac.).
gentryi *J.A. Steyermark in Ann. Missouri Bot. Gard.,* 74(1): 89 (1987)—Venezuela.

PIPTATHERUM (Gramin.).
frutescens *G. Heynhold, Nomencl. Bot. Hort.:* 524, 629 (1841), nom. superfl.—Greece.
humile *(Bor) S. Kumar & M.B. Raizada in Indian J. Forest.,* 9(2): 177 (1986): Oryzopsis humilis.

PIRIQUETA (Turnerac.).
mexicana *P.A. Fryxell & S.D. Koch in Aliso,* 11(4): 558 (1987)—Mexico.

PISONIA (Nyctaginac.).
sechellarum *F. Friedmann in Bull. Mus. Nation. Hist. Nat., B, Adansonia,* Ser. 4, 8(4): 384 (1986 publ. 1987)—Seychelles Is.
wagneriana *F.R. Fosberg in Phytologia,* 62(3): 177 (1987)—Hawaiian Is.

PITCAIRNIA (Bromeliac.).
altoatratoensis *G.S. Varadarajan & E. Forero in Syst. Bot.,* 12(3): 412 (1987)—Colombia.
atrorubens
var. pallidobracteata *Gross & W. Rauh in Abh. Akad. Wiss. Lit. Mainz, Math.-Naturwiss. Klasse, Trop. Subtrop. Pflanzenwelt,* 58: 9 (1986), as 'pallido-bracteata'—Panama.
brachysperma
var. snowii *G.S. Varadarajan & E. Forero in Syst. Bot.,* 12(3): 413 (1987)—Colombia.
hammelii *H.E. Luther in J. Bromeliad Soc.,* 37(3): 128 (1987)—Panama.
jaramilloi *G.S. Varadarajan & E. Forero in Syst. Bot.,* 12(3): 410 (1987)—Colombia.
killipiana
var. viridis *G.S. Varadarajan & E. Forero in Syst. Bot.,* 12(3): 413 (1987)—Colombia.
kressii *H.E. Luther in J. Bromeliad Soc.,* 37(5): 214 (1987)—Panama.
lymansmithiana *H.E. Luther in J. Bromeliad Soc.,* 37(5): 212 (1987)—Panama.
matogrossensis *E. Pereira & E.M.C. Leme in Bradea,* 4(34): 269 (1986)—Brazil.
occidentalis
var. alversonii *(L.B. Smith & Read) G.S. Varadarajan & E. Forero in Syst. Bot.,* 12(3): 413 (1987): P. alversonii.
undulatosepala *W. Rauh in Abh. Akad. Wiss. Lit. Mainz, Math.-Naturwiss. Klasse, Trop. Subtrop. Pflanzenwelt,* 58: 16 (1986), as 'undulato-sepala'—Mexico.

PITHECELLOBIUM (Legumin.).
campylacanthus *L. Rico & M. Sousa S. in Ann. Missouri Bot. Gard.,* 73(4): 722 (1986 publ. 1987)—Mexico, Honduras, Nicaragua.

PITYOPSIS (Compos.).
aspera
var. adenolepis *(Fern.) J.C. Semple & F.D. Bowers in Univ. Waterloo Biol. Ser.,* 29: 18 (1985): *Chrysopsis* adenolepis.
graminifolia
var. latifolia *(Fern.) J.C. Semple & F.D. Bowers in Univ. Waterloo Biol. Ser.,* 29: 28 (1985): Chrysopsis graminifolia var. latifolia.
var. tenuifolia *(Torr.) J.C. Semple & F.D. Bowers in Univ. Waterloo Biol. Ser.,* 29: 24 (1985): *Inula graminifolia* β tenuifolia.

PLANTAGO (Plantaginac.).
holosteum
var. littoralis *(Rouy) M. Kerguélen in Lejeunia,* 120: 141 (1987): P. recurvata var. littoralis.
humilis
subsp. insularis *(Gren. & Godron) M. Kerguélen & J. Lambinon in Lejeunia,* 120: 141 (1987): P. subulata var. insularis.

PLATANTHERA (Orchidac.).
lancifolia *(J. Rohlena) A.W. Hill in Index Kewensis,* Supp. 6: 159 (1916–1920 publ. 1926): P. montana var. lancifolia.

PLATYSTELE (Orchidac.).
rauhii *C.A. Luer in Orchidee,* 38(2): 61 (1987)—Peru.

PLECTRANTHUS (Labiat.).
acaulis *R.K. Brummitt & J.H. Seyani in Kew Bull.,* 42(3): 695 (1987)—Malawi, Zambia.
salubenii *R.K. Brummitt & J.H. Seyani in Kew Bull.,* 42(3): 690 (1987)—Malawi, Zambia, Tanzania.
viphyensis *R.K. Brummitt & J.H. Seyani in Kew Bull.,* 42(3): 692 (1987)—Malawi.
zebrarum *R.K. Brummitt & J.H. Seyani in Kew Bull.,* 42(3): 694 (1987)—Malawi, Zambia.

PLEIOTAXIS (Compos.).
bampsiana *S. Lisowski in Bull. Jard. Bot. Nation. Belg.,* 57(1–2): 247 (1987)—Zaïre.
duvigneaudii *S. Lisowski in Bull. Jard. Bot. Nation. Belg.,* 57(1–2): 249 (1987)—Zaïre.
jeffreyana *S. Lisowski in Bull. Jard. Bot. Nation. Belg.,* 57(1–2): 250 (1987)—Zaïre.
lawalreana *S. Lisowski in Bull. Jard. Bot. Nation. Belg.,* 57(1–2): 251 (1987)—Zaïre.
lejolyana *S. Lisowski in Bull. Jard. Bot. Nation. Belg.,* 57(1–2): 252 (1987)—Zaïre.
perfoliata *S. Lisowski in Bull. Jard. Bot. Nation. Belg.,* 57(1–2): 253 (1987)—Zaïre.
petitiana *S. Lisowski in Bull. Jard. Bot. Nation. Belg.,* 57(1–2): 254 (1987)—Zaïre.
robynsiana *S. Lisowski in Bull. Jard. Bot. Nation. Belg.,* 57(1–2): 255 (1987)—Zaïre.
upembensis *S. Lisowski in Bull. Jard. Bot. Nation. Belg.,* 57(1–2): 256 (1987)—Zaïre.

PLEUROPTEROPYRUM (Polygonac.).
hagei *(Royle ex Bab.) A.H. Munshi & G.N. Javeid, Syst. Stud. Polygonac. Kashmir Himalaya (J. Econ. Taxon. Bot. Addit. Ser. 2):* 79 (1986): Polygonum hagei.
polystachyum *(Wallich ex Meissn.) A.H. Munshi & G.N. Javeid, Syst. Stud. Polygonac. Kashmir Himalaya (J. Econ. Taxon. Bot. Addit. Ser. 2):* 78 (1986): Polygonum polystachyum.
rumicifolium *(Royle ex Bab.) A.H. Munshi & G.N. Javeid, Syst. Stud. Polygonac. Kashmir Himalaya (J. Econ. Taxon. Bot. Addit. Ser. 2):* 79 (1986): Polygonum rumicifolium.

PLEUROPTEROPYRUM :–
tortuosum *(Don) A.H. Munshi & G.N. Javeid, Syst. Stud. Polygonac. Kashmir Himalaya (J. Econ. Taxon. Bot. Addit. Ser. 2):* 80 (1986): Polygonum tortuosum.

PLEUROSPERMUM (Umbellif.).
longicarpum *Shan & Z.H. Pan in Fl. Xizangica,* 3: 426 (1986), nom. nov.: Pterocyclus wolffianus *Fedde ex Wolff.*

PLEUROTHALLIS (Orchidac.).
sect. Abortivae *C.A. Luer, Ic. Pleurothallid. III - Syst. Pleurothallis (Monogr. Syst. Bot. Missouri Bot. Gard. 20):* 57 (1986).
subgen. Acianthera *(Scheidw.) C.A. Luer, Ic. Pleurothallid. III - Syst. Pleurothallis (Monogr. Syst. Bot. Missouri Bot. Gard. 20):* 12 (1986): gen. Acianthera.
subsect. Acroniae *C.A. Luer, Ic. Pleurothallid. III - Syst. Pleurothallis (Monogr. Syst. Bot. Missouri Bot. Gard. 20):* 61 (1986), nom. nov.: gen. Acronia *Presl.*
subgen. Aenigma *C.A. Luer, Ic. Pleurothallid. III - Syst. Pleurothallis (Monogr. Syst. Bot. Missouri Bot. Gard. 20):* 26 (1986).
sect. Aenigmata *C.A. Luer, Ic. Pleurothallid. III - Syst. Pleurothallis (Monogr. Syst. Bot. Missouri Bot. Gard. 20):* 26 (1986).
subgen. Ancipitia *C.A. Luer, Ic. Pleurothallid. III - Syst. Pleurothallis (Monogr. Syst. Bot. Missouri Bot. Gard. 20):* 29 (1986).
subgen. Andreettaea *(Luer) C.A. Luer, Ic. Pleurothallid. III - Syst. Pleurothallis (Monogr. Syst. Bot. Missouri Bot. Gard. 20):* 31 (1986): gen. Andreettaea.
subgen. Apoda-Prorepentia *C.A. Luer, Ic. Pleurothallid. III - Syst. Pleurothallis (Monogr. Syst. Bot. Missouri Bot. Gard. 20):* 31 (1986), nom. nov.: P. sect. Apodae-Prorepentes *Lindl.*
subsect. Apodae-Caespitosae *(Lindl.) C.A. Luer, Ic. Pleurothallid. III - Syst. Pleurothallis (Monogr. Syst. Bot. Missouri Bot. Gard. 20):* 84 (1986): P. sect. Apodae-Caespitosae.
subgen. Arthrosia *C.A. Luer, Ic. Pleurothallid. III - Syst. Pleurothallis (Monogr. Syst. Bot. Missouri Bot. Gard. 20):* 34 (1986).
subgen. Crocodeilanthe *(Rchb. f. & Warsc.) C.A. Luer, Ic. Pleurothallid. III - Syst. Pleurothallis (Monogr. Syst. Bot. Missouri Bot. Gard. 20):* 34 (1986): gen. Crocodeilanthe.
sect. Cryptophoranthae *C.A. Luer, Ic. Pleurothallid. III - Syst. Pleurothallis (Monogr. Syst. Bot. Missouri Bot. Gard. 20):* 16 (1986), nom. nov.: gen Cryptophoranthus *Barb. Rodr.*
sect. Cucumeres *C.A. Luer, Ic. Pleurothallid. III - Syst. Pleurothallis (Monogr. Syst. Bot. Missouri Bot. Gard. 20):* 81 (1986).

PLEUROTHALLIS :-

subgen. Dracontia *C.A. Luer, Ic. Pleurothallid. III - Syst. Pleurothallis (Monogr. Syst. Bot. Missouri Bot. Gard. 20):* 38 (1986).

subgen. Dresslera *C.A. Luer, Ic. Pleurothallid. III - Syst. Pleurothallis (Monogr. Syst. Bot. Missouri Bot. Gard. 20):* 38 (1986).

subgen. Elongatia *C.A. Luer, Ic. Pleurothallid. III - Syst. Pleurothallis (Monogr. Syst. Bot. Missouri Bot. Gard. 20):* 41 (1986), nom. nov.: P. sect. Elongatae *Lindl.*

subgen. Empusella *C.A. Luer, Ic. Pleurothallid. III - Syst. Pleurothallis (Monogr. Syst. Bot. Missouri Bot. Gard. 20):* 41 (1986).

subgen. Kraenzlinella *(Kuntze) C.A. Luer, Ic. Pleurothallid. III - Syst. Pleurothallis (Monogr. Syst. Bot. Missouri Bot. Gard. 20):* 44 (1986): gen. Kraenzlinella.

subsect. Longicaulae *(Barb. Rodr.) C.A. Luer, Ic. Pleurothallid. III - Syst. Pleurothallis (Monogr. Syst. Bot. Missouri Bot. Gard. 20):* 86 (1986): Lepanthes sect. Longicaulae.

subsect. Macrophyllae-Racemosae *(Lindl.) C.A. Luer, Ic. Pleurothallid. III - Syst. Pleurothallis (Monogr. Syst. Bot. Missouri Bot. Gard. 20):* 61 (1986): P. sect. Macrophyllae-Racemosae.

subgen. Masdevalliantha *C.A. Luer, Ic. Pleurothallid. III - Syst. Pleurothallis (Monogr. Syst. Bot. Missouri Bot. Gard. 20):* 44 (1986).

sect. Mentosae *C.A. Luer, Ic. Pleurothallid. III - Syst. Pleurothallis (Monogr. Syst. Bot. Missouri Bot. Gard. 20):* 86 (1986).

subgen. Mirabilia *C.A. Luer, Ic. Pleurothallid. III - Syst. Pleurothallis (Monogr. Syst. Bot. Missouri Bot. Gard. 20):* 47 (1986).

subgen. Mirandia *C.A. Luer, Ic. Pleurothallid. III - Syst. Pleurothallis (Monogr. Syst. Bot. Missouri Bot. Gard. 20):* 47 (1986).

sect. Muscariae *C.A. Luer, Ic. Pleurothallid. III - Syst. Pleurothallis (Monogr. Syst. Bot. Missouri Bot. Gard. 20):* 89 (1986).

subgen. Mystax *C.A. Luer, Ic. Pleurothallid. III - Syst. Pleurothallis (Monogr. Syst. Bot. Missouri Bot. Gard. 20):* 50 (1986).

subsect. Pectinatae *C.A. Luer, Ic. Pleurothallid. III - Syst. Pleurothallis (Monogr. Syst. Bot. Missouri Bot. Gard. 20):* 20 (1986).

sect. Phloeophilae *C.A. Luer, Ic. Pleurothallid. III - Syst. Pleurothallis (Monogr. Syst. Bot. Missouri Bot. Gard. 20):* 17 (1986), nom. nov.: gen. Phloeophila *Hoehne & Schltr.*

subgen. Physosiphon *(Lindl.) C.A. Luer, Ic. Pleurothallid. III - Syst. Pleurothallis (Monogr. Syst. Bot. Missouri Bot. Gard. 20):* 50 (1986): gen. Physosiphon.

subgen. Physothallis *(Garay) C.A. Luer, Ic. Pleurothallid. III - Syst. Pleurothallis (Monogr. Syst. Bot. Missouri Bot. Gard. 20):* 53 (1986): gen. Physothallis.

subgen. Pleurobotryum *(Barb. Rodr.) C.A. Luer, Ic. Pleurothallid. III - Syst. Pleurothallis (Monogr. Syst. Bot. Missouri Bot. Gard. 20):* 53 (1986): gen. Pleurobotryum.

PLEUROTHALLIS :-

subgen. Pseudoctomeria *(Krzl.) C.A. Luer, Ic. Pleurothallid. III - Syst. Pleurothallis (Monogr. Syst. Bot. Missouri Bot. Gard. 20):* 67 (1986): gen. Pseudoctomeria.

subgen. Restrepioidia *C.A. Luer, Ic. Pleurothallid. III - Syst. Pleurothallis (Monogr. Syst. Bot. Missouri Bot. Gard. 20):* 70 (1986), nom. nov.: P. sect. Restrepioideae *Lindl.*

subgen. Rhynchopera *(Kl.) C.A. Luer, Ic. Pleurothallid. III - Syst. Pleurothallis (Monogr. Syst. Bot. Missouri Bot. Gard. 20):* 70 (1986): gen. Rhynchopera.

subgen. Rubellia *C.A. Luer, Ic. Pleurothallid. III - Syst. Pleurothallis (Monogr. Syst. Bot. Missouri Bot. Gard. 20):* 73 (1986).

subgen. Sarracenella *(Luer) C.A. Luer, Ic. Pleurothallid. III - Syst. Pleurothallis (Monogr. Syst. Bot. Missouri Bot. Gard. 20):* 73 (1986): gen. Sarracenella.

subgen. Scopula *C.A. Luer, Ic. Pleurothallid. III - Syst. Pleurothallis (Monogr. Syst. Bot. Missouri Bot. Gard. 20):* 76 (1986), nom. nov.: gen. Colombiana *Ospina.*

subsect. Sicariae *(Lindl.) C.A. Luer, Ic. Pleurothallid. III - Syst. Pleurothallis (Monogr. Syst. Bot. Missouri Bot. Gard. 20):* 20 (1986): P. sect. Sicariae.

subgen. Talpinaria *(Karst.) C.A. Luer, Ic. Pleurothallid. III - Syst. Pleurothallis (Monogr. Syst. Bot. Missouri Bot. Gard. 20):* 96 (1986): gen. Talpinaria.

sect. Tomentosae *C.A. Luer, Ic. Pleurothallid. III - Syst. Pleurothallis (Monogr. Syst. Bot. Missouri Bot. Gard. 20):* 22 (1986).

sect. Tribuloides *C.A. Luer, Ic. Pleurothallid. III - Syst. Pleurothallis (Monogr. Syst. Bot. Missouri Bot. Gard. 20):* 91 (1986).

sect. Tricarinatae *C.A. Luer, Ic. Pleurothallid. III - Syst. Pleurothallis (Monogr. Syst. Bot. Missouri Bot. Gard. 20):* 22 (1986), without latin descr.

sect. Tripteranthae *C.A. Luer, Ic. Pleurothallid. III - Syst. Pleurothallis (Monogr. Syst. Bot. Missouri Bot. Gard. 20):* 94 (1986).

sect. Truncatae *C.A. Luer, Ic. Pleurothallid. III - Syst. Pleurothallis (Monogr. Syst. Bot. Missouri Bot. Gard. 20):* 67 (1986).

sect. Unciferae *C.A. Luer, Ic. Pleurothallid. III - Syst. Pleurothallis (Monogr. Syst. Bot. Missouri Bot. Gard. 20):* 94 (1986).

sect. Vestigipetalae *C.A. Luer, Ic. Pleurothallid. III - Syst. Pleurothallis (Monogr. Syst. Bot. Missouri Bot. Gard. 20):* 26 (1986).

subgen. Xenion *C.A. Luer, Ic. Pleurothallid. III - Syst. Pleurothallis (Monogr. Syst. Bot. Missouri Bot. Gard. 20):* 96 (1986).

asaroides *(Krzl.) C.A. Luer, Ic. Pleurothallid. III - Syst. Pleurothallis (Monogr. Syst. Bot. Missouri Bot. Gard. 20):* 73 (1986): Physosiphon asaroides.

carrenoi *G. Carnevali F.C. & I. Ramírez de Carnevali in Ernstia,* 44: 18 (1987)—Venezuela.

PLEUROTHALLIS :–
cylindrica *(Luer) C.A. Luer, Ic. Pleurothallid. III - Syst. Pleurothallis (Monogr. Syst. Bot. Missouri Bot. Gard. 20):* 53 (1986): Physothallis cylindrica.
minimifolia *C.A. Luer, Ic. Pleurothallid. III - Syst. Pleurothallis (Monogr. Syst. Bot. Missouri Bot. Gard. 20):* 16 (1986), nom. nov.: Cryptophoranthus minimus *Cogn.*
muricata *T. Hashimoto in Bull. Nation. Sci. Mus., B (Tokyo),* 13(1): 35 (1987)—Peru.
neobradei *C.A. Luer, Ic. Pleurothallid. III - Syst. Pleurothallis (Monogr. Syst. Bot. Missouri Bot. Gard. 20):* 17 (1986), nom. nov.: Physosiphon bradei *Schltr.*
neoharlingii *C.A. Luer, Ic. Pleurothallid. III - Syst. Pleurothallis (Monogr. Syst. Bot. Missouri Bot. Gard. 20):* 53 (1986), nom. nov.: Physothallis harlingii *Garay.*
neojordanensis *C.A. Luer, Ic. Pleurothallid. III - Syst. Pleurothallis (Monogr. Syst. Bot. Missouri Bot. Gard. 20):* 16 (1986), nom. nov.: Cryptophoranthus jordanensis *Brade.*
neojuergensii *C.A. Luer, Ic. Pleurothallid. III - Syst. Pleurothallis (Monogr. Syst. Bot. Missouri Bot. Gard. 20):* 16 (1986), nom. nov.: Cryptophoranthus juergensii *Schltr.*
ocellus *(Luer) C.A. Luer, Ic. Pleurothallid. III - Syst. Pleurothallis (Monogr. Syst. Bot. Missouri Bot. Gard. 20):* 31 (1986): Andreettaea ocellus.
paulensis *(Hoehne & Schltr.) C.A. Luer, Ic. Pleurothallid. III - Syst. Pleurothallis (Monogr. Syst. Bot. Missouri Bot. Gard. 20):* 17 (1986): Phloeophila paulensis.
punctatiflora *C.A. Luer, Ic. Pleurothallid. III - Syst. Pleurothallis (Monogr. Syst. Bot. Missouri Bot. Gard. 20):* 16 (1986), nom. nov.: Cryptophoranthus punctatus *Barb. Rodr.*
sarracenia *C.A. Luer, Ic. Pleurothallid. III - Syst. Pleurothallis (Monogr. Syst. Bot. Missouri Bot. Gard. 20):* 73 (1986), nom. nov.: Physosiphon pubescens *Barb. Rodr.*
sieberi *C.A. Luer, Ic. Pleurothallid. III - Syst. Pleurothallis (Monogr. Syst. Bot. Missouri Bot. Gard. 20):* 34 (1986), nom. nov.: Sprecklinia floribunda *Lindl.*
similis *(Schltr.) C.A. Luer, Ic. Pleurothallid. III - Syst. Pleurothallis (Monogr. Syst. Bot. Missouri Bot. Gard. 20):* 17 (1986): Cryptophoranthus similis.
spicata *(Dutra) C.A. Luer, Ic. Pleurothallid. III - Syst. Pleurothallis (Monogr. Syst. Bot. Missouri Bot. Gard. 20):* 17 (1986): Cryptophoranthus spicatus.

PLOWMANIA A.T. Hunziker & R. Subils in Kurtziana, 18: 127 (1986). *SOLANACEAE.*
nyctaginoides *(Standley) A.T. Hunziker & R. Subils in Kurtziana,* 18: 127 (1986): Brunfelsia nyctaginoides.

PLUCHEA (Compos.).
rufescens *(DC.) A.J. Scott in Kew Bull.,* 42(4): 854 (1987): Monarrhenus rufescens.

POA (Gramin.).
acicularifolia
subsp. ophitalis *E. Edgar in New Zealand J. Bot.,* 24(3): 444 (1986)—New Zealand.
alpina
subsp. atlantica *(Trabut) A.M. Romo in Treb. Inst. Bot. Barcelona,* 11: 40 (1987): P. alpina var. atlantica.
anceps
subsp. polyphylla *(Hackel) E. Edgar in New Zealand J. Bot.,* 24(3): 452 (1986): P. polyphylla.
angustifolia
var. laxispicula *D.F. Cui in Acta Bot. Bor.-Occid. Sin.,* 7(2): 102 (1987)—Xinjiang.
arjinsanensis *D.F. Cui in Acta Bot. Bor.-Occid. Sin.,* 5(2): 87 (1987)—Xinjiang.
aucklandica
subsp. campbellensis *(Petrie) E. Edgar in New Zealand J. Bot.,* 24(3): 467 (1986): P. campbellensis.
subsp. rakiura *E. Edgar in New Zealand J. Bot.,* 24(3): 467 (1986)—New Zealand.
biflora ***Ferat** ex C.C. Gillet & J.-H. Magne, Nouv. Fl. Française,* éd. 3, 538 (1873), non Retz (1789)—Spain.
celsa *E. Edgar in New Zealand J. Bot.,* 24(3): 463 (1986)—New Zealand.
cita *E. Edgar in New Zealand J. Bot.,* 24(3): 446 (1986), nom. nov.: P. caespitosa *Sprengel.*
eragrostis *(L.) Brotero, Fl. Lusit.,* 1: 103 (1804), non L. (1753): Briza eragrostis.
festucoides *N.R. Cui in Acta Bot. Bor.-Occid. Sin.,* 7(2): 95 (1987)—Xinjiang.
subsp. kunluensis *N.R. Cui in Acta Bot. Bor.-Occid. Sin.,* 7(2): 97 (1987)—Xinjiang.
florida *N.R. Cui in Acta Bot. Bor.-Occid. Sin.,* 7(2): 91 (1987)—Xinjiang.
glauca
f. pallida *(Lange) J. Cayouette in Nat. Canad.,* 113(4): 332 (1986): P. glauca γ pallida.
hesperia *E. Edgar in New Zealand J. Bot.,* 24(3): 442 (1986)—New Zealand.
intrusa *E. Edgar in New Zealand J. Bot.,* 24(3): 463 (1986)—New Zealand.
lejeunii *(Dumort.) B.D. Jackson in Index Kewensis,* 2: 573 (1753–1885 publ. 1894): P. pratensis * (subsp.) lejeunii.
macroanthera *D.F. Cui in Acta Bot. Bor.-Occid. Sin.,* 7(2): 97 (1987)—Xinjiang.
subsp. meilitzyka *D.F. Cui in Acta Bot. Bor.-Occid. Sin.,* 7(2): 100 (1987)—Xinjiang.
maia *E. Edgar in New Zealand J. Bot.,* 24(3): 470 (1986)—New Zealand.
major *D.F. Cui in Acta Bot. Bor.-Occid. Sin.,* 7(2): 83 (1987)—Xinjiang.
membranigluma *D.F. Cui in Acta Bot. Bor.-Occid. Sin.,* 7(2): 89 (1987)—Xinjiang.
nemoralis
var. parca *N.R. Cui in Acta Bot. Bor.-Occid. Sin.,* 7(2): 103 (1987)—Xinjiang.
parvissima *Gui ex D.F. Cui in Acta Bot. Bor.-Occid. Sin.,* 7(2): 85 (1987)—Xinjiang.

POA :–
perligularis *H. Scholz in Willdenowia,* 16(2): 404 (1987), nom. nov.: P. bulbosa subsp. perligulata *H. Scholz.*
poiphagorum
var. hunczilapensis *Keng ex D.F. Cui in Acta Bot. Bor.-Occid. Sin.,* 7(2): 100 (1987)—Xinjiang.
scabristemmed *N.R. Cui in Acta Bot. Bor.-Occid. Sin.,* 7(2): 93 (1987)—Xinjiang.
senex *E. Edgar in New Zealand J. Bot.,* 24(3): 477 (1986)—New Zealand.
sublimis *E. Edgar in New Zealand J. Bot.,* 24(3): 465 (1986)—New Zealand.
subvestita *(Hackel) E. Edgar in New Zealand J. Bot.,* 24(3): 436 (1986): P. novaezelandiae var. subvestita.
sudicola *E. Edgar in New Zealand J. Bot.,* 24(3): 437 (1986)—New Zealand.
tonsa *E. Edgar in New Zealand J. Bot.,* 24(3): 477 (1986)—New Zealand.

POCHOTA (Bombacac.).
quinata *(Jacq.) W.D. Stevens in Taxon* 36(2): 463 (1987): Bombax quinatum.
sessilis *(Benth.) W.D. Stevens in Taxon* 36(2): 463 (1987): Pachira sessilis.

PODOCARPIUM (Legumin.).
ser. Tuberifera *C. Chen & X.J. Cui in Acta Bot. Yunnanica,* 9(3): 307 (1987).
fallax *(Schindl.) C. Chen & X.J. Cui in Acta Bot. Yunnanica,* 9(3): 306 (1987): Desmodium fallax.
var. densum *C. Chen & X.J. Cui in Acta Bot. Yunnanica,* 9(3): 306 (1987)—China.
menglaense *C. Chen & X.J. Cui in Acta Bot. Yunnanica,* 9(3): 305 (1987)—China.

POECILOSTACHYS (Gramin.).
oplismenoides *(Hack.) W.D. Clayton in Kew Bull.,* 42(2): 403 (1987): Panicum oplismenoides.

POIKILACANTHUS (Acanthac.).
glandulosus *(Nees) L. Ariza Espinar in Kurtziana,* 17: 159 (1984): Orthotactus glandulosus.
harlingii *D.C. Wasshausen in Op. Bot.,* 92: 275 (1987)—Ecuador.

POIRETIA (Legumin.).
bahiana *C. Müller in Rev. Brasil. Bot.,* 9(1): 23 (1986 publ. 1987)—Brazil.
crenata *C. Müller in Rev. Brasil. Bot.,* 9(1): 23 (1986 publ. 1987)—Brazil.
elegans *C. Müller in Rev. Brasil. Bot.,* 9(1): 26 (1986 publ. 1987)—Brazil.
marginata *C. Müller in Rev. Brasil. Bot.,* 9(1): 26 (1986 publ. 1987)—Brazil.
mattogrossensis *C. Müller in Rev. Brasil. Bot.,* 9(1): 26 (1986 publ. 1987)—Brazil.

POLYCARPAEA (Caryophyllac.).
pulvinata *M.G. Gilbert in Kew Bull.,* 42(3): 703 (1987)—Somali Republic.

POLYCARPAEA :–
rupicola *J.-P. Lebrun & A.L. Stork in Bull. Mus. Nation. Hist. Nat., B, Adansonia,* Ser. 4, 9(1): 39 (1987)—Cameroun.
ventiversa *M.G. Gilbert in Kew Bull.,* 42(3): 704 (1987)—Somali Republic.

POLYCARPON (Caryophyllac.).
tetraphyllum
f. depauperatum *(Merino) F. Amich García in An. Jard. Bot. Madrid,* 43(2): 467 (1986 publ. 1987): P. tetraphyllum var. depauperatum.
f. ovalifolium *(Merino) F. Amich García in An. Jard. Bot. Madrid,* 43(2): 467 (1986 publ. 1987): P. tetraphyllum var. ovalifolium.

POLYGALA (Polygalac.).
comosa
var. hybrida *(DC.) D.A. Petelin in Sosud. Rast. Sovet. Dal'nego Vostoka,* 2: 192 (1987): P. hybrida.
lactiflora *J. Paiva & R.K. Brummitt in An. Jard. Bot. Madrid,* 43(1): 24 (1986)—Malawi.
nicaeensis
subsp. pannonica *(Pócs) Melzer ex T. Wraber & M. Palma in Biol. Vestn.,* 30(2): 119 (1982): P. nicaeensis var. pannonica.

POLYGONATUM (Convallariac.).
adnatum *S.Y. Liang in Acta Phytotax. Sin.,* 25(1): 65 (1987)—China.
longipedunculatum *S.Y. Liang in Acta Phytotax. Sin.,* 25(1): 64 (1987)—China.
vietnamicum *L.I. Abramova in Novosti Sist. Vÿssh. Rast.,* 23: 47 (1986)—Vietnam.

POLYGONUM (Polygonac.).
aviculare
var. heterophyllum *(Lindm.) A.H. Munshi & G.N. Javeid, Syst. Stud. Polygonac. Kashmir Himalaya (J. Econ. Taxon. Bot. Addit. Ser. 2):* 55 (1986): P. heterophyllum.
balansae
subsp. battandieri *(Maire & Sennen) W. Greuter in Willdenowia,* 16(2): 451 (1987): P. aviculare subsp. battandieri.
subsp. induratum *(Asch. & Barbey) W. Greuter in Willdenowia,* 16(2): 451 (1987): P. induratum.
subsp. rhizoxylon *(Pau & Font Quer) W. Greuter in Willdenowia,* 16(2): 451 (1987): P. rhizoxylon.
baltistanicum *M.N. Chaudhri in Pakistan Syst.,* 3(1): 30 (1987)—Pakistan.
bellardii
var. densa *J. Pastor & A. Aparicio in Lagascalia,* 14(1): 133 (1986)—Spain.
ebracteatum *A.H. Munshi & G.N. Javeid, Syst. Stud. Polygonac. Kashmir Himalaya (J. Econ. Taxon. Bot. Addit. Ser. 2):* 56 (1986)—India.

POLYGONUM :–
ganderbalense *A.H. Munshi & G.N. Javeid, Syst. Stud. Polygonac. Kashmir Himalaya (J. Econ. Taxon. Bot. Addit. Ser. 2)*: 55 (1986), as 'ganderbalensis'—India.
hassanabdalicum *M.N. Chaudhri in Pakistan Syst.*, 3(1): 28 (1987)—Pakistan.
mirajabii *M.N. Chaudhri in Pakistan Syst.*, 3(1): 28 (1987)—Pakistan.
senegalense
subsp. albotomentosum *(R.A. Grah.) G. Germishuizen in Bothalia,* 16(2): 233 (1986): P. senegalense f. albotomentosum.
viviparum
var. ramiflora *L. Villar in An. Jard. Bot. Madrid,* 44(1): 185 (1987)—Spain.

POLYSTACHYA (Orchidac.).
fabriana *D. Geerinck in Bull. Jard. Bot. Nation. Belg.*, 56(3–4): 492 (1986)—Rwanda.
stewartiana *D. Geerinck in Bull. Jard. Bot. Nation. Belg.*, 56(3–4): 493 (1986)—Zaïre.

PONERORCHIS (Orchidac.).
renzii *S. Deva & H.B. Naithani, Orchid Fl. N.W. Himalaya:* 199 (1986)—India.

POPULUS (Salicac.).
cana *T.Y. Sun in J. Nanjing Forest. Univ.*, 1986(4): 111 (1986)—China.
intramongolica *T.Y. Sun & E.W. Ma in J. Nanjing Forest. Univ.*, 1986(4): 109 (1986)—China.
platyphylla
var. flaviflora *T.Y. Sun in J. Nanjing Forest. Univ.*, 1986(4): 114 (1986)—China.
var. glauca *T.Y. Sun & Z.F. Chen in J. Nanjing Forest. Univ.*, 1986(4): 115 (1986)—China.
pseudosimonii
var. patula *T.Y. Sun in J. Nanjing Forest. Univ.*, 1986(4): 113 (1986)—China.
qiongdaoensis *T. Hong & P. Luo in Bull. Bot. Res. North-East. Forest. Inst.*, 7(3): 67 (1987)—Hainan.
simonii
var. breviamenta *T.Y. Sun in J. Nanjing Forest. Univ.*, 1986(4): 116 (1986)—China.
var. griseoalba *T.Y. Sun in J. Nanjing Forest. Univ.*, 1986(4): 116 (1986)—China.
var. ovata *T.Y. Sun in J. Nanjing Forest. Univ.*, 1986(4): 115 (1986)—China.

POTAMOGETON (Potamogetonac.).
×haynesii *C.B. Hellquist & G.E. Crow in Brittonia,* 38(4): 415 (1986)—U.S.A. (Michigan, Minnesota, Vermont), Canada (Ontario). (P. strictifolius × zosteriformis).
×kyushuensis *Y. Kadono & G. Wiegleb in J. Jap. Bot.*, 62(3): 76 (1987)—Japan. (P. maackianus × oxyphyllus).
×lanceolatifolius *(Tiselius) C.D. Preston in Watsonia,* 16(4): 437 (1987): P. gramineus f. lanceolatifolius.

POTAMOGETON :–
octandrus
var. asiaticus *(A. Benn.) N.N. Tsvelev in Sosud. Rast. Sovet. Dal'nego Vostoka,* 2: 323 (1987): P. asiaticus.
var. limosellifolius *(Maxim. ex Korsh.) N.N. Tsvelev in Sosud. Rast. Sovet. Dal'nego Vostoka,* 2: 323 (1987): P. limosellifolius.
×yamagataensis *Y. Kadono & G. Wiegleb in J. Jap. Bot.*, 62(3): 73 (1987)—Japan.

POTENTILLA (Rosac.).
ser. Asperrimae *V.A. Chevtaeva in Izv. Akad. Nauk Tadzh. SSR, Biol. Nauk,* 1(90): 16 (1983).
ser. Bungeanae *V.A. Chevtaeva in Izv. Akad. Nauk Tadzh. SSR, Biol. Nauk,* 1(90): 17 (1983).
subgen. Dasiphora *(Rafin.) G. Panigrahi & B.K. Dikshit in Bull. Bot. Surv. India,* 27(1–4): 179 (1985 publ. 1987): gen. Dasiphora.
ser. Desertae *V.A. Chevtaeva in Izv. Akad. Nauk Tadzh. SSR, Biol. Nauk,* 3(84): 22 (1981).
sect. Desertae *(Czevt.) V.A. Chevtaeva in Izv. Akad. Nauk Tadzh. SSR, Biol. Nauk,* 1(90): 17 (1983), without basionym ref.: Basionym not stated.
sect. Duchesneae *(J.E. Smith) B.K. Dikshit & G. Panigrahi in Bull. Bot. Surv. India,* 27(1–4): 181 (1985 publ. 1987): gen. Duchesnea.
ser. Heterosepalae *V.A. Chevtaeva in Izv. Akad. Nauk Tadzh. SSR, Biol. Nauk,* 1(90): 16 (1983).
ser. Intermediae *V.A. Chevtaeva in Izv. Akad. Nauk Tadzh. SSR, Biol. Nauk,* 1(90): 16 (1983).
sect. Lipskyanae *V.A. Chevtaeva in Izv. Akad. Nauk Tadzh. SSR, Biol. Nauk,* 3(84): 23 (1981).
ser. Nevadensis *V.A. Chevtaeva in Izv. Akad. Nauk Tadzh. SSR, Biol. Nauk,* 1(90): 18 (1983).
ser. Norvegicae *V.A. Chevtaeva in Izv. Akad. Nauk Tadzh. SSR, Biol. Nauk,* 1(90): 16 (1983).
ser. Pentandrae *V.A. Chevtaeva in Izv. Akad. Nauk Tadzh. SSR, Biol. Nauk,* 1(90): 16 (1983).
ser. Persicae *V.A. Chevtaeva in Izv. Akad. Nauk Tadzh. SSR, Biol. Nauk,* 1(90): 18 (1983).
ser. Rivales *V.A. Chevtaeva in Izv. Akad. Nauk Tadzh. SSR, Biol. Nauk,* 1(90): 15 (1983).
agrimonioides
var. capitata *J. Soják in Preslia,* 59(4): 302 (1987)—U.S.S.R. (Middle Asia).
var. intercedens *J. Soják in Preslia,* 59(4): 302 (1987)—U.S.S.R. (Middle Asia).
var. kobdoensis *J. Soják in Preslia,* 59(4): 302 (1987)—Mongolia.
var. malacotricha *(Juz.) J. Soják in Preslia,* 59(4): 302 (1987): P. malacotricha.

POTENTILLA :-
var. micans *J. Soják in Preslia,* 59(4): 302 (1987)—U.S.S.R. (Siberia).
var. strigosella *J. Soják in Preslia,* 59(4): 302 (1987)—U.S.S.R. (Caucasus).
angelliae *N.H. Holmgren in Brittonia,* 39(3): 342 (1987)—U.S.A. (Utah).
anglica
subsp. nesogenes *(Briq.) Gamisans ex M. Kerguélen in Lejeunia,* 120: 145 (1987): P. procumbens subsp. nesogenes.
conferta
var. laxiuscula *J. Soják in Preslia,* 59(4): 303 (1987)—U.S.S.R. (Middle Asia).
cottamii *N.H. Holmgren in Brittonia,* 39(3): 340 (1987)—U.S.A. (Nevada, Utah).
drosocalyx *J. Soják in Bot. Jahrb.,* 109(1): 33 (1987)—China, Xizang.
ferganensis *J. Soják in Bot. Jahrb.,* 109(1): 35 (1987)—U.S.S.R. (Middle Asia).
ghazniensis *J. Soják in Bot. Jahrb.,* 109(1): 42 (1987)—Afghanistan.
kansuensis *J. Soják in Bot. Jahrb.,* 109(1): 34 (1987)—China.
macropoda *J. Soják in Bot. Jahrb.,* 109(1): 43 (1987)—U.S.S.R. (Middle Asia).
nanshanica *J. Soják in Bot. Jahrb.,* 109(1): 28 (1987)—China.
omissa *J. Soják in Bot. Jahrb.,* 109(1): 39 (1987)—U.S.S.R. (Siberia, Soviet Far East, Middle Asia).
pamiroalaica
var. quasisericea *J. Soják in Bot. Jahrb.,* 109(1): 32 (1987)—India.
panigrahiana *B.K. Dikshit in Bull. Bot. Surv. India,* 27(1–4): 107 (1985 publ. 1987)—India, Sikkim, Nepal, Bhutan, Xizang.
penniphylla *J. Soják in Bot. Jahrb.,* 109(1): 38 (1987)—U.S.S.R. (Middle Asia).
pensylvanica
var. brevipila *J. Soják in Preslia,* 59(4): 301 (1987)—Canada (Yukon).
pulvinaris
subsp. argenteus *P. Hartvig & A. Strid in Bot. Jahrb.,* 108(2–3): 335 (1987)—Turkey.
supina
subsp. arabica *J. Soják in Preslia,* 59(3): 272 (1987)—Iraq.
subsp. costata *J. Soják in Preslia,* 59(3): 272 (1987)—U.S.S.R. (Siberia).
tschimganica *J. Soják in Bot. Jahrb.,* 109(1): 45 (1987)—U.S.S.R. (Middle Asia).
turkestanica *J. Soják in Bot. Jahrb.,* 109(1): 26 (1987)—Xinjiang, Mongolia.
vilijuga *J. Soják in Bot. Jahrb.,* 109(1): 30 (1987)—India.
williamsii *J. Soják in Bot. Jahrb.,* 109(1): 25 (1987)—Nepal.

POUZOLZIA (Urticac.).
peteri *I. Friis in Kew Bull.,* 42(3): 666 (1987), nom. nov.: P. cordata *A. Peter.*

PRANGOS (Umbellif.).
lachnantha *(Korov.) M.G. Pimenov & E.V. Klyuĭkov in Byull. Mosk. Obshch. Ispўt. Prir., Biol.,* 91(6): 110 (1986): Schrenkia lachnantha.

PRENANTHES (Compos.).
angustiloba *C. Shih in Acta Phytotax. Sin.,* 25(3): 193 (1987)—China.
leptantha *C. Shih in Acta Phytotax. Sin.,* 25(3): 193 (1987)—China.
pyramidalis *C. Shih in Acta Phytotax. Sin.,* 25(3): 191 (1987)—China.
racemiformis *C. Shih in Acta Phytotax. Sin.,* 25(3): 191 (1987)—China.

PRESTOEA (Palmae).
humilis *A. Henderson & J.A. Steyermark in Brittonia,* 38(4): 311 (1986)—Venezuela.
trichoclada *(Burret) H. Balslev & A. Henderson in Principes,* 31(1): 11 (1987): Euterpe trichoclada.

PRIMULA (Primulac.).
advena
var. euprepes *(W.W. Sm.) Chen & C.M. Hu in Fl. Xizangica,* 3: 799 (1986), without basionym ref.: P. maximowiczii var. euprepes.
anvilensis *S. Kelso in Syst. Bot.,* 12(1): 10 (1987)—Alaska.

PRIONOSTEMMA (Celastrac.).
delagoensis
var. ritschardii *(Wilczek) N. Hallé in Fl. Gabon,* 29: 232 (1986): Loeseneriella ritschardii.

PRISTIMERA (Celastrac.).
andongensis
var. cinerascens *N. Hallé in Fl. Gabon,* 29: 211 (1986)—Cameroun, Gabon, Zaïre.
var. volkensii *(Loesener) N. Hallé in Fl. Gabon,* 29: 211 (1986): Hippocratea volkensii.
biholongii *N. Hallé in Fl. Gabon,* 29: 206 (1986)—Cameroun.
luteoviridis
var. kribiana *N. Hallé in Fl. Gabon,* 29: 202 (1986)—Cameroun.

PRITZELAGO (Crucif.).
alpina
subsp. auerswaldii *(Willkomm) J. Soják in Čas. Nár. Muz. (Prague),* 153(3): 145 (1984 publ. 1985): Hutchinsia auerswaldii.
subsp. font-queri *(Sauvage) J. Soják in Čas. Nár. Muz. (Prague),* 153(3): 145 (1984 publ. 1985): Hutchinsia font-queri.
subsp. polatschekii *(Laínz) J. Soják in Čas. Nár. Muz. (Prague),* 153(3): 145 (1984 publ. 1985): Hutchinsia alpina subsp. polatschekii.

PROSOPIS (Legumin.).
koelziana
var. puberula *J. Léonard in Bull. Jard. Bot. Nation. Belg.,* 56(3–4): 485 (1986)—Saudi Arabia, South Yemen.

PROSTANTHERA (Labiat.).
carrickiana *B.J. Conn in Muelleria,* 6(5): 371 (1987)—Australia (W. Australia).

PROTASPARAGUS (Asparagac.).
virgatus *(Baker) A.A. Obermeyer in Fl. Australia,* 45: 494 (1987): Asparagus virgatus.

PROTEA (Proteac.).
angolensis
var. roseola *S.M. Chisumpa & R.K. Brummitt in Kew Bull.,* 42(4): 823 (1987)—Zambia.
argyrea
subsp. zambiana *S.M. Chisumpa & R.K. Brummitt in Kew Bull.,* 42(4): 834 (1987)—Zambia.
baumii
subsp. robusta *S.M. Chisumpa & R.K. Brummitt in Kew Bull.,* 42(4): 839 (1987)—Zambia, Tanzania, Zaïre, Angola.
caffra
subsp. gazensis *(Beard) S.M. Chisumpa & R.K. Brummitt in Kew Bull.,* 42(4): 830 (1987): P. gazensis.
subsp. kilimandscharica *(Engl.) S.M. Chisumpa & R.K. Brummitt in Kew Bull.,* 42(4): 830 (1987): P. kilimandscharica.
subsp. mafingensis *S.M. Chisumpa & R.K. Brummitt in Kew Bull.,* 42(4): 830 (1987)—Zambia, Malawi.
subsp. nyasae *(Rendle) S.M. Chisumpa & R.K. Brummitt in Kew Bull.,* 42(4): 830 (1987): P. nyasae.
heckmanniana
subsp. angustifolia *(Hauman) S.M. Chisumpa & R.K. Brummitt in Kew Bull.,* 42(4): 844 (1987): P. heckmanniana var. angustifolia.
kibarensis
subsp. cuspidata *(Beard) S.M. Chisumpa & R.K. Brummitt in Kew Bull.,* 42(4): 845 (1987): P. cuspidata.
madiensis
subsp. occidentalis *(Beard) S.M. Chisumpa & R.K. Brummitt in Kew Bull.,* 42(4): 819 (1987): P. occidentalis.
matonchiana *S.M. Chisumpa & R.K. Brummitt in Kew Bull.,* 42(4): 846 (1987)—Zambia.
micans
subsp. lemairei *(De Wild.) S.M. Chisumpa & R.K. Brummitt in Kew Bull.,* 42(4): 842 (1987): P. lemairei.
subsp. makutuensis *S.M. Chisumpa & R.K. Brummitt in Kew Bull.,* 42(4): 843 (1987)—Zambia.

PROTEA :–
subsp. suffruticosa *(Beard) S.M. Chisumpa & R.K. Brummitt in Kew Bull.,* 42(4): 843 (1987): P. suffruticosa.
subsp. trichophylla *(Engl. & Gilg) S.M. Chisumpa & R.K. Brummitt in Kew Bull.,* 42(4): 842 (1987): P. trichophylla.
paludosa
subsp. kolweziensis *S.M. Chisumpa & R.K. Brummitt in Kew Bull.,* 42(4): 850 (1987)—Zaïre.
subsp. secundifolia *(Hauman) S.M. Chisumpa & R.K. Brummitt in Kew Bull.,* 42(4): 849 (1987): P. secundifolia.
petiolaris
subsp. elegans *S.M. Chisumpa & R.K. Brummitt in Kew Bull.,* 42(4): 825 (1987)—Tanzania, Zaïre, Zambia, Malawi, Mozambique, Zimbabwe.
poggei
subsp. haemantha *(Engl. & Gilg) S.M. Chisumpa & R.K. Brummitt in Kew Bull.,* 42(4): 851 (1987): P. haemantha.
subsp. heliophila *S.M. Chisumpa & R.K. Brummitt in Kew Bull.,* 42(4): 852 (1987)—Angola.
subsp. mwinilungensis *S.M. Chisumpa & R.K. Brummitt in Kew Bull.,* 42(4): 852 (1987)—Zambia.

PRUNUS (Rosac.).
linearipetalus *Y.N. Lee in Korean J. Bot.,* 25(4): 176 (1982)—Korea.
mira
subsp. nepalensis *(Serafimov) H. Hara in Enum. Fl. Pl. Nepal,* 2: 142 (1979): Persica mira subsp. nepalensis.

PSAMMISIA (Ericac.).
amazonica *J.L. Luteyn in Op. Bot.,* 92: 117 (1987)—Peru.
aurantiaca *J.L. Luteyn in Op. Bot.,* 92: 121 (1987)—Ecuador.
incana *J.L. Luteyn in Op. Bot.,* 92: 118 (1987)—Ecuador.
pterocalyx *J.L. Luteyn in Op. Bot.,* 92: 117 (1987)—Colombia.

PSAMMOGETON (Umbellif.).
lamondiae *L. Engstrand & K.H. Rechinger in Fl. Iran.,* 162: 150 (1987)—Pakistan.
ranunculifolius *(Boiss.) L. Engstrand in Fl. Iran.,* 162: 149 (1987): Pimpinella ranunculifolia.
registanicus *K.H. Rechinger in Fl. Iran.,* 162: 148 (1987)—Afghanistan.
ternatus *(Rech. f.) L. Engstrand in Fl. Iran.,* 162: 151 (1987): Pimpinella ternata.

PSEUDAIDIA D.D. Tirvengadum in Bull. Mus. Nation. Hist. Nat., B, Adansonia, Ser. 4, 8(3): 286 (1986). *RUBIACEAE.*
speciosa *(Bedd.) D.D. Tirvengadum in Bull. Mus. Nation. Hist. Nat., B, Adansonia,* Ser. 4, 8(3): 287 (1986): Griffithia speciosa.

×PSEUDELYMUS M.E. Barkworth & D.R. Dewey in Great Basin Nat., 43(4): 570 (1983). *GRAMINEAE.* (Elymus × Pseudoroegneria).
saxicola *(Scribner & J.G. Smith) M.E. Barkworth & D.R. Dewey in Great Basin Nat.*, 43(4): 570 (1983): Elymus saxicolus.

PSEUDOSAGOTIA R. de S. Secco in Bol. Mus. Paraense Emílio Goeldi, Bot., 2(1): 23 (1985). *EUPHORBIACEAE.*
brevipetiolata *R. de S. Secco in Bol. Mus. Paraense Emílio Goeldi, Bot.*, 2(1): 24 (1985)—Venezuela.

PSEUDOVANILLA L.A. Garay in Bot. Mus. Leafl. Harvard Univ., 30(4): 234 (1986). *ORCHIDACEAE.*
affinis *(J.J. Sm.) L.A. Garay in Bot. Mus. Leafl. Harvard Univ.*, 30(4): 235 (1986): Galeola affinis.
anomala *(Ames & L.O. Wms.) L.A. Garay in Bot. Mus. Leafl. Harvard Univ.*, 30(4): 235 (1986): Vanilla anomala.
foliata *(F. Muell.) L.A. Garay in Bot. Mus. Leafl. Harvard Univ.*, 30(4): 235 (1986): Ledgeria foliata.
gracilis *(Schltr.) L.A. Garay in Bot. Mus. Leafl. Harvard Univ.*, 30(4): 235 (1986): Galeola gracilis.
philippinensis *(Ames) L.A. Garay in Bot. Mus. Leafl. Harvard Univ.*, 30(4): 236 (1986): Galeola philippinensis.
ponapensis *(Kaneh. & Yamam.) L.A. Garay in Bot. Mus. Leafl. Harvard Univ.*, 30(4): 236 (1986): Vanilla ponapensis.
ternatensis *(J.J. Sm.) L.A. Garay in Bot. Mus. Leafl. Harvard Univ.*, 30(4): 236 (1986): Galeola ternatensis.
vanilloides *(Schltr.) L.A. Garay in Bot. Mus. Leafl. Harvard Univ.*, 30(4): 236 (1986): Galeola vanilloides.

PSEUDUVARIA (Annonac.).
hylandii *L.W. Jessup in Austrobaileya*, 2(4): 308 (1987)—Australia (Queensland).
mulgraveana *L.W. Jessup in Austrobaileya*, 2(4): 310 (1987)—Australia (Queensland).
var. glabrescens *L.W. Jessup in Austrobaileya*, 2(4): 312 (1987)—Australia (Queensland).
villosa *L.W. Jessup in Austrobaileya*, 2(4): 312 (1987)—Australia (Queensland).

PSILANTHUS (Rubiac.).
subgen. Afrocoffea *(Moens) D.M. Bridson in Kew Bull.*, 42(2): 453 (1987): Coffea subgen. Afrocoffea.
lebrunianus *(Germain & Kesler) Leroy ex D.M. Bridson in Kew Bull.*, 42(2): 456 (1987): Coffea lebruniana.

PSITTACANTHUS (Loranthac.).
barlowii *J. Kuijt in Fl. Ecuador*, 24(32C): 185 (1986)—Ecuador.

PSITTACANTHUS :–
pangui *J. Kuijt in Fl. Ecuador*, 24(32C): 177 (1986)—Ecuador.
truncatus *J. Kuijt in Fl. Ecuador*, 24(32C): 174 (1986)—Ecuador.
wiensii *J. Kuijt in Fl. Ecuador*, 24(32C): 182 (1986)—Ecuador.

PSYCHOTRIA (Rubiac.).
faxlucens *D.H. Lorence & J.D. Dwyer in Bol. Soc. Bot. Méx.*, 47: 50 (1987)—Mexico.
guanchezii *J.A. Steyermark in Ann. Missouri Bot. Gard.*, 74(1): 109 (1987)—Venezuela.
izabalensis
subsp. oaxacana *D.H. Lorence in Bol. Soc. Bot. Méx.*, 47: 53 (1987)—Mexico.
luteotuba *D.H. Lorence in Bol. Soc. Bot. Méx.*, 47: 55 (1987)—Mexico.
megalantha *D.H. Lorence in Bol. Soc. Bot. Méx.*, 47: 58 (1987)—Mexico.
ronaldii *J.A. Steyermark in Ann. Missouri Bot. Gard.*, 74(1): 111 (1987)—Venezuela.
sohmeri *M. Kiehn in Pl. Syst. Evol.*, 154(3–4): 215 (1986), nom. nov.: P. fosbergii *Sohmer.*
sousae *D.H. Lorence & J.D. Dwyer in Bol. Soc. Bot. Méx.*, 47: 59 (1987)—Mexico.
veracruzensis *D.H. Lorence & J.D. Dwyer in Bol. Soc. Bot. Méx.*, 47: 61 (1987)—Mexico.

PTERACANTHUS (Acanthac.).
hygrophiloides *(C.B. Clarke ex W.W. Smith) H.W. Li in Fl. Xizangica*, 4: 415 (1985): Strobilanthes hygrophiloides.
versicolor *(Diels) H.W. Li in Fl. Xizangica*, 4: 415 (1985): Strobilanthes versicolor.

PTEROCEPHALIDIUM G. López González in An. Jard. Bot. Madrid, 43(2): 251 (1986 publ. 1987). *DIPSACACEAE.*
diandrum *(Lag.) G. López González in An. Jard. Bot. Madrid*, 43(2): 252 (1986 publ. 1987): Scabiosa diandra.

PTEROCYMBIUM R.Brown, Pterocymbium: 219 (4 June 1844). *STERCULIACEAE.* †
javanicum *R. Brown, Pterocymbium:* 219 (4 June 1844)—Java. †

PTEROSTYLIS (Orchidac.).
bicolor *M.A. Clements & D.L. Jones in Proc. Roy. Soc. Queensl.*, 98: 126 (1987)—Australia (New South Wales, Queensland).
bicornis *D.L. Jones & M.A. Clements in Proc. Roy. Soc. Queensl.*, 98: 124 (1987)—Australia (Queensland).
dolichochila *D.L. Jones & M.A. Clements in Fl. S. Australia*, 4: 2117 (1986)—Australia (S. Australia, Victoria).
erythroconcha *M.A. Clements & D.L. Jones in Fl. S. Australia*, 4: 2118 (1986)—Australia (S. Australia).
excelsa *M.A. Clements in Fl. S. Australia*, 4: 2119 (1986)—Australia (S. Australia, Victoria).
ovata *M.A. Clements in Fl. S. Australia*, 4: 2124 (1986)—Australia (S. Australia).

PTEROSTYLIS :–
xerophila *M.A. Clements in Fl. S. Australia,* 4: 2130 (1986)—Australia (S. Australia, Victoria).

PTERYGOTA (Sterculiac.).
alata *(Roxb.) R. Brown Pterocymbium:* 234 (4 June 1844): Sterculia alata. †

PTYCHOSPERMA (Palmae).
hentyi *F.B. Essig in Principes,* 31(3): 113 (1987)—New Guinea (Papua New Guinea).

PUCCINELLIA (Gramin.).
anisoclada
subsp. melderisiana *Kit Tan in Pl. Syst. Evol.,* 154(1–2): 127 (1986)—Turkey.

PULTENAEA (Legumin.).
linearifolia *A. Strid in Pl. Syst. Evol.,* 155(1–4): 344 (1987)—Australia (W. Australia).
rigida
var. ovata *J.Z. Weber in Fl. S. Australia,* 2: 681 (1986)—Australia (S. Australia).

PURSHIA (Rosac.).
alba *(Goodding) E.I. Golubkova in Bot. Zhurn.,* 72(6): 776 (1987): Cowania alba.
pinkavae *C.G. Schaack in Phytologia,* 63(4): 301 (1987)—U.S.A. (Arizona).

PUTTERLICKIA (Celastrac.).
retrospinosa *A.E. van Wyk & S.C. Mostert in S. Afr. J. Bot.,* 53(4): 267 (1987)—South Africa (Natal, Cape Province).

PUYA (Bromeliac.).
werneriana *R.W. Read & L.B. Smith in Abh. Akad. Wiss. Lit. Mainz, Math.-Naturwiss. Klasse, Trop. Subtrop. Pflanzenwelt,* 58: 5 (1986)—Peru.

PYRETHRUM (Compos.).
sovetkinae *S.S. Kovalevskaya in Novosti Sist. Vyssh. Rast.,* 23: 241 (1986)—U.S.S.R. (Middle Asia).

PYROSTRIA (Rubiac.).
affinis *(Robyns) D.M. Bridson in Kew Bull.,* 42(3): 628 (1987): Dinocanthium affine.
bispathacea *(Mildbr.) D.M. Bridson in Kew Bull.,* 42(3): 629 (1987): Plectronia bispathacea.
hystrix *(Bremek.) D.M. Bridson in Kew Bull.,* 42(3): 629 (1987): Dinocanthium hystrix.
lobulata *D.M. Bridson in Kew Bull.,* 42(3): 627 (1987)—Rwanda, Tanzania, Zambia.
phyllanthoidea *(Baill.) D.M. Bridson in Kew Bull.,* 42(3): 629 (1987): Canthium phyllanthoideum.
socotrana *(Radcliffe-Smith) D.M. Bridson in Kew Bull.,* 42(3): 625 (1987): Canthium socotranum.

Q

QUERCUS (Fagac.).
×beckyae *B.D. Gaynor in Mich. Bot.,* 26(2): 65 (1987)—U.S.A. (Michigan). (Q. macrocarpa × prinoides).
glauca
var. podocupula *C.F. Shen in C.F. Shen & T.S. Liu, Taxon. & Pollen Morph. Fagaceae Taiwan:* 53 (1984)—Taiwan.
liaoi *C.F. Shen in C.F. Shen & T.S. Liu, Taxon. & Pollen Morph. Fagaceae Taiwan:* 55 (1984)—Taiwan.
longinux
var. kuoi *(Liao) C.F. Shen in C.F. Shen & T.S. Liu, Taxon. & Pollen Morph. Fagaceae Taiwan:* 56 (1984): Cyclobalanopsis longinux var. kuoi.
salicina
var. asiatica *C.F. Shen in C.F. Shen & T.S. Liu, Taxon. & Pollen Morph. Fagaceae Taiwan:* 63 (1984)—Taiwan.
×wagneri *B.D. Gaynor in Mich. Bot.,* 26(2): 67 (1987)—U.S.A. (Michigan). (Q. bicolor × prinoides).

QUIINA (Quiinac.).
colonensis *(W.G. D'Arcy) W.G. D'Arcy, Fl. Panama Checklist & Index, 2 (Monogr. Syst. Bot. Missouri Bot. Gard. 18):* 534, 337 (1987): Lacunaria colonensis.

R

RABDOSIA (Labiat.).
megathyrsoides *H.W. Li in Fl. Xizangica,* 4: 221 (1985)—Xizang.

RACOSPERMA (Legumin.).
acradenia *(F. Muell.) L. Pedley in Austrobaileya,* 2(4): 344 (1987): Acacia acradenia.
adenogonium *L. Pedley in Austrobaileya,* 2(4): 316 (1987), nom. nov.: Acacia deltoidea var. pungens *Bentham.*
adsurgens *(Maiden & Blakely) L. Pedley in Austrobaileya,* 2(4): 344 (1987): Acacia adsurgens.
aduncum *(Cunn. ex G. Don) L. Pedley in Austrobaileya,* 2(4): 344 (1987): Acacia adunca.
allenianum *(Maiden) L. Pedley in Austrobaileya,* 2(4): 344 (1987): Acacia alleniana.
amblygonum *(Cunn. ex Benth.) L. Pedley in Austrobaileya,* 2(4): 344 (1987): Acacia amblygona.

RACOSPERMA :–

ammophilum *(Pedley) L. Pedley in Austrobaileya,* 2(4): 344 (1987): Acacia ammophila.
ancistrocarpum *(Maiden & Blakely) L. Pedley in Austrobaileya,* 2(4): 344 (1987): Acacia ancistrocarpa.
aneurum *(F. Muell. ex Benth.) L. Pedley in Austrobaileya,* 2(4): 344 (1987): Acacia aneura.
angustum *(Maiden & Blakely) L. Pedley in Austrobaileya,* 2(4): 344 (1987): Acacia angusta.
apreptum *(Pedley) L. Pedley in Austrobaileya,* 2(4): 344 (1987): Acacia aprepta.
argyraeum *(Tindale) L. Pedley in Austrobaileya,* 2(4): 344 (1987): Acacia argyraea.
argyrodendron *(Domin) L. Pedley in Austrobaileya,* 2(4): 345 (1987): Acacia argyrodendron.
armillatum *L. Pedley in Austrobaileya,* 2(4): 325 (1987)—Australia (Queensland).
armitii *(F. Muell. ex Maiden) L. Pedley in Austrobaileya,* 2(4): 345 (1987): Acacia armitii.
asperulaceum *(F. Muell.) L. Pedley in Austrobaileya,* 2(4): 345 (1987): Acacia asperulacea.
attenuatum *(Maiden & Blakely) L. Pedley in Austrobaileya,* 2(4): 345 (1987): Acacia attenuata.
aulacocarpum *(Cunn. ex Benth.) L. Pedley in Austrobaileya,* 2(4): 345 (1987): Acacia aulacocarpa.
var. fruticosum *(C. White) L. Pedley in Austrobaileya,* 2(4): 345 (1987): Acacia aulacocarpa var. fruticosa.
baeuerlenii *(Maiden & R. Baker) L. Pedley in Austrobaileya,* 2(4): 345 (1987): Acacia baeuerlenii.
baileyanum *(F. Muell.) L. Pedley in Austrobaileya,* 2(4): 345 (1987): Acacia baileyana.
bakeri *(Maiden) L. Pedley in Austrobaileya,* 2(4): 345 (1987): Acacia bakeri.
bancroftii *(Maiden) L. Pedley in Austrobaileya,* 2(4): 345 (1987): Acacia bancroftii.
baueri *(Benth.) L. Pedley in Austrobaileya,* 2(4): 345 (1987): Acacia baueri.
betchei *(Maiden & Blakely) L. Pedley in Austrobaileya,* 2(4): 345 (1987): Acacia betchei.
binervatum *(DC.) L. Pedley in Austrobaileya,* 2(4): 345 (1987): Acacia binervata.
blakei *(Pedley) L. Pedley in Austrobaileya,* 2(4): 345 (1987): Acacia blakei.
subsp. diphyllum *(Tindale) L. Pedley in Austrobaileya,* 2(4): 345 (1987): Acacia diphylla.
brachycarpum *(Pedley) L. Pedley in Austrobaileya,* 2(4): 345 (1987): Acacia brachycarpa.
brachystachyum *(Benth.) L. Pedley in Austrobaileya,* 2(4): 345 (1987): Acacia brachystachya.

RACOSPERMA :–

brassii *(Pedley) L. Pedley in Austrobaileya,* 2(4): 345 (1987): Acacia brassii.
brownei *(Poiret) L. Pedley in Austrobaileya,* 2(4): 346 (1987): Mimosa brownei.
brunioides *(Cunn. ex G. Don) L. Pedley in Austrobaileya,* 2(4): 346 (1987): Acacia brunioides.
subsp. graniticum *(Pedley) L. Pedley in Austrobaileya,* 2(4): 346 (1987): Acacia brunioides subsp. granitica.
burbidgeae *(Pedley) L. Pedley in Austrobaileya,* 2(4): 346 (1987): Acacia burbidgeae.
burrowii *(Maiden) L. Pedley in Austrobaileya,* 2(4): 346 (1987): Acacia burrowii.
buxifolium *(Cunn.) L. Pedley in Austrobaileya,* 2(4): 346 (1987): Acacia buxifolia.
subsp. pubiflorum *(Pedley) L. Pedley in Austrobaileya,* 2(4): 346 (1987): Acacia buxifolia subsp. pubiflora.
calanthum *(Pedley) L. Pedley in Austrobaileya,* 2(4): 346 (1987): Acacia calantha.
calcicola *(Forde & Ising) L. Pedley in Austrobaileya,* 2(4): 346 (1987): Acacia calcicola.
calyculatum *(Cunn. ex Benth.) L. Pedley in Austrobaileya,* 2(4): 346 (1987): Acacia calyculata.
cambagei *(R. Baker) L. Pedley in Austrobaileya,* 2(4): 346 (1987): Acacia cambagei.
canum *(Maiden) L. Pedley in Austrobaileya,* 2(4): 346 (1987): Acacia cana.
caroleae *(Pedley) L. Pedley in Austrobaileya,* 2(4): 346 (1987): Acacia caroleae.
catenulatum *(C. White) L. Pedley in Austrobaileya,* 2(4): 346 (1987): Acacia catenulata.
centrinervium *(Maiden & Blakely) L. Pedley in Austrobaileya,* 2(4): 346 (1987): Acacia centrinervia.
chinchillense *(Tindale) L. Pedley in Austrobaileya,* 2(4): 346 (1987): Acacia chinchillensis.
chippendalei *(Pedley) L. Pedley in Austrobaileya,* 2(4): 346 (1987): Acacia chippendalei.
chisholmii *(Bailey) L. Pedley in Austrobaileya,* 2(4): 346 (1987): Acacia chisholmii.
cincinnatum *(F. Muell.) L. Pedley in Austrobaileya,* 2(4): 347 (1987): Acacia cincinnata.
complanatum *(Cunn. ex Benth.) L. Pedley in Austrobaileya,* 2(4): 347 (1987): Acacia complanata.
confertum *(Cunn. ex Benth.) L. Pedley in Austrobaileya,* 2(4): 347 (1987): Acacia conferta.
conjunctifolium *(F. Muell.) L. Pedley in Austrobaileya,* 2(4): 347 (1987): Acacia conjunctifolia.
cowleanum *(Tate) L. Pedley in Austrobaileya,* 2(4): 347 (1987): Acacia cowleana.

RACOSPERMA :-

crassicarpum *(Cunn. ex Benth.) L. Pedley in Austrobaileya,* 2(4): 347 (1987): Acacia crassicarpa.

crassum *(Pedley) L. Pedley in Austrobaileya,* 2(4): 347 (1987): Acacia crassa.

subsp. longicomum *(Pedley) L. Pedley in Austrobaileya,* 2(4): 347 (1987): Acacia crassa subsp. longicoma.

cretatum *(Pedley) L. Pedley in Austrobaileya,* 2(4): 347 (1987): Acacia cretata.

cultriforme *(Cunn. ex G. Don) L. Pedley in Austrobaileya,* 2(4): 347 (1987): Acacia cultriformis.

curranii *(Maiden) L. Pedley in Austrobaileya,* 2(4): 347 (1987): Acacia curranii.

cyperophyllum *(F. Muell. ex Benth.) L. Pedley in Austrobaileya,* 2(4): 347 (1987): Acacia cyperophylla.

dawsonii *(R. Baker) L. Pedley in Austrobaileya,* 2(4): 347 (1987): Acacia dawsonii.

dealbatum *(Link) L. Pedley in Austrobaileya,* 2(4): 358 (1987): Acacia dealbata.

debile *(Tindale) L. Pedley in Austrobaileya,* 2(4): 347 (1987): Acacia debilis.

decorum *(H.G. Reichb.) L. Pedley in Austrobaileya,* 2(4): 347 (1987): Acacia decora.

decurrens *(Willd.) L. Pedley in Austrobaileya,* 2(4): 358 (1987): Acacia decurrens.

deltoideum *(Cunn. ex G. Don) L. Pedley in Austrobaileya,* 2(4): 315 (1987): Acacia deltoidea.

deuteroneurum *(Pedley) L. Pedley in Austrobaileya,* 2(4): 347 (1987): Acacia deuteroneura.

dictyophlebum *(F. Muell.) L. Pedley in Austrobaileya,* 2(4): 347 (1987): Acacia dictyophleba.

dietrichianum *(F. Muell.) L. Pedley in Austrobaileya,* 2(4): 347 (1987): Acacia dietrichiana.

difficile *(Maiden) L. Pedley in Austrobaileya,* 2(4): 348 (1987): Acacia difficilis.

dimidiatum *(Benth.) L. Pedley in Austrobaileya,* 2(4): 348 (1987): Acacia dimidiata.

drepanocarpum *(F. Muell.) L. Pedley in Austrobaileya,* 2(4): 348 (1987): Acacia drepanocarpa.

subsp. latifolium *(Pedley) L. Pedley in Austrobaileya,* 2(4): 348 (1987): Acacia drepanocarpa subsp. latifolia.

elatum *(Cunn. ex Benth.) L. Pedley in Austrobaileya,* 2(4): 358 (1987): Acacia elata.

eremophiloides *(Pedley & P. Forster) L. Pedley in Austrobaileya,* 2(4): 348 (1987): Acacia eremophiloides.

estrophiolatum *(F. Muell.) L. Pedley in Austrobaileya,* 2(4): 348 (1987): Acacia estrophiolata.

everistii *(Pedley) L. Pedley in Austrobaileya,* 2(4): 348 (1987): Acacia everistii.

excelsum *(Benth.) L. Pedley in Austrobaileya,* 2(4): 348 (1987): Acacia excelsa.

RACOSPERMA :-

subsp. angustum *(Pedley) L. Pedley in Austrobaileya,* 2(4): 348 (1987): Acacia excelsa subsp. angusta.

falciforme *(DC.) L. Pedley in Austrobaileya,* 2(4): 348 (1987): Acacia falciformis.

fasciculiferum *(F. Muell. ex Benth.) L. Pedley in Austrobaileya,* 2(4): 348 (1987): Acacia fasciculifera.

filicifolium *(Cheel & Welch) L. Pedley in Austrobaileya,* 2(4): 348 (1987): Acacia filicifolia.

fimbriatum *(Cunn. ex G. Don) L. Pedley in Austrobaileya,* 2(4): 348 (1987): Acacia fimbriata.

flavescens *(Cunn. ex Benth.) L. Pedley in Austrobaileya,* 2(4): 348 (1987): Acacia flavescens.

fleckeri *(Pedley) L. Pedley in Austrobaileya,* 2(4): 348 (1987): Acacia fleckeri.

flexifolium *(Cunn. ex Benth.) L. Pedley in Austrobaileya,* 2(4): 348 (1987): Acacia flexifolia.

floribundum *(Vent.) L. Pedley in Austrobaileya,* 2(4): 348 (1987): Mimosa floribunda.

floydii *(Tindale) L. Pedley in Austrobaileya,* 2(4): 348 (1987): Acacia floydii.

froggattii *(Maiden) L. Pedley in Austrobaileya,* 2(4): 318 (1987): Acacia froggattii.

galioides *(Benth.) L. Pedley in Austrobaileya,* 2(4): 349 (1987): Acacia galioides.

georginae *(Bailey) L. Pedley in Austrobaileya,* 2(4): 349 (1987): Acacia georginae.

gittinsii *(Pedley) L. Pedley in Austrobaileya,* 2(4): 349 (1987): Acacia gittinsii.

glaucocarpum *(Maiden & Blakely) L. Pedley in Austrobaileya,* 2(4): 349 (1987): Acacia glaucocarpa.

gnidium *(Benth.) L. Pedley in Austrobaileya,* 2(4): 349 (1987): Acacia gnidium.

gonocladum *(F. Muell.) L. Pedley in Austrobaileya,* 2(4): 349 (1987): Acacia gonoclada.

grandifolium *(Pedley) L. Pedley in Austrobaileya,* 2(4): 349 (1987): Acacia grandifolia.

graniticum *(Maiden) L. Pedley in Austrobaileya,* 2(4): 349 (1987): Acacia granitica.

gunnii *(Benth.) L. Pedley in Austrobaileya,* 2(4): 349 (1987): Acacia gunnii.

guymeri *(Tindale) L. Pedley in Austrobaileya,* 2(4): 349 (1987): Acacia guymeri.

hakeoides *(Cunn. ex Benth.) L. Pedley in Austrobaileya,* 2(4): 349 (1987): Acacia hakeoides.

hammondii *(Maiden) L. Pedley in Austrobaileya,* 2(4): 349 (1987): Acacia hammondii.

handonis *(Pedley) L. Pedley in Austrobaileya,* 2(4): 349 (1987): Acacia handonis.

harpophyllum *(F. Muell. ex Benth.) L. Pedley in Austrobaileya,* 2(4): 349 (1987): Acacia harpophylla.

RACOSPERMA :–

hemignostum *(F. Muell.) L. Pedley in Austrobaileya,* 2(4): 349 (1987): Acacia hemignosta.
hemsleyi *(Maiden) L. Pedley in Austrobaileya,* 2(4): 349 (1987): Acacia hemsleyi.
hillianum *(Maiden) L. Pedley in Austrobaileya,* 2(4): 349 (1987): Acacia hilliana.
hockingsii *(Pedley) L. Pedley in Austrobaileya,* 2(4): 349 (1987): Acacia hockingsii.
holosericeum(Cunn. ex G. Don) L. Pedley in Austrobaileya, 2(4): 349 (1987): Acacia holosericea.
holotrichum *(Pedley) L. Pedley in Austrobaileya,* 2(4): 350 (1987): Acacia holotricha.
homalocladum *(F. Muell.) L. Pedley in Austrobaileya,* 2(4): 350 (1987): Acacia homaloclada.
hubbardianum *(Pedley) L. Pedley in Austrobaileya,* 2(4): 350 (1987): Acacia hubbardiana.
humifusum *(Cunn. ex Benth.) L. Pedley in Austrobaileya,* 2(4): 350 (1987): Acacia humifusa.
hyaloneurum *(Pedley) L. Pedley in Austrobaileya,* 2(4): 350 (1987): Acacia hyaloneura.
implexum *(Benth.) L. Pedley in Austrobaileya,* 2(4): 350 (1987): Acacia implexa.
irroratum *(Sieber ex Sprengel) L. Pedley in Austrobaileya,* 2(4): 350 (1987): Acacia irrorata.
islanum *(Pedley) L. Pedley in Austrobaileya,* 2(4): 350 (1987): Acacia islana.
ixiophyllum *(Benth.) L. Pedley in Austrobaileya,* 2(4): 350 (1987): Acacia ixiophylla.
ixodes *(Pedley) L. Pedley in Austrobaileya,* 2(4): 350 (1987): Acacia ixodes.
jackesianum *(Pedley) L. Pedley in Austrobaileya,* 2(4): 350 (1987): Acacia jackesiana.
johnsonii *(Pedley) L. Pedley in Austrobaileya,* 2(4): 350 (1987): Acacia johnsonii.
jucundum *(Maiden & Blakely) L. Pedley in Austrobaileya,* 2(4): 350 (1987): Acacia jucunda.
juliferum
  subsp. gilbertense *(Pedley) L. Pedley in Austrobaileya,* 2(4): 350 (1987): Acacia julifera subsp. gilbertensis.
juncifolium *(Benth.) L. Pedley in Austrobaileya,* 2(4): 350 (1987): Acacia juncifolia.
kempeanum *(F. Muell.) L. Pedley in Austrobaileya,* 2(4): 350 (1987): Acacia kempeana.
laccatum *(Pedley) L. Pedley in Austrobaileya,* 2(4): 350 (1987): Acacia laccata.
lanigerum *(Cunn.) L. Pedley in Austrobaileya,* 2(4): 350 (1987): Acacia lanigera.
latifolium *(Benth.) L. Pedley in Austrobaileya,* 2(4): 350 (1987): Acacia latifolia.
latisepalum *(Pedley) L. Pedley in Austrobaileya,* 2(4): 351 (1987): Acacia latisepala.
lautum *(Pedley) L. Pedley in Austrobaileya,* 2(4): 351 (1987): Acacia lauta.

RACOSPERMA :–

lazaridis *(Pedley) L. Pedley in Austrobaileya,* 2(4): 351 (1987): Acacia lazaridis.
legnotum *(Pedley) L. Pedley in Austrobaileya,* 2(4): 351 (1987): Acacia legnota.
leichhardtii *(Benth.) L. Pedley in Austrobaileya,* 2(4): 351 (1987): Acacia leichhardtii.
leiocalyx *(Domin) L. Pedley in Austrobaileya,* 2(4): 351 (1987): Acacia glaucescens var. leiocalyx.
leptocarpum *(Cunn. ex Benth.) L. Pedley in Austrobaileya,* 2(4): 351 (1987): Acacia leptocarpa.
leptolobum *(Pedley) L. Pedley in Austrobaileya,* 2(4): 351 (1987): Acacia leptoloba.
leptostachyum *(Benth.) L. Pedley in Austrobaileya,* 2(4): 351 (1987): Acacia leptostachya.
leucocladum *(Tindale) L. Pedley in Austrobaileya,* 2(4): 351 (1987): Acacia leucoclada.
  subsp. argentifolium *(Tindale) L. Pedley in Austrobaileya,* 2(4): 351 (1987): Acacia leucoclada subsp. argentifolia.
ligulatum *(Cunn. ex Benth.) L. Pedley in Austrobaileya,* 2(4): 351 (1987): Acacia ligulata.
limbatum *(F. Muell.) L. Pedley in Austrobaileya,* 2(4): 351 (1987): Acacia limbata.
lineatum *(Cunn. ex G. Don) L. Pedley in Austrobaileya,* 2(4): 351 (1987): Acacia lineata.
longipedunculatum *(Pedley) L. Pedley in Austrobaileya,* 2(4): 351 (1987): Acacia longipedunculata.
longispicatum *(Benth.) L. Pedley in Austrobaileya,* 2(4): 351 (1987): Acacia longispicata.
  subsp. velutinum *(Pedley) L. Pedley in Austrobaileya,* 2(4): 351 (1987): Acacia longispicata subsp. velutina.
longissimum *(H. Wendl.) L. Pedley in Austrobaileya,* 2(4): 352 (1987): Acacia longissima.
lorolobum *(Tindale) L. Pedley in Austrobaileya,* 2(4): 352 (1987): Acacia loroloba.
lysiphloium *(F. Muell.) L. Pedley in Austrobaileya,* 2(4): 352 (1987): Acacia lysiphloia.
macradenium *(Benth.) L. Pedley in Austrobaileya,* 2(4): 352 (1987): Acacia macradenia.
maidenii *(F. Muell.) L. Pedley in Austrobaileya,* 2(4): 352 (1987): Acacia maidenii.
mangium *(Willd.) L. Pedley in Austrobaileya,* 2(4): 352 (1987): Acacia mangium.
maranoense *(Pedley) L. Pedley in Austrobaileya,* 2(4): 352 (1987): Acacia maranoensis.
mearnsii *(De Wild.) L. Pedley in Austrobaileya,* 2(4): 358 (1987): Acacia mearnsii.
megalanthum *(F. Muell.) L. Pedley in Austrobaileya,* 2(4): 352 (1987): Acacia megalantha.
meiospermum *L. Pedley in Austrobaileya,* 2(4): 321 (1987)—Australia (Queensland).

RACOSPERMA :-

melleodorum *(Pedley) L. Pedley in Austrobaileya,* 2(4): 352 (1987): Acacia melleodora.

melvillei *(Pedley) L. Pedley in Austrobaileya,* 2(4): 352 (1987): Acacia melvillei.

microcephalum *(Pedley) L. Pedley in Austrobaileya,* 2(4): 352 (1987): Acacia microcephala.

microspermum *(Pedley) L. Pedley in Austrobaileya,* 2(4): 352 (1987): Acacia microsperma.

mimulum *(Pedley) L. Pedley in Austrobaileya,* 2(4): 352 (1987): Acacia mimula.

montanum *(Benth.) L. Pedley in Austrobaileya,* 2(4): 352 (1987): Acacia montana.

monticola *(J. Black) L. Pedley in Austrobaileya,* 2(4): 352 (1987): Acacia monticola.

muellerianum *(Maiden & R. Baker) L. Pedley in Austrobaileya,* 2(4): 352 (1987): Acacia muelleriana.

multisiliquum *(Benth.) L. Pedley in Austrobaileya,* 2(4): 352 (1987): Acacia simsii var. multisiliqua.

murrayanum *(F. Muell. ex Benth.) L. Pedley in Austrobaileya,* 2(4): 352 (1987): Acacia murrayana.

neriifolium *(Cunn. ex Benth.) L. Pedley in Austrobaileya,* 2(4): 353 (1987): Acacia neriifolia.

subsp. pustulum *(Maiden & Blakely) L. Pedley in Austrobaileya,* 2(4): 353 (1987): Acacia pustula.

nesophilum *(Pedley) L. Pedley in Austrobaileya,* 2(4): 353 (1987): Acacia nesophila.

nuperrimum *(E.G. Baker) L. Pedley in Austrobaileya,* 2(4): 353 (1987): Acacia nuperrima.

subsp. cassiterum *(Pedley) L. Pedley in Austrobaileya,* 2(4): 353 (1987): Acacia nuperrima subsp. cassitera.

obtusifolium *(Cunn.) L. Pedley in Austrobaileya,* 2(4): 353 (1987): Acacia obtusifolia.

oligophlebum *(Pedley) L. Pedley in Austrobaileya,* 2(4): 353 (1987): Acacia oligophleba.

omalophyllum *(Cunn. ex Benth.) L. Pedley in Austrobaileya,* 2(4): 353 (1987): Acacia omalophylla.

ommatospermum *L. Pedley in Austrobaileya,* 2(4): 327 (1987)—Australia (Queensland).

orites *(Pedley) L. Pedley in Austrobaileya,* 2(4): 353 (1987): Acacia orites.

orthocarpum *(F. Muell.) L. Pedley in Austrobaileya,* 2(4): 353 (1987): Acacia orthocarpa.

oshanesii *(F. Muell. & Maiden) L. Pedley in Austrobaileya,* 2(4): 353 (1987): Acacia oshanesii.

oswaldii *(F. Muell.) L. Pedley in Austrobaileya,* 2(4): 353 (1987): Acacia oswaldii.

paniculatum *L. Pedley in Austrobaileya,* 2(4): 324 (1987)—Australia (Queensland).

parramattense *(Tindale) L. Pedley in Austrobaileya,* 2(4): 358 (1987): Acacia parramattensis.

RACOSPERMA :-

pendulum *(Cunn. ex G. Don) L. Pedley in Austrobaileya,* 2(4): 353 (1987): Acacia pendula.

penninerve

var. longiracemosum *(Domin) L. Pedley in Austrobaileya,* 2(4): 353 (1987): Acacia penninervis var. longiracemosa.

perangustum *(C. White) L. Pedley in Austrobaileya,* 2(4): 353 (1987): Acacia fimbriata var. perangusta.

phlebocarpum *(F. Muell.) L. Pedley in Austrobaileya,* 2(4): 353 (1987): Acacia phlebocarpa.

platycarpum *(F. Muell.) L. Pedley in Austrobaileya,* 2(4): 354 (1987): Acacia platycarpa.

plectocarpum *(Cunn. ex Benth.) L. Pedley in Austrobaileya,* 2(4): 354 (1987): Acacia plectocarpa.

subsp. tanumbirinense *(Maiden) L. Pedley in Austrobaileya,* 2(4): 354 (1987): Acacia tanumbirinensis.

podalyriifolium *(Cunn. ex G. Don) L. Pedley in Austrobaileya,* 2(4): 354 (1987): Acacia podalyriifolia.

polifolium *(Pedley) L. Pedley in Austrobaileya,* 2(4): 354 (1987): Acacia polifolia.

polyadenium *L. Pedley in Austrobaileya,* 2(4): 322 (1987)—Australia (Queensland).

polybotryum *(Benth.) L. Pedley in Austrobaileya,* 2(4): 354 (1987): Acacia polybotrya.

polystachyum *(Cunn. ex Benth.) L. Pedley in Austrobaileya,* 2(4): 354 (1987): Acacia polystachya.

pravifolium *(F. Muell.) L. Pedley in Austrobaileya,* 2(4): 354 (1987): Acacia pravifolia.

pravissimum *(F. Muell. ex Benth.) L. Pedley in Austrobaileya,* 2(4): 359 (1987): Acacia pravissima.

pruinosum *(Cunn. ex Benth.) L. Pedley in Austrobaileya,* 2(4): 354 (1987): Acacia pruinosa.

pubicostum *(C. White) L. Pedley in Austrobaileya,* 2(4): 354 (1987): Acacia pubicosta.

pubifolium *(Pedley) L. Pedley in Austrobaileya,* 2(4): 354 (1987): Acacia pubifolia.

pubirhachis *(Pedley) L. Pedley in Austrobaileya,* 2(4): 354 (1987): Acacia pubirhachis.

purpureipetalum *(Bailey) L. Pedley in Austrobaileya,* 2(4): 354 (1987): Acacia purpureapetala.

quadrilaterale *(DC.) L. Pedley in Austrobaileya,* 2(4): 354 (1987): Acacia quadrilateralis.

ramiflorum *(Domin) L. Pedley in Austrobaileya,* 2(4): 354 (1987): Acacia ramiflora.

ramulosum *(W.V. Fitzg.) L. Pedley in Austrobaileya,* 2(4): 354 (1987): Acacia ramulosa.

resinicostatum *(Pedley) L. Pedley in Austrobaileya,* 2(4): 354 (1987): Acacia resinicostata.

RACOSPERMA :–
retiveneum *(F. Muell.) L. Pedley in Austrobaileya,* 2(4): 354 (1987): Acacia retivenea.
rhodoxylon *(Maiden) L. Pedley in Austrobaileya,* 2(4): 354 (1987): Acacia rhodoxylon.
rigens *(Cunn. ex G. Don) L. Pedley in Austrobaileya,* 2(4): 355 (1987): Acacia rigens.
rothii *(Bailey) L. Pedley in Austrobaileya,* 2(4): 355 (1987): Acacia rothii.
rubidum *(Cunn.) L. Pedley in Austrobaileya,* 2(4): 355 (1987): Acacia rubida.
ruppii *(Maiden & Betche) L. Pedley in Austrobaileya,* 2(4): 355 (1987): Acacia ruppii.
salicinum *(Lindley) L. Pedley in Austrobaileya,* 2(4): 355 (1987): Acacia salicina.
salignum *(Labill.) L. Pedley in Austrobaileya,* 2(4): 355 (1987): Mimosa saligna.
saxicola *(Pedley) L. Pedley in Austrobaileya,* 2(4): 355 (1987): Acacia saxicola.
semilunatum *(Maiden & Blakely) L. Pedley in Austrobaileya,* 2(4): 355 (1987): Acacia semilunata.
semirigidum *(Maiden & Blakely) L. Pedley in Austrobaileya,* 2(4): 355 (1987): Acacia semirigida.
shirleyi *(Maiden) L. Pedley in Austrobaileya,* 2(4): 355 (1987): Acacia shirleyi.
simsii *(Benth.) L. Pedley in Austrobaileya,* 2(4): 355 (1987): Acacia simsii.
spanium *(Pedley) L. Pedley in Austrobaileya,* 2(4): 355 (1987): Acacia spania.
sparsiflorum *(Maiden) L. Pedley in Austrobaileya,* 2(4): 355 (1987): Acacia sparsiflora.
spectabile *(Cunn. ex Benth.) L. Pedley in Austrobaileya,* 2(4): 355 (1987): Acacia spectabilis.
spirorbe *(Labill.) L. Pedley in Austrobaileya,* 2(4): 355 (1987): Acacia spirorbis.
subsp. solandri *(Benth.) L. Pedley in Austrobaileya,* 2(4): 355 (1987): Acacia solandri.
spondylophyllum *(F. Muell.) L. Pedley in Austrobaileya,* 2(4): 355 (1987): Acacia spondylophylla.
stenophyllum *(Cunn. ex Benth.) L. Pedley in Austrobaileya,* 2(4): 355 (1987): Acacia stenophylla.
stipuligerum *(F. Muell.) L. Pedley in Austrobaileya,* 2(4): 356 (1987): Acacia stipuligera.
subsp. glabrifolium *(Maiden & Blakely) L. Pedley in Austrobaileya,* 2(4): 356 (1987): Acacia stipuligera var. glabrifolia.
stipulosum *(F. Muell.) L. Pedley in Austrobaileya,* 2(4): 317 (1987): Acacia stipulosa.
storyi *(Tindale) L. Pedley in Austrobaileya,* 2(4): 356 (1987): Acacia storyi.
stowardii *(Maiden) L. Pedley in Austrobaileya,* 2(4): 356 (1987): Acacia stowardii.

RACOSPERMA :–
striatifolium *(Pedley) L. Pedley in Austrobaileya,* 2(4): 356 (1987): Acacia striatifolia.
strongylophyllum *(F. Muell.) L. Pedley in Austrobaileya,* 2(4): 356 (1987): Acacia strongylophylla.
sublanatum *(Benth.) L. Pedley in Austrobaileya,* 2(4): 318 (1987): Acacia sublanata.
tenuinerve *(Pedley) L. Pedley in Austrobaileya,* 2(4): 356 (1987): Acacia tenuinervis.
tenuissimum *(F. Muell.) L. Pedley in Austrobaileya,* 2(4): 356 (1987): Acacia tenuissima.
tephrinum *(Pedley) L. Pedley in Austrobaileya,* 2(4): 356 (1987): Acacia tephrina.
tetragonophyllum *(F. Muell.) L. Pedley in Austrobaileya,* 2(4): 356 (1987): Acacia tetragonophylla.
tindaleae *(Pedley) L. Pedley in Austrobaileya,* 2(4): 356 (1987): Acacia tindaleae.
torulosum *(Benth.) L. Pedley in Austrobaileya,* 2(4): 356 (1987): Acacia torulosa.
translucens *(Cunn. ex Hook.) L. Pedley in Austrobaileya,* 2(4): 356 (1987): Acacia translucens.
tripterum *(Benth.) L. Pedley in Austrobaileya,* 2(4): 356 (1987): Acacia triptera.
tropicum *(Maiden & Blakely) L. Pedley in Austrobaileya,* 2(4): 356 (1987): Acacia cunninghamii var. tropica.
ulicifolium *(Salisb.) L. Pedley in Austrobaileya,* 2(4): 356 (1987): Mimosa ulicifolia.
umbellatum *(Cunn. ex Benth.) L. Pedley in Austrobaileya,* 2(4): 356 (1987): Acacia umbellata.
unciferum *(Benth.) L. Pedley in Austrobaileya,* 2(4): 356 (1987): Acacia uncifera.
uncinatum *(Lindley) L. Pedley in Austrobaileya,* 2(4): 357 (1987): Acacia uncinata.
venulosum *(Benth.) L. Pedley in Austrobaileya,* 2(4): 357 (1987): Acacia venulosa.
vernicifluum *(Cunn.) L. Pedley in Austrobaileya,* 2(4): 357 (1987): Acacia verniciflua.
verticillatum
var. latifolium *L. Pedley in Austrobaileya,* 2(4): 358 (1987), nom. nov.: Acacia verticillata var. latifolia *Benth.*
victoriae *(Benth.) L. Pedley in Austrobaileya,* 2(4): 357 (1987): Acacia victoriae.
viscidulum *(Benth.) L. Pedley in Austrobaileya,* 2(4): 357 (1987): Acacia viscidula.
wardellii *(Tindale) L. Pedley in Austrobaileya,* 2(4): 357 (1987): Acacia wardellii.
whitei *(Maiden) L. Pedley in Austrobaileya,* 2(4): 357 (1987): Acacia whitei.
wickhamii *(Benth.) L. Pedley in Austrobaileya,* 2(4): 357 (1987): Acacia wickhamii.

RAIMUNDOCHLOA A.M. Molina in Parodiana, 4(2): 402 (1986 publ. 1987), nom. nov.: gen. Parodiochloa A.M. Molina (Gramin.).

RAIMUNDOCHLOA :-
trachyantha *(Phil.) A.M. Molina in Parodiana,* 4(2): 402 (1986 publ. 1987), without exact basionym page: Koeleria trachyantha.

RANDIA (Rubiac.).
hidalgensis *D.H. Lorence in Bol. Soc. Bot. Méx.,* 47: 38 (1987)—Mexico.
lonicerioides *J.D. Dwyer & D.H. Lorence in Bol. Soc. Bot. Méx.,* 47: 40 (1987)—Mexico, Costa Rica.
matudae *D.H. Lorence & J.D. Dwyer in Bol. Soc. Bot. Méx.,* 47: 42 (1987), nom. nov.: Casasia chiapensis *Miranda.*
pterocarpa *D.H. Lorence & J.D. Dwyer in Bol. Soc. Bot. Méx.,* 47: 43 (1987)—Mexico.
retroflexa *D.H. Lorence & M. Nee in Brittonia,* 39(3): 371 (1987)—Mexico.
vazquezii *D.H. Lorence & J.D. Dwyer in Bol. Soc. Bot. Méx.,* 47: 46 (1987)—Mexico.

RANUNCULUS (Ranunculac.).
abaensis *W.T. Wang in Acta Phytotax. Sin.,* 25(1): 36 (1987)—China.
auricomus
subsp. eversii *E. Julin in Mem. Soc. Fauna Flora Fenn.,* 62(4): 120 (1986)—Sweden.
balangshanicus *W.T. Wang in Bull. Bot. Res. North-East. Forest. Inst.,* 7(2): 105 (1987)—China.
cassubicus
subsp. kemerovensis *G. Kvist in Ann. Bot. Fenn.,* 24(1): 79 (1987)—U.S.S.R. (Siberia).
caucasicus
subsp. pavlii *A.G. Elenevskiĭ & T.G. Derviz-Sokolova in Novosti Sist. Vȳssh. Rast.,* 23: 56 (1986)—U.S.S.R. (Crimea).
dissectus
subsp. aragazii *(Grossh.) Yu.I. Bulanȳĭ & T.G. Derviz-Sokolova in Byull. Mosk. Obshch. Ispȳt. Prir., Biol.,* 91(6): 107 (1986): R. aragazii.
subsp. szowitsianus *(Boiss.) A.G. Elenevskiĭ & T.G. Derviz-Sokolova in Byull. Mosk. Obshch. Ispȳt. Prir., Biol.,* 91(6): 107 (1986): R. szowitsianus.
fallax
subsp. altior *(Marklund) G. Kvist & J. Jalas in Ann. Bot. Fenn.,* 24(2): 222 (1987): R. altior.
hirtellus
var. parciflorus *R.A. Qureshi & M.N. Chaudhri in Pakistan Syst.,* 3(1): 12 (1987)—Pakistan.
kohistanensis *R.A. Qureshi & M.N. Chaudhri in Pakistan Syst.,* 3(1): 16 (1987), nom. nov.: R. saxatilis *R.A. Qureshi & M.N. Chaudhri.*
kuepferi *W. Greuter & Burdet in Willdenowia,* 16(2): 452 (1987), nom. nov.: R. plantagineus subsp. occidentalis *Küpfer.*
laetus
subsp. chitralicus *R.A. Qureshi & M.N. Chaudhri in Pakistan Syst.,* 3(1): 14 (1987)—Pakistan.

RANUNCULUS :-
var. latifolius *R.A. Qureshi & M.N. Chaudhri in Pakistan Syst.,* 3(1): 14 (1987)—Pakistan.
×levenensis *Druce ex R.J. Gornall in Watsonia,* 16(4): 383 (1987)—Great Britain, Ireland, Norway, Sweden, U.S.S.R. (European Russia), Germany, Italy, Switzerland, ?Faeroes. (R. flammula × reptans).
malinovskii *A.G. Elenevskiĭ & T.G. Derviz-Sokolova in Novosti Sist. Vȳssh. Rast.,* 23: 59 (1986)—U.S.S.R. (Ukraine).
nephelogenes
var. geniculatus *(Hand.-Mazz.) W.T. Wang in Bull. Bot. Res. North-East. Forest. Inst.,* 7(2): 109 (1987): R. pulchellus var. geniculatus.
var. longicaulis *(C.A. Mey.) W.T. Wang in Bull. Bot. Res. North-East. Forest. Inst.,* 7(2): 110 (1987), as '(Trautv.)': R. longicaulis.
var. pubescens *W.T. Wang in Bull. Bot. Res. North-East. Forest. Inst.,* 7(2): 110 (1987)—China, Xizang.
quelpaertensis
var. albiflorus *Y.N. Lee in Korean J. Bot.,* 25(4): 177 (1982)—Korea.
serpens
subsp. polyanthemoides *(Boreau) M. Kerguélen & J. Lambinon in Lejeunia,* 120: 149 (1987): R. polyanthemoides.
subsp. polyanthemophyllus *(W. Koch & H. Hess) M. Kerguélen in Lejeunia,* 120: 149 (1987): R. polyanthemophyllus.
triangularis *W.T. Wang in Acta Phytotax. Sin.,* 25(1): 37 (1987)—China.
trichophyllus
var. glaber *(Tamura) H. Hara in Enum. Fl. Pl. Nepal,* 2: 20 (11 June 1979), isonym; see M. Tamura *in Sci. Rep. Osaka Univ.,* 27(2): 114 (1978 publ. 31 March 1979): Batrachium trichophyllum var. glabrum.
zhungdianensis *W.T. Wang in Bull. Bot. Res. North-East. Forest. Inst.,* 7(2): 104 (1987)—China.

RAPHIONACME (Asclepiadac.).
elsana *H.J.T. Venter & R.L. Verhoeven in S. Afr. J. Bot.,* 53(2): 177 (1987)—South Africa (Natal).

RAUVOLFIA (Apocynac.).
×ivanovii *M.M. Granda & V.R. Fuentes in Rev. Jard. Bot. Nacion. Univ. Habana,* 6(2): 42 (1985 publ. 1986)—Cultivation. (R. ligustrina × viridis).

RAVENEA (Palmae).
moorei *J. Dransfield & N.W. Uhl in Principes,* 30(4): 159 (1986)—Comoro Archipelago.

REAUMURIA (Tamaricac.).
babataghi *V.P. Bochantsev in Bot. Zhurn.,* 72(6): 814 (1987)—U.S.S.R. (Middle Asia).

REBUTIA (Cactac.).
cylindrica *(Donald & A. Lau) J.D. Donald in Bradleya,* 5: 93 (1987): Sulcorebutia cylindrica.
fidaiana *(Backeberg) D. Hunt in Bradleya,* 5: 94 (1987): Echinocactus fidaianus.
mentosa *(F. Ritter) J.D. Donald in Bradleya,* 5: 93 (1987): Sulcorebutia mentosa.
neocumingii *(Backeberg) D. Hunt in Bradleya,* 5: 94 (1987): Weingartia neocumingii.
neumanniana *(Werdermann) D. Hunt in Bradleya,* 5: 94 (1987): Weingartia neumanniana.

REGELIA (Myrtac.).
punicea *(N. Byrnes) B.A. Barlow in Brunonia,* 9(1): 95 (1987): Melaleuca punicea.

REGISTANIELLA K.H. Rechinger in Fl. Iran., 162: 152 (1987). *UMBELLIFERAE.*
hapaxlegomena *K.H. Rechinger in Fl. Iran.,* 162: 152 (1987)—Afghanistan.

REINHARDTIA (Palmae).
paiewonskiana *R.W. Read, T.A. Zanoni & M. Mejía in Brittonia,* 39(1): 20 (1987)—Hispaniola (Dominican Republic).

REMIJIA (Rubiac.).
marahuacensis *J.A. Steyermark in Ann. Missouri Bot. Gard.,* 74(1): 111 (1987)—Venezuela.

REMUSATIA (Arac.).
ornata *(Schott) H. Li & Q.F. Guo in Acta Phytotax. Sin.,* 25(5): 414 (1987): Gonatanthus ornatus.

RENEALMIA (Zingiberac.).
cylindrica *P.J.M. Maas & H. Maas in Notes Roy. Bot. Gard. Edinburgh,* 44(2): 242 (1987)—Colombia, Panama, Ecuador.
pirrensis *P.J.M. Maas & H. Maas in Notes Roy. Bot. Gard. Edinburgh,* 44(2): 237 (1987)—Panama.
purpurea *P.J.M. Maas & H. Maas in Notes Roy. Bot. Gard. Edinburgh,* 44(2): 247 (1987)—Peru.
variegata *P.J.M. Maas & H. Maas in Notes Roy. Bot. Gard. Edinburgh,* 44(2): 241 (1987)—Ecuador, Panama, Colombia.

RHINACTINIDIA (Compos.).
eremophila *(Bunge) Novopokr. ex V.P. Bochantsev in Novosti Sist. Vÿssh. Rast.,* 23: 180 (1986): Aster eremophilus.
subsp. grubovii *V.P. Bochantsev in Novosti Sist. Vÿssh. Rast.,* 23: 180 (1986)—Mongolia.
limoniifolia *(Less.) Novopokr. ex V.P. Bochantsev in Novosti Sist. Vÿssh. Rast.,* 23: 179 (1986): Rhinactina limoniifolia.
novopokrovskyi *(Krasch. & Iljin) Novopokr. ex V.P. Bochantsev in Novosti Sist. Vÿssh. Rast.,* 23: 181 (1986): Aster novopokrovskyi.

RHINACTINIDIA :-
popovii *(Botsch.) V.P. Bochantsev in Novosti Sist. Vÿssh. Rast.,* 23: 179 (1986): Aster popovii.

RHIPSALIS (Cactac.).
baccifera
subsp. erythrocarpa *(Schumann) W. Barthlott in Bradleya,* 5: 100 (1987): R. erythrocarpa.
subsp. horrida *(Baker) W. Barthlott in Bradleya,* 5: 100 (1987): R. horrida.
subsp. mauritiana *(DC.) W. Barthlott in Bradleya,* 5: 100 (1987): R. cassytha var. mauritiana.
occidentalis *W. Barthlott & W. Rauh in Kakt. And. Sukk.,* 38(1): 17 (1987)—Peru, Ecuador.

RHODIOLA (Crassulac.).
chrysanthemifolia
subsp. liciae *(R.-Hamet) H. Ohba in Acta Phytotax. Geobot.,* 38: 216 (1987): Sedum liciae.

RHODODENDRON (Ericac.).
ser. Adenogyna *W. Spethmann in Pl. Syst. Evol.,* 157(1–2): 30 (1987), nom. nov.: R. subser. Adenogynum *Tagg.*
subsect. Brachycalyces *W. Spethmann in Pl. Syst. Evol.,* 157(1–2): 28 (1987), nom. nov.: R. sect. Brachycalyx *Sweet.*
sect. Bullatorhodion *W. Spethmann in Pl. Syst. Evol.,* 157(1–2): 25 (1987), nom. nov.: R. ser. Edgeworthii *Hutch.*
ser. Calophyta *W. Spethmann in Pl. Syst. Evol.,* 157(1–2): 29 (1987), nom. nov.: R. subser. Calophytum *Tagg.*
subsect. Canadensia *(Rehd.) W. Spethmann in Pl. Syst. Evol.,* 157(1–2): 28 (1987), nom. nov.: R. subser. Canadense *Rehd.*
subsect. Caucasica *W. Spethmann in Pl. Syst. Evol.,* 157(1–2): 30 (1987), nom. nov.: R. subser. Caucasicum *Tagg.*
subsect. Ciliicalyces *W. Spethmann in Pl. Syst. Evol.,* 157(1–2): 26 (1987), nom. nov.: R. subser. Ciliicalyx *Hutch.*
subsect. Davidia *W. Spethmann in Pl. Syst. Evol.,* 157(1–2): 29 (1987), without type.
ser. Davidia *W. Spethmann in Pl. Syst. Evol.,* 157(1–2): 29 (1987), nom. nov.: R. subser. Davidii *Tagg.*
sect. Destaminorhodion *W. Spethmann in Pl. Syst. Evol.,* 157(1–2): 27 (1987).
ser. Griffithiana *W. Spethmann in Pl. Syst. Evol.,* 157(1–2): 30 (1987), nom. nov.: R. subser. Griffithianum *Tagg.*
sect. Lactanthes *W. Spethmann in Pl. Syst. Evol.,* 157(1–2): 30 (1987).
subsect. Lutea *W. Spethmann in Pl. Syst. Evol.,* 157(1–2): 28 (1987), nom. nov.: R. subser. Luteum *Rehd.*
subgen. Maddenodendron *W. Spethmann in Pl. Syst. Evol.,* 157(1–2): 25 (1987).
sect. Maddenodendron *W. Spethmann in Pl. Syst. Evol.,* 157(1–2): 26 (1987).

RHODODENDRON :–

subsect. Megacalyces *W. Spethmann in Pl. Syst. Evol.*, 157(1–2): 26 (1987), nom. nov.: R. subser. Megacalyx *Hutch.*

sect. Microrhodion *W. Spethmann in Pl. Syst. Evol.*, 157(1–2): 27 (1987), nom. nov.: R. ser. Micranthum *Hutch.*

sect. Moupinensorhodion *W. Spethmann in Pl. Syst. Evol.*, 157(1–2): 25 (1987), nom. nov.: R. ser. Moupinense *Hutch.*

subsect. Obtusa *W. Spethmann in Pl. Syst. Evol.*, 157(1–2): 28 (1987), nom. nov.: R. subser. Obtusum *Rehd.*

ser. Orbicularia *W. Spethmann in Pl. Syst. Evol.*, 157(1–2): 29 (1987), nom. nov.: R. subser. Orbiculare *Tagg.*

subsect. Oreodoxa *W. Spethmann in Pl. Syst. Evol.*, 157(1–2): 29 (1987), without type.

ser. Oreodoxa *(Tagg) W. Spethmann in Pl. Syst. Evol.*, 157(1–2): 29 (1987): R. subser. Oreodoxa.

subsect. Pogonantha *W. Spethmann in Pl. Syst. Evol.*, 157(1–2): 27 (1987), nom. nov., without replaced synonym ref.: R. sect. Pogonanthum *G. Don.*

sect. Pseudorhodion *W. Spethmann in Pl. Syst. Evol.*, 157(1–2): 26 (1987).

ser. Roxieana *W. Spethmann in Pl. Syst. Evol.*, 157(1–2): 30 (1987), nom. nov.: R. subser. Roxieanum *Tagg.*

subsect. Sanguinea *W. Spethmann in Pl. Syst. Evol.*, 157(1–2): 30 (1987), nom. nov.: R. subser. Sanguineum *Tagg.*

subsect. Setosa *W. Spethmann in Pl. Syst. Evol.*, 157(1–2): 27 (1987), nom. nov.: R. sect. Setosum *Philipson & Philipson.*

subsect. Tashiroia *W. Spethmann in Pl. Syst. Evol.*, 157(1–2): 28 (1987), nom. nov.: R. subser. Tashiroi *Rehd.*

ser. Wasonia *W. Spethmann in Pl. Syst. Evol.*, 157(1–2): 31 (1987), nom. nov.: R. subser. Wasonii *Tagg.*

subsect. Yunnanensia *W. Spethmann in Pl. Syst. Evol.*, 157(1–2): 27 (1987).

amakusaense *(Takada ex Yamazaki) T. Yamazaki in J. Jap. Bot.*, 62(3): 72 (1987): R. viscistylum var. amakusaense.

brachycarpum

f. fauriae *(Franch.) G. Murata in Acta Phytotax. Geobot.*, 38: 56 (1987): R. fauriae.

hyugaense *(Yamazaki) T. Yamazaki in J. Jap. Bot.*, 62(3): 72 (1987): R. viscistylum var. hyugaense.

japonoheptamerum *S. Kitamura in Acta Phytotax. Geobot.*, 38: 382 (1987), nom. nov.: Hymenanthes japonica *Blume.*

var. hondoense *(Nakai) S. Kitamura in Acta Phytotax. Geobot.*, 38: 382 (1987): R. metternichii var. hondoense.

kangdingense *Z.J. Zhao in Bull. Bot. Res. North-East. Forest. Inst.*, 7(3): 61 (1987)—China.

leiboense *Z.J. Zhao in Bull. Bot. Res. North-East. Forest. Inst.*, 7(3): 57 (1987)—China.

RHODODENDRON :–

mainlingense *S.H. Huang & R.C. Fang in Fl. Xizangica*, 3: 674 (1986)—Xizang.

mianningense *Z.J. Zhao in Bull. Bot. Res. North-East. Forest. Inst.*, 7(3): 59 (1987)—China.

nanjianense *K.M. Feng & Z.H. Yang in Acta Bot. Yunnanica*, 9(1): 55 (1987)—China.

nyingchiense *R.C. Fang & S.H. Huang in Fl. Xizangica*, 3: 674 (1986)—Xizang.

periclymenoides

f. album *(Ait.) C.F. Reed in Phytologia*, 63(5): 411 (1987): Azalea nudiflora δ alba.

f. glandiferum *(Porter) C.F. Reed in Phytologia*, 63(5): 411 (1987): Azalea nudiflora var. glandifera.

principis

var. vellereum *(Hutch. ex Tagg.) T.L. Ming in Fl. Xizangica*, 3: 621 (1986): R. vellereum.

reticulatum

var. bifolium *T. Yamazaki in J. Jap. Bot.*, 62(9): 288 (1987)—Japan.

rubropunctatum *T.L. Ming in Fl. Xizangica*, 3: 603 (1986), as 'rubro-punctatum'—Xizang.

saxatile *B.Y. Ding & Y.Y. Fang in Bull. Bot. Res. North-East. Forest. Inst.*, 7(2): 29 (1987)—China.

taishunense *B.Y. Ding & Y.Y. Fang in Bull. Bot. Res. North-East. Forest. Inst.*, 7(2): 27 (1987)—China.

tenuifolium *R.C. Fang & S.H. Huang in Fl. Xizangica*, 3: 647 (1986)—Xizang.

xichangense *Z.J. Zhao in Bull. Bot. Res. North-East. Forest. Inst.*, 7(3): 60 (1987)—China.

yakumontanum *(Yamazaki) T. Yamazaki in J. Jap. Bot.*, 62(3): 72 (1987): R. nudipes subsp. yakumontanum.

yedoense

var. hallaisanense *(Léveillé) T. Yamazaki in J. Jap. Bot.*, 62(9): 260 (1987): R. hallaisanense.

RHODOPHIALA (Amaryllidac.).

herbertiana *(Lindley) A.T. Hunziker in Lorentzia*, 5: 13 (1985): Phycella herbertiana.

phycelloides *(Herbert) A.T. Hunziker in Lorentzia*, 5: 15 (1985): Habranthus phycelloides.

RHUS (Anacardiac.).

radicans

subsp. negundo *(Greene) J. McNeill in Nat. Canad.*, 108(3): 238 (1981): Toxicodendron negundo.

subsp. rydbergii *(Small ex Rydb.) J. McNeill in Nat. Canad.*, 108(3): 238 (1981): R. rydbergii.

RHYNCHOSIA (Legumin.).

emarginata *G. Germishuizen in Bothalia*, 17(2): 181 (1987)—South Africa (Cape Province).

RHYNCHOSPORA (Cyperac.).

calderana *D.A. Simpson in Kew Bull.*, 42(4): 891 (1987)—Brazil.

RHYNCHOSPORA :–
ierensis *C.D. Adams in Phytologia,* 63(4): 298 (1987)—Trinidad.
pirrensis *W.W. Thomas in Brittonia,* 38(4): 314 (1986)—Panama.

RHYNCHOSTYLIS (Orchidac.).
gigantea
subsp. violacea *(Lindl.) E.A. Christenson in Indian Orchid J.,* 1(4):154 (1985); *et in Selbyana,* 9(1): 169 (1986): Vanda violacea.

RITCHIEA (Capparac.).
simplicifolia
var. caloneura *(Gilg) L.E. Kers in Fl. Cameroun,* 29: 102 (1986): R. caloneura.

ROEMERIA (Papaverac.).
refracta
subsp. occidentalis *J.W. Kadereit in Flora (Germany),* 179(2): 138 (1987)—Iran, Turkey, U.S.S.R. (Middle Asia).

ROLLANDIA (Campanulac.).
alba *H. St. John & W. Takeuchi in Phytologia,* 63(5): 367 (1987)—Hawaiian Is.
obatae *H. St. John in Phytologia,* 63(5): 367 (1987)—Hawaiian Is.

ROMULEA (Iridac.).
unifolia *M.P. de Vos in S. Afr. J. Bot.,* 53(3): 247 (1987)—South Africa (Cape Province).

RONDELETIA (Rubiac.).
anguillensis *R.A. Howard & E.A. Kellogg in J. Arnold Arbor.,* 68(1): 127 (1987)—Leeward Is.
arida *A. Borhidi & M. Fernandez Zequeira in Acta Bot. Hung.,* 33(1–2): 105 (1987)—Cuba.
bissei *A. Borhidi & M. Fernandez Zequeira in Acta Bot. Hung.,* 33(1–2): 107 (1987)—Cuba.
bracteosa *A. Borhidi & M. Fernandez Zequeira in Acta Bot. Hung.,* 33(1–2): 109 (1987)—Cuba.
steirophylloides *A. Borhidi & M. Fernandez Zequeira in Acta Bot. Hung.,* 33(1–2): 112 (1987)—Cuba.
tubulosa *A. Borhidi & M. Fernandez Zequeira in Acta Bot. Hung.,* 33(1–2): 113 (1987)—Cuba.

RONNBERGIA (Bromeliac.).
carvalhoi *G. Martinelli & E.M.C. Leme in J. Bromeliad Soc.,* 37(2): 79 (1987)—Brazil.

RORIPPA (Crucif.).
pseudoislandica *H.J. Chowdhery & R.R. Rao in Indian J. Forest.,* 8(2): 150 (1985)—India.
pyrenaica
subsp. hispanica *(Boiss. & Reut.) M. Kerguélen in Lejeunia,* 120: 150 (1987): Nasturtium hispanicum.
viaria *V.I. Dorofeev in Bot. Zhurn.,* 71(11): 1559 (1986)—U.S.S.R. (European Russia).

ROSA (Rosac.).
dumalis
f. dentatistipulata *R. Popek in Fragm. Flor. Geobot.,* 29(3–4): 346 (1983 publ. 1986)—Poland.
f. nervipilosa *R. Popek in Fragm. Flor. Geobot.,* 29(3–4): 347 (1983 publ. 1986)—Poland.
f. subglabristyla *R. Popek in Fragm. Flor. Geobot.,* 29(3–4): 347 (1983 publ. 1986)—Poland.
kostrakiewiczii *R. Popek in Fragm. Flor. Geobot.,* 29(3–4): 350 (1983 publ. 1986): R. pubescens subsp. kostrakiewiczii.
obtusifolia
var. dunensis *C. Arnaiz, J.-M. Gehu & J. Gehu-Franck in Bull. Soc. Bot. Nord France,* 33(3–4): 72 (1980 publ. 1981), without type—France.
platyacantha
subsp. crassipes *(Vass.) I.O. Buzunova in Novosti Sist. Vÿssh. Rast.,* 23: 77 (1986): R. platyacantha var. crassipes.
rubiginosa
var. dunensis *C. Arnaiz, J.-M. Gehu & J. Gehu-Franck in Bull. Soc. Bot. Nord France,* 33(3–4): 69 (1980 publ. 1981), without type—France.

ROSMARINUS (Labiat.).
×lavandulaceus
var. clementei *J.L. Rosúa in Lagascalia,* 14(2): 185 (1986)—Spain.
var. noeanus *(Maire) J.L. Rosúa in Lagascalia,* 14(2): 184 (1986): R. × noeanus.
var. subtomentosus *(Maire & Weiller) J.L. Rosúa in Lagascalia,* 14(2): 185 (1986): R. officinalis var. subtomentosus.
var. trogloditarum *(Maire & Weiller) J.L. Rosúa in Lagascalia,* 14(2): 185 (1986): R. officinalis var. trogloditarum.

ROSULARIA (Crassulac.).
blepharophylla *U. Eggli in Kakt. And. Sukk.,* 38(6): 134 (1987)—Turkey.
sempervivum
subsp. amanensis *U. Eggli in Kakt. And. Sukk.,* 38(6): 135 (1987)—Turkey.
subsp. glaucophylla *U. Eggli in Kakt. And. Sukk.,* 38(6): 136 (1987)—Turkey.
serpentinica
var. gigantea *U. Eggli in Kakt. And. Sukk.,* 386: 137 (1987)—Turkey.

ROTALA (Lythrac.).
rosea *(Poir.) C.D.K. Cook ex H. Hara in Enum. Fl. Pl. Nepal,* 2: 173 (11 June 1979): Ammannia rosea.

RUBUS (Rosac.).
subgen. Micranthobatus *(Fritsch) C. Kalkman in Blumea,* 32(2): 324 (1987): R. sect. Micranthobatus.
cooperi *D.G. Long in Notes Roy. Bot. Gard. Edinburgh,* 44(2): 259 (1987)—Bhutan.

RUBUS :–
cordiformis *C. Kalkman in Blumea*, 32(2): 331 (1987)—New Guinea (Papua New Guinea).
haesitans *H.O. Martensen & Walsemann in Svensk Bot. Tidskr.*, 81(4): 265 (1987)—Germany, Denmark.
illecebrosus
var. yakusimensis *(Masamune) Hatusima ex H. Ohba in J. Fac. Sci. Univ. Tokyo, Bot.*, 14(2): 89 (1987): R. yakusimensis.
juennensis *G.H. Leute & W. Maurer in Carinthia II*, 173/93: 237 (1983)—Austria.
niveus
var. micranthus *(D. Don) H. Hara in Enum. Fl. Pl. Nepal*, 2: 146 (1979): R. micranthus.
palmatus
var. yakumontanus *(Masam.) Hatusima ex H. Ohba in J. Fac. Sci. Univ. Tokyo, Bot.*, 14(2): 90 (1987): R. yakumontanus.
royenii *C. Kalkman in Blumea*, 32(2): 333 (1987)—New Guinea (Papua New Guinea, W. Irian), New Britain.
var. hispidus *C. Kalkman in Blumea*, 32(2): 336 (1987)—New Guinea (Papua New Guinea), New Britain.
var. ikilimbu *(P. Royen) C. Kalkman in Blumea*, 32(2): 336 (1987): R. diclinis var. ikilimbu.
sikkimensis
var. canescens *Hara ex D.G. Long in Notes Roy. Bot. Gard. Edinburgh*, 44(2): 262 (1987)—Bhutan.
widderi *W. Maurer in Phyton (Austria)*, 26(2): 227 (1987)—Austria.

RUDBECKIA (Compos.).
scabrifolia *L.E. Brown in Phytologia*, 61(6): 367 (1986)—U.S.A. (Texas).

RUELLIA (Acanthac.).
chartacea *(T. Anders.) D.C. Wasshausen in Op. Bot.*, 92: 265 (1987): Aphelandra chartacea.
golfodulcensis *L.H. Durkee in Fieldiana, Bot.*, n.s., 18: 75 (1986)—Costa Rica.
palustris *L.H. Durkee in Fieldiana, Bot.*, n.s., 18: 77 (1986)—Costa Rica.

RUMEX (Polygonac.).
gangotrianus *B.S. Aswal & S.K. Srivastava in Candollea*, 42(1): 389 (1987)—India.

RUPIPHILA M.G. Pimenov & T.V. Lavrova in Byull. Mosk. Obshch. Ispȳt. Prir., Biol., 91(2): 97 (1986). *UMBELLIFERAE.*
tachiroei *(Franch. & Savat.) M.G. Pimenov & T.V. Lavrova in Byull. Mosk. Obshch. Ispȳt. Prir., Biol.*, 91(2): 97 (1986): Seseli tachiroei.

RYTIGYNIA (Rubiac.).
subgen. Fadogiopsis *B. Verdcourt in Kew Bull.*, 42(1): 185 (1987).
subgen. Sali *B. Verdcourt in Kew Bull.*, 42(1): 185 (1987).

RYTIGYNIA :–
acuminatissima
subsp. pedunculata *B. Verdcourt in Kew Bull.*, 42(1): 154 (1987)—Burundi.
adenodonta
var. reticulata *(Robyns) B. Verdcourt in Kew Bull.*, 42(1): 167 (1987): R. reticulata.
bridsoniae
subsp. kahuzica *B. Verdcourt in Kew Bull.*, 42(1): 160 (1987)—Zaïre.
bugoyensis
subsp. glabriflora *B. Verdcourt in Kew Bull.*, 42(1): 184 (1987)—Tanzania, Malawi.
caudatissima *B. Verdcourt in Kew Bull.*, 42(1): 180 (1987)—Tanzania.
celastroides *(Baill.) B. Verdcourt in Kew Bull.*, 42(1): 171 (1987): Canthium celastroides.
var. australis *B. Verdcourt in Kew Bull.*, 42(1): 172 (1987)—Mozambique.
var. nuda *B. Verdcourt in Kew Bull.*, 42(1): 171 (1987)—Zanzibar, Tanzania.
griseovelutina *B. Verdcourt in Kew Bull.*, 42(1): 183 (1987)—Tanzania.
hirsutiflora *B. Verdcourt in Kew Bull.*, 42(1): 183 (1987)—Tanzania.
ignobilis *B. Verdcourt in Kew Bull.*, 42(1): 164 (1987)—Tanzania.
induta *(Bullock) B. Verdcourt & D.M. Bridson in Kew Bull.*, 42(1): 169 (1987): Canthium indutum.
lichenoxenos
subsp. glabrituba *B. Verdcourt in Kew Bull.*, 42(1): 169 (1987)—Tanzania.
longicaudata *B. Verdcourt in Kew Bull.*, 42(1): 180 (1987)—Tanzania.
macrura *B. Verdcourt in Kew Bull.*, 42(1): 178 (1987)—Zimbabwe, Malawi, Mozambique.
monantha
var. lusakati *B. Verdcourt in Kew Bull.*, 42(1): 174 (1987)—Tanzania.
mrimaensis *B. Verdcourt in Kew Bull.*, 42(1): 170 (1987)—Kenya.
neglecta
var. vatkeana *(Hiern) B. Verdcourt in Kew Bull.*, 42(1): 158 (1987): Canthium vatkeanum.
parvifolia *B. Verdcourt in Kew Bull.*, 42(1): 172 (1987)—Kenya.
pawekiae *B. Verdcourt in Kew Bull.*, 42(1): 168 (1987), as 'pawekae'—Malawi.
pergracilis *B. Verdcourt in Kew Bull.*, 42(1): 157 (1987)—Tanzania.
pseudolongicaudata *B. Verdcourt in Kew Bull.*, 42(1): 182 (1987)—Tanzania.
pubescens *B. Verdcourt in Kew Bull.*, 42(1): 168 (1987)—Tanzania.
rubiginosa
subsp. cymigera *(Brem.) B. Verdcourt in Kew Bull.*, 42(1): 175 (1987): Hutchinsonia cymigera.
saliensis *B. Verdcourt in Kew Bull.*, 42(1): 185 (1987)—Tanzania.
uhligii *(K. Schum. & K. Krause) B. Verdcourt in Kew Bull.*, 42(1): 165 (1987): Vangueria uhligii.

S

SABAZIA (Compos.).
gonzaleziae *B.L. Turner in Phytologia,* 63(4): 307 (1987), as 'gonzalezae'—Mexico.

SAGERETIA (Rhamnac.).
thea
var. bornmuelleri *(Schneid.) M.C. Johnston ex H. Hara in Enum. Fl. Pl. Nepal,* 2: 92 (1979): S. brandrethiana var. bornmuelleri.

SALACCA (Palmae).
graciliflora *J.P. Mogea in Fed. Mus. J. (Kuala Lumpur),* 29: 6 (1984)—Malaya.
minuta *J.P. Mogea in Fed. Mus. J. (Kuala Lumpur),* 29: 11 (1984)—Malaya.
multiflora *J.P. Mogea in Fed. Mus. J. (Kuala Lumpur),* 29: 13 (1984)—Malaya.
ramosiana *J.P. Mogea in Principes,* 30(4): 161 (1986)—Borneo (Sabah), Philippine Is.

SALACIA (Celastrac.).
alwynii *A.M.W. Mennega in Proc. Kon. Nederl. Akad. Wetensch., C,* 87(1): 84 (1984)—Peru, Venezuela.
annettae *N. Hallé in Fl. Gabon,* 29: 134 (1986)—Gabon.
belingana *N. Hallé in Fl. Gabon,* 29: 148 (1986)—Gabon.
bullata *A.M.W. Mennega in Proc. Kon. Nederl. Akad. Wetensch., C,* 87(1): 81 (1984)—Brazil.
capitulata *N. Hallé in Fl. Gabon,* 29: 116 (1986)—Gabon, Cameroun, Zaïre.
cerasifera
var. wendjiensis *(Wilczek) N. Hallé in Fl. Gabon,* 29: 110 (1986): S. wendjiensis.
coronata *N. Hallé in Fl. Gabon,* 29: 154 (1986)—Gabon.
dimidia *N. Hallé in Fl. Gabon,* 29: 102 (1986)—Cameroun.
diplasia *N. Hallé in Fl. Gabon,* 29: 101 (1986)—Gabon.
elegans
var. inurbana *N. Hallé in Fl. Gabon,* 29: 153 (1986)—Gabon.
erecta
var. dewildemaniana *(Wilczek) N. Hallé in Fl. Gabon,* 29: 115 (1986): S. dewildemaniana.
var. kabweensis *(Wilczek) N. Hallé in Fl. Gabon,* 29: 115 (1986): S. kabweensis.
var. leonardii *(Wilczek) N. Hallé in Fl. Gabon,* 29: 115 (1986): S. leonardii.
ferrifodina *N. Hallé in Fl. Gabon,* 29: 104 (1986)—Gabon.
germainii
var. chlorion *(Hallé) N. Hallé in Fl. Gabon,* 29: 109 (1986): S. chlorion.
lehmbachii
var. aurantiaca *(Hallé) N. Hallé in Fl. Gabon.,* 29: 74 (1986): S. uregaensis var. aurantiaca.

SALACIA :–
var. leonensis *(Hutchinson & M.B. Moss) N. Hallé in Fl. Gabon,* 29: 73 (1986): S. leonensis.
var. manus-lacertae *N. Hallé in Fl. Gabon,* 29: 73 (1986)—Gabon.
var. pes-ranulae *N. Hallé in Fl. Gabon,* 29: 75 (1986)—Cameroun.
var. uregaensis *(Wilczek) N. Hallé in Fl. Gabon,* 29: 74 (1986): S. uregaensis.
letouzeyana *N. Hallé in Fl. Gabon,* 29: 86 (1986)—Cameroun, Nigeria.
loloensis
var. sibangana *N. Hallé in Fl. Gabon,* 29: 82 (1986)—Gabon, Cameroun, Nigeria.
longipes *(Oliver) N. Hallé in Fl. Gabon,* 29: 65 (1986): Hippocratea ? longipes.
var. camerunensis *(Loesener) N. Hallé in Fl. Gabon,* 29: 66 (1986): S. ? camerunensis.
var. longipetiolata *(Loesener) N. Hallé in Fl. Gabon,* 29: 69 (1986): S. camerunensis var. longipetiolata.
mamba *N. Hallé in Fl. Gabon,* 29: 145 (1986)—Gabon, Cameroun, Congo - Brazzaville.
nitida
var. bipindensis *(Loesener) N. Hallé in Fl. Gabon,* 29: 144 (1986): S. bipindensis.
paradoxa *A.M.W. Mennega in Proc. Kon. Nederl. Akad. Wetensch., C,* 87(1): 87 (1984)—Brazil.
pierrei *N. Hallé in Fl. Gabon,* 29: 128 (1986)—Gabon.
preussii
var. louisii *(Wilczek) N. Hallé in Fl. Gabon,* 29: 52 (1986): S. louisii.
staudtiana
var. cerasiocarpa *(Wilczek) N. Hallé in Fl. Gabon,* 29: 63 (1986): S. tshopoensis var. cerasiocarpa.
var. scaphodisca *N. Hallé in Fl. Gabon,* 29: 64 (1986)—Gabon, Congo - Brazzaville.
var. tshopoensis *(De Wildeman) N. Hallé in Fl. Gabon,* 29: 62 (1986): S. tshopoensis.
subicterica *N. Hallé in Fl. Gabon,* 29: 84 (1986)—Gabon.
tuberculata
var. ngaziensis *(Louis ex Wilczek) N. Hallé in Fl. Gabon,* 29: 123 (1986): S. ngaziensis.
villiersii *N. Hallé in Fl. Gabon,* 29: 123 (1986)—Gabon.
whytei
var. toussaintii *(Wilczek) N. Hallé in Fl. Gabon,* 29: 100 (1986): S. toussaintii.
var. vermoeseniana *(Wilczek) N. Hallé in Fl. Gabon,* 29: 96 (1986): S. vermoeseniana.

SALICORNIA (Chenopodiac.).
borealis *S.L. Wolff & R.L. Jefferies in Canad. J. Bot.,* 65(7): 1425 (1987)—Canada (Manitoba).
maritima *S.L. Wolff & R.L. Jefferies in Canad. J. Bot.,* 65(7): 1424 (1987)—Canada (Quebec, New Brunswick, Nova Scotia).

SALICORNIA :–
oliveri
var. nidiformis *(König) M. Kerguélen in Lejeunia,* 120: 153 (1987): S. stricta subsp. nidiformis.

SALIX (Salicac.).
nothosect. Amylix *T.E. Díaz Gonzalez & F. Llamas in Acta Bot. Malacitana,* 12: 126 (1987). (S. sect. Amygdaline × sect. Salix).
nothosect. Amytrix *T.E. Díaz Gonzalez & F. Llamas in Acta Bot. Malacitana,* 12: 133 (1987). (S. sect. Amygdalinae × sect. Vetrix).
nothosect. Calix *T.E. Díz Gonzalez & F. Llamas in Acta Bot. Malacitana,* 12: 133 (1987). (S. sect. Canae × sect. Helix).
nothosect. Camen *T.E. Díaz Gonzalez & F. Llamas in Acta Bot. Malacitana,* 12: 132 (1987). (S. sect. Canae × sect. Vimen).
nothosect. Catrix *T.E. Díaz Gonzalez & F. Llamas in Acta Bot. Malacitana,* 12: 131 (1987). (S. sect. Canae × sect. Vetrix).
nothosect. Hetrix *T.E. Díaz Gonzalez & F. Llamas in Acta Bot. Malacitana,* 12: 132 (1987). (S. sect. Helix × sect. Vetrix).
nothosubgen. Velix *T.E. Díaz Gonzalez & F. Llamas in Acta Bot. Malacitana,* 12: 133 (1987). (S. subgen. Salix × subgen. Vetrix).
nothosect. Vimetrix *T.E. Díaz Gonzalez & F. Llamas in Acta Bot. Malacitana,* 12: 130 (1987). (S. sect. Vetrix × sect. Vimen).
annulifera
var. glabra *P.Y. Mao & W.Z. Li in Acta Bot. Yunnanica,* 9(1): 49 (1987)—China.
contortiapiculata *P.Y. Mao & W.Z. Li in Acta Bot. Yunnanica,* 9(1): 43 (1987)—China.
daguanensis *P.Y. Mao & P.X. He in Acta Bot. Yunnanica,* 9(1): 45 (1987)—China.
×expectata *Rivas-Martínez, T.E. Díaz Gonzalez, J.A. Prieto, J. Loidi & A. Penas in Acta Bot. Malacitana,* 12: 128 (1987)—Spain. (S. atrocinerea × cantabrica).
fengiana
var. gymnocarpa *P.Y. Mao & W.Z. Li in Acta Bot. Yunnanica,* 9(1): 53 (1987)—China.
floccosa
var. leoigyna *P.Y. Mao & W.Z. Li in Acta Bot. Yunnanica,* 9(1): 49 (1987)—China.
glarearum *P.Y. Mao & W.Z. Li in Acta Bot. Yunnanica,* 9(1): 49 (1987), as 'glareorum'—China.
×longissima *T.E. Díaz Gonzalez & J. Andrés in Acta Bot. Malacitana,* 12: 132 (1987)—Spain. (S. elaeagnos subsp. angustifolia × viminalis).
magnifica
var. microphylla *P.Y. Mao in Acta Bot. Yunnanica,* 9(1): 45 (1987)—China.
mucronata
subsp. capensis *(Thunb.) K.L. Immelman in Bothalia,* 17(2): 173 (1987): S. capensis.
subsp. hirsuta *(Thunb.) K.L. Immelman in Bothalia,* 17(2): 173 (1987): S. hirsuta.

SALIX :–
subsp. wilmsii *(Seemen) K.L. Immelman in Bothalia,* 17(2): 176 (1987): S. wilmsii.
subsp. woodii *(Seemen) K.L. Immelman in Bothalia,* 17(2): 176 (1987), without basionym page no. which is same as Beibl. no.: S. woodii.
×multidentata *T.E. Díaz Gonzalez & F. Llamas in Acta Bot. Malacitana,* 12: 134 (1987)—Spain. (S. atrocinerea × triandra subsp. discolor).
oreinoma
var. wawashanica *P.Y. Mao & P.X. He in Acta Bot. Yunnanica,* 9(1): 47 (1987)—China.
×pormensis *T.E. Díaz Gonzalez & F. Llamas in Acta Bot. Malacitana,* 12: 129 (1987)—Spain. (S. cantabrica × caprea).
×pseudoelaeagnos *T.E. Díaz Gonzalez & F. Llamas in Acta Bot. Malacitana,* 12: 133 (1987)—Spain. (S. elaeagnos subsp. angustifolia × purpurea subsp. lambertiana).
×pseudosalvifolia *T.E. Díaz Gonzalez & E. Puente in Acta Bot. Malacitana,* 12: 131 (1987)—Spain. (S. elaeagnos subsp. angustifolia × salvifolia).
×rijosa *Rivas-Martínez, T.E. Díaz Gonzalez, J.A.F. Prieto, J. Loidi & A. Penas in Acta Bot. Malacitana,* 12: 134 (1987)—Spain. (S. cantabrica × triandra subsp. discolor).
rivulicola *P.Y. Mao & W.Z. Li in Acta Bot. Yunnanica,* 9(1): 47 (1987)—China.
scopulicola *P.Y. Mao & W.Z. Li in Acta Bot. Yunnanica,* 9(1): 51 (1987)—China.
×viridifolia *T.E. Díaz Gonzalez & A. Penas in Acta Bot. Malacitana,* 12: 130 (1987)—Spain. (S. salvifolia × viminalis).

SALSOLA (Chenopodiac.).
griffithii *(Bunge) H. Freitag & K. Khani in Pl. Syst. Evol.,* 155(1–4): 50 (1987): Noaea griffithii.

SALVIA (Labiat.).
blancoana
subsp. amethystea *(Emb. & Maire) R. Figuerola in An. Jard. Bot. Madrid,* 44(1): 174 (1987): S. aucheri var. amethystea.
var. aurasiaca *(Maire) R. Figuerola in An. Jard. Bot. Madrid,* 44(1): 174 (1987): S. aucheri var. aurasiaca.
var. lagascana *(Webb) R. Figuerola in An. Jard. Bot. Madrid,* 44(1): 174 (1987): S. lavandulifolia var. lagascana.
subsp. mesatlantica *(Maire) R. Figuerola in An. Jard. Bot. Madrid,* 44(1): 174 (1987): S. aucheri var. mesatlantica.
var. vellerea *(Cuatrec.) R. Figuerola in An. Jard. Bot. Madrid,* 44(1): 174 (1987): S. officinalis var. vellerea.
eizi-matudae *T.P. Ramamoorthy in Taxon,* 36(3): 588 (1987), nom. nov.: S. matudae *Epling.*

SALVIA :–
lavandulifolia
var. adenostachys *(O. Bolòs & Vigo) R. Figuerola in An. Jard. Bot. Madrid,* 44(1): 174 (1987): S. officinalis f. adenostachys.
subsp. approximata *(Pau) R. Figuerola in An. Jard. Bot. Madrid,* 43(1): 178 (1986): S. approximata.
var. trichostachya *(Font Quer ex O. Bolòs & Vigo) R. Figuerola in An. Jard. Bot. Madrid,* 44(1): 174 (1987): S. officinalis f. trichostachya.
manantlanensis *T.P. Ramamoorthy in Bull. Mus. Nation. Hist. Nat., B, Adansonia,* Ser. 4, 9(2): 173 (1987)—Mexico.

SAMBUCUS (Caprifoliac.).
subgen. Ebulus *(Spach) M.L. Samutina in Bot. Zhurn.,* 71(11): 1558 (1986): S. sect. Ebulus.
ser. Latipinnae *Pojark. ex V.A. Nedoluzhko in Sosud. Rast. Sovet. Dal'nego Vostoka,* 2: 293 (1987).
ser. Racemosae *V.A. Nedoluzhko in Sosud. Rast. Sovet. Dal'nego Vostoka,* 2: 294 (1987).
sect. Wightia *M.L. Samutina in Bot. Zhurn.,* 71(11): 1558 (1986).
nigra
f. porphyrophylla *E.C. Nelson in Plantsman,* 8(3): 190 (1986)—Cultivation ex Great Britain.
pubens
var. melanocarpa *(A. Gray) M.L. Samutina in Novosti Sist. Vyssh. Rast.,* 23: 173 (1986): S. melanocarpa.
sibirica
subsp. miquelii *(Nakai) M.L. Samutina in Novosti Sist. Vyssh. Rast.,* 23: 169 (1986): S. racemosa var. miquelii.
sieboldiana
var. pinnatisecta *G.Y. Luo & P.H. Huang in Bull. Bot. Res. North-East. Forest. Inst.,* 7(2): 147 (1987)—China.
williamsii
subsp. coreana *(Nakai) M.L. Samutina in Novosti Sist. Vyssh. Rast.,* 23: 176 (1986): S. sieboldiana var. coreana.

SANGUISORBA (Rosac.).
officinalis
subsp. longifolia *(Bertol.) K.M. Purohit & G. Panigrahi in Bull. Bot. Soc. Bengal,* 36(1–2): 34 (1982 publ. 1986): S. longifolia.

SANICULA (Umbellif.).
haleakalae *H. St. John in Bot. Jahrb.,* 108(1): 52 (1986)—Hawaiian Is.
lobata *(Hillebr. ex Wolff) H. St. John in Bot. Jahrb.,* 108(1): 57 (1986): S. sandwicensis var. lobata.
mariversa *K.M. Nagata & S.M. Gon in Syst. Bot.,* 12(3): 406 (1987)—Hawaiian Is.

SARCOCHILUS (Orchidac.).
gildasii *N. Hallé in Bull. Mus. Nation. Hist. Nat., B, Adansonia,* Ser. 4, 8(3): 234 (1986)—New Caledonia.
hillii
var. thycolus *N. Hallé in Bull. Mus. Nation. Hist. Nat., B, Adansonia,* Ser. 4, 8(3): 232 (1986)—New Caledonia.

SARCOGLYPHIS (Orchidac.).
comberi *(J.J. Wood) J.J. Wood in Orchid Rev.,* 95(1123): 167 (1987): Cleisostoma comberi.

SARCOMELICOPE (Rutac.).
petiolaris *(A. Gray) A.C. Smith in Fl. Vitiensis Nova,* 3: 508 (1985): Acronychia petiolaris.

SASA (Gramin.).
oblongula *C.H. Hu in J. Bamboo Res.,* 6(4): 18 (1987)—China.

SASAELLA (Gramin.).
caudiceps
var. psilovaginula *S. Suzuki in J. Jap. Bot.,* 62(9): 278 (1987)—Japan.

SATUREJA (Labiat.).
laxiflora
subsp. zuvandica *(D. Kapanadze) D.A. Kapanadze in Fl. Gruzii, ed. 2,* 11: 188 (1987): S. zuvandica.

SATYRIA (Ericac.).
boliviana *J.L. Luteyn in Op. Bot.,* 92: 124 (1987)—Bolivia.

SAUROPUS (Euphorbiac.).
sect. Synostemon *(F. v. Muell.) P.T. Li in Acta Phytotax. Sin.,* 25(2): 137 (1987): gen. Synostemon.
reticulatus *X.L. Mo ex P.T. Li in Acta Phytotax. Sin.,* 25(2): 133 (1987)—China.
tsiangii *P.T. Li in Acta Phytotax. Sin.,* 25(2): 135 (1987)—China.
yanhuianus *P.T. Li in Acta Phytotax. Sin.,* 25(2): 134 (1987)—China.

SAUSSUREA (Compos.).
gossypiphora
var. conaensis *S.W. Liu in Fl. Xizangica,* 4: 879 (1985)—Xizang.
grandiceps *S.W. Liu in Fl. Xizangica,* 4: 874 (1985)—Xizang.
gyacaensis *S.W. Liu in Fl. Xizangica,* 4: 877 (1985)—Xizang.
nuda
subsp. densa *(Hook.) G.W. Douglas in Canad. J. Bot.,* 64(11): 2727 (1986): S. alpina var. densa.
pubifolia *S.W. Liu in Fl. Xizangica,* 4: 870 (1985)—Xizang.
var. lhasaensis *S.W. Liu in Fl. Xizangica,* 4: 871 (1985)—Xizang.
tridactyla
var. maiduoganla *S.W. Liu in Fl. Xizangica,* 4: 879 (1985)—Xizang.

SAUVAGESIA (Ochnac.).
falcisepala *C. Sastre in Phytologia,* 59(5): 313 (21 February 1986); *et in Bull. Mus. Nation. Hist. Nat., B, Adansonia, Ser. 4,* 8(1): 13 (8 July 1986); *see Bull. Mus. Nation. Hist. Nat., B, Adansonia, Ser. 4,* 9(2): 124 (1987)— Venezuela. †
guianensis
subsp. araracuarensis *(Sastre) J.A. Steyermark in Ann. Missouri Bot. Gard.,* 74(1): 101 (1987): S. guianensis var. araracuarensis.
subsp. gleasoniana *(Lasser) J.A. Steyermark in Ann. Missouri Bot. Gard.,* 74(1): 101 (1987): Leitgebia gleasoniana.
subsp. guaiquinimensis *J.A. Steyermark in Ann. Missouri Bot. Gard.,* 74(1): 101 (1987)—Venezuela.
subsp. sipapoensis *J.A. Steyermark in Ann. Missouri Bot. Gard.,* 74(1): 101 (1987)—Venezuela.
marahuacensis *J.A. Steyermark in Ann. Missouri Bot. Gard.,* 74(1): 98 (1987)—Venezuela.

SAXIFRAGA (Saxifragac.).
×akinfievii
nothovar. oettingenii *(Galuško & Kudrjašova) R. Horný & K. Mirko Webr, Lomikameny sekce Porophyllum,* 1: 125 (1985): S. × oettingenii.
amabilis *H. Ohba & M. Wakabayashi in J. Jap. Bot.,* 62(6): 162 (1987)—Nepal.
carpetana
subsp. atlantica *(Boiss. & Reuter) A.M. Romo in Treb. Inst. Bot. Barcelona,* 11: 16 (1987): S. carpetana var. atlantica.
decipiens
subsp. sternbergii *(Willd.) M. Kerguélen in Lejeunia,* 120: 155 (1987): S. sternbergii.
harai *H. Ohba & M. Wakabayashi in J. Jap. Bot.,* 62(6): 163 (1987)—Nepal.
mallae *H. Ohba & M. Wakabayashi in J. Jap. Bot.,* 62(6): 165 (1987)—Nepal.
marginata
var. bubakii *(Rohlena) R. Horný in R. Horný & K. Mirko Webr, Lomikameny sekce Porophyllum,* 1: 96 (1985): S. rocheliana var. bubakii.
mucronulata
subsp. sikkimensis *(Hultén) H. Hara in Enum. Fl. Pl. Nepal,* 2: 153 (1979): S. flagellaris subsp. sikkimensis.
obtusa *(Sprague) R. Horný & K. Mirko Webr, Lomikameny sekce Porophyllum,* 1: 106 (1985): S. scardica var. obtusa.
pentadactylis
subsp. almanzorii *P. Vargas in An. Jard. Bot. Madrid,* 43(2): 457 (1986 publ. 1987)—Spain.
×prudhommei *P. Aubin in Bull. Mens. Soc. Linn. Lyon,* 56(1): 15 (1987)—France. (S. granulata × intricata).

SAXIFRAGA :–
spruneri
var. deorum *(Pénzes) R. Horný & K. Mirko Webr, Lomikameny sekce Porophyllum,* 1: 117 (1985): S. spruneri subsp. deorum.
stenophylla
subsp. hoffmeisteri *(Klotzsch) H. Hara in Enum. Fl. Pl. Nepal,* 2: 155 (1979): S. hoffmeisteri.

SCABIOSA (Dipsacac.).
tschiliensis
var. superba *(Grün.) S.Y. He in Fl. Reipubl. Popul. Sin.,* 73(1): 81 (1986): S. superba.

SCAEVOLA (Goodeniac.).
albida
var. pallida *(R. Br.) R.C. Carolin in Fl. S. Australia,* 3: 1410 (1986): S. pallida.
angustata *R.C. Carolin in Fl. S. Australia,* 3: 1410 (1986)—Australia (S. Australia).
glabrata *R.C. Carolin in Fl. S. Australia,* 3: 1412 (1986), nom. nov.: S. ovalifolia β glabra *R. Br.*
parvibarbata *R.C. Carolin in Fl. S. Australia,* 3: 1414 (1986)—Australia (N. Terr., Queensland, New South Wales, W. Australia, S. Australia).

SCALIGERIA (Umbellif.).
gongylotaxis *K.H. Rechinger in Fl. Iran.,* 162: 226 (1987)—Afghanistan.

SCAPHOCHLAMYS (Zingiberac.).
argentea *R.M. Smith in Notes Roy. Bot. Gard. Edinburgh,* 44(2): 209 (1987)—Borneo (Sarawak).
petiolata *(K. Schum.) R.M. Smith in Notes Roy. Bot. Gard. Edinburgh,* 44(2): 210 (1987): Haplochorema petiolatum.
reticosa *(Ridley) R.M. Smith in Notes Roy. Bot. Gard. Edinburgh,* 44(2): 209 (1987): Gastrochilus reticosus.

SCELOCHILUS (Orchidac.).
rubriflorus *K. Senghas in Orchidee,* 38(3): 120 (1987)—Peru.

SCHEFFLERA (Araliac.).
feriarum *D.G. Frodin in Proc. Acad. Nat. Sci. Philadelphia,* 138(2): 422 (1986)—Philippine Is.
heteroclada *D.G. Frodin in Proc. Acad. Nat. Sci. Philadelphia,* 138(2): 416 (1986)—Philippine Is.
merrittii *D.G. Frodin in Proc. Acad. Nat. Sci. Philadelphia,* 138(2): 418 (1986)—Philippine Is.
minutipetiolata *D.G. Frodin in Proc. Acad. Nat. Sci. Philadelphia,* 138(2): 410 (1986)—Philippine Is.
pawoiensis *D.G. Frodin in Proc. Acad. Nat. Sci. Philadelphia,* 138(2): 417 (1986)—Philippine Is.
procumbens *(Hemsley) F. Friedmann in Bull. Mus. Nation. Hist. Nat., B, Adansonia,* Ser. 4, 8(3): 254 (1986): Geopanax procumbens.

SCHISTOSTEMON (Humiriac.).
sylvaticum *D. Sabatier in Proc. Kon. Nederl. Akad. Wetensch., C,* 90(2): 206 (1987)—French Guiana.

SCHIZACHYRIUM (Gramin.).
scoparium
var. polycladum *(Scribn. & Ball) C.F. Reed in Phytologia,* 63(5): 410 (1987): Andropogon scoparius var. polycladus.

SCHIZOSTIGMA G.A.W. Arnott in Madras J. Lit. Sci., 12: 50 (1840). *CUCURBITACEAE.*

SCHOEPFIA (Olacac.).
paradoxa *(Bisse & Berazaín) R. Berazaín Iturralde in Rev. Jard. Bot. Nacion. Univ. Habana,* 6(2): 10 (1985 publ. 1986), without exact basionym page: Pseudogonocalyx paradoxa.

SCHRADERA (Rubiac.).
marahuacensis *J.A. Steyermark in Ann. Missouri Bot. Gard.,* 74(1): 114 (1987)—Venezuela.

SCILLA (Hyacinthac.).
spetana *Z. Kereszty in Symb. Bot. Upsal.,* 27(2): 111 (1986 publ. 1987)—Hungary.
vindobonensis
subsp. borhidiana *Z. Kereszty in Symb. Bot. Upsal.,* 27(2): 111 (1986 publ. 1987)—Hungary.
var. transdanubialis *Z. Kereszty in Symb. Bot. Upsal.,* 27(2): 111 (1986 publ. 1987)—Hungary.

SCIRPUS (Cyperac.).
oligosetus *A.E. Kozhevnikov in Bot. Zhurn.,* 72(9): 1255 (1987)—U.S.S.R. (Soviet Far East), ?China.
rongchengensis *F.Z. Li in Acta Phytotax. Sin.,* 25(1): 71 (1987)—China.

SCLERIA (Cyperac.).
orchardii *C.D. Adams in Phytologia,* 63(4): 299 (1987)—Trinidad.

SCLEROCACTUS (Cactac.).
erectocentrus *(J. Coulter) N.P. Taylor in Bradleya,* 5: 94 (1987): Echinocactus erectocentrus.
intertextus *(Engelmann) N.P. Taylor in Bradleya,* 5: 94 (1987): Echinocactus intertextus.
var. dasyacanthus *(Engelmann) N.P. Taylor in Bradleya,* 5: 94 (1987): Echinocactus intertextus var. dasyacanthus.
johnsonii *(Engelmann) N.P. Taylor in Bradleya,* 5: 94 (1987): Echinocactus johnsonii.
mariposensis *(Hester) N.P. Taylor in Bradleya,* 5: 94 (1987): Echinomastus mariposensis.
papyracanthus *(Engelmann) N.P. Taylor in Bradleya,* 5: 94 (1987): Mammillaria papyracantha.

SCLEROCACTUS :-
scheeri *(Salm-Dyck) N.P. Taylor in Bradleya,* 5: 94 (1987): Echinocactus scheeri.
schlesseri *K. Heil & S.L. Welsh in Great Basin Nat.,* 46(4): 677 (1986)—U.S.A. (Nevada).
uncinatus *(Galeotti) N.P. Taylor in Bradleya,* 5: 94 (1987): Echinocactus uncinatus.
var. crassihamatus *(F.A.C. Weber) N.P. Taylor in Bradleya,* 5: 94 (1987): Echinocactus crassihamatus.
var. wrightii *(Engelmann) N.P. Taylor in Bradleya,* 5: 94 (1987): Echinocactus uncinatus var. wrightii.
unguispinus *(Engelmann) N.P. Taylor in Bradleya,* 5: 94 (1987): Echinocactus unguispinus.
var. durangensis *(Runge) N.P. Taylor in Bradleya,* 5: 94 (1987): Echinocactus durangensis.
warnockii *(L. Benson) N.P. Taylor in Bradleya,* 5: 94 (1987): Neolloydia warnockii.

SCLEROCHLOA (Gramin.).
triticum *J. Woods, Tourist's Fl.:* 411 (1850)—Locality not stated.

SCLEROLAENA (Chenopodiac.).
fontinalis *P.G. Wilson in Fl. S. Australia,* 1: 302 (1986)—Australia (S. Australia).

SCORZONERA (Compos.).
subgen. Piptopogon *(C.A. Meyer ex Turcz.) C. Díaz de la Guardia & G. Blanca in An. Jard. Bot. Madrid,* 43(2): 330 (1986 publ. 1987): S. sect. Piptopogon.
albicans
f. longifolia *A.M. Negrillo Galindo in Stud. Bot. (Salamanca),* 5: 115 (1986)—Spain.
angustifolia
var. minor *(Willk.) C. Díaz de la Guardia & G. Blanca in An. Jard. Bot. Madrid,* 43(2): 338 (1986 publ. 1987): S. graminifolia var. minor.
czerepanovii *R.V. Kamelin in Bot. Zhurn.,* 71(12): 1680 (1986), nom. nov., without replaced synonym ref.: S. lanata *(L.) Hoffm.*
laciniata
var. subulata *(DC.) C. Díaz de la Guardia & G. Blanca in An. Jard. Bot. Madrid,* 43(2): 349 (1986 publ. 1987): Podospermum subulatum.
pamirica *C. Shih in Acta Phytotax. Sin.,* 25(1): 48 (1987)—Xinjiang.
sandrasica *P. Hartvig & A. Strid in Bot. Jahrb.,* 108(2–3): 311 (1987)—Turkey.

SCUTELLARIA (Labiat.).
kiusiana
var. kuromidakensis *T. Yahara in J. Fac. Sci. Univ. Tokyo, Bot.,* 14(2): 108 (1987)—Japan.
molanguitensis *P. Hiriart Valencia in Bol. Soc. Bot. Méx.,* 46: 43 (1984 publ. 1986)—Mexico.

SCUTELLARIA :–
orientalis
subsp. tortumensis *Kit Tan & F. Sorger in Pl. Syst. Evol.*, 155(1–4): 98 (1987)—Turkey.
rubropunctata
var. yakusimensis *(Masamune) T. Yahara in J. Fac. Sci. Univ. Tokyo, Bot.*, 14(2): 108 (1987): S. indica var. yakusimensis.
rupestris
subsp. caroli-henrici *R. von Bothmer in Pl. Syst. Evol.*, 155(1–4): 242 (1987)—Greece.
subsp. olympica *R. von Bothmer in Pl. Syst. Evol.*, 155(1–4): 245 (1987)—Greece.
subsp. rechingeri *R. von Bothmer in Pl. Syst. Evol.*, 155(1–4): 246 (1987)—Greece.
zaprjagaevii *T.F. Kochkareva & E.P. Zhogoleva in Izv. Akad. Nauk Tadzh. SSR, Biol. Nauk*, 1(102): 15 (1986)—U.S.S.R. (Middle Asia).

SEBAEA (Gentianac.).
condensata *J. Klackenberg in Bull. Mus. Nation. Hist. Nat., B, Adansonia*, Ser. 4, 9(2): 133 (1987)—Madagascar.

SEDUM (Crassulac.).
subgen. Chuhsingianum *K.T. Fu in Bull. Bot. Res. North-East. Forest. Inst.*, 7(1): 67 (1987).
abyssinicum *(A. Rich.) R. Hamet in Bull. Soc. Bot. France*, 59: 139 (1912): Sempervivum abyssinicum. †
celatum
f. calcaratum *K.T. Fu in Bull. Bot. Res. North-East. Forest. Inst.*, 7(1): 69 (1987)—China.
chuhsingense *K.T. Fu in Bull. Bot. Res. North-East. Forest. Inst.*, 7(1): 68 (1987)—China.
×hommelsii *H. 't Hart in Bot. Jahrb.*, 109(1): 8 (1987)—France.
×lorenzoi *H. 't Hart in Bot. Jahrb.*, 109(1): 10 (1987)—France.
verticillatum
f. bulbiferum *N. Yonezawa in J. Phytogeogr. Taxon.*, 34(2): 69 (1986)—Japan.

SELENICEREUS (Cactac.).
wittii *(Schum.) G.D. Rowley in Excelsa*, 12: 36 (1986): Cereus wittii.

SEMENOVIA (Umbellif.).
heracleifolia *(Wolff) I.C. Hedge & J.M. Lamond in Fl. Iran.*, 162: 489 (1987): Peucedanum heracleifolium.
lasiocarpa
subsp. incana *(Rech. f. & H. Riedl) R. Alava in Fl. Iran.*, 162: 488 (1987): Platytaenia lasiocarpa subsp. incana.
subsp. thomsonii *(C.B. Clarke) R. Alava in Fl. Iran.*, 162: 488 (1987): Heracleum thomsonii.
macrocarpa *(Rech. f. & H. Riedl) R. Alava in Fl. Iran.*, 162: 481 (1987): Platytaenia macrocarpa.

SEMENOVIA :–
millefolia *(Diels) V.M. Vinogradova & R.V. Kamelin in Novosti Sist. Vyssh. Rast.*, 23: 96 (1986): Heracleum millefolium
montana *R.V. Kamelin & V.M. Vinogradova in Novosti Sist. Vyssh. Rast.*, 23: 97 (1986)—China.
radiata *(Rech. f. & H. Riedl) R. Alava in Fl. Iran.*, 162: 484 (1987): Platytaenia lasiocarpa subsp. radiata.
subscaposa *(Rech. f.) R. Alava in Fl. Iran.*, 162: 484 (1987): Zozimia subscaposa.

SEMPERVIVUM (Crassulac.).
brevipetalum *Kit Tan & F. Sorger in Pl. Syst. Evol.*, 154(1–2): 120 (1986)—Turkey.
globiferum
subsp. aghricum *Kit Tan & F. Sorger in Pl. Syst. Evol.*, 154(1–2): 121 (1986)—Turkey.

SENECIO (Compos.).
andrzejowskyi *N.N. Tsvelev in Novosti Sist. Vyssh. Rast.*, 23: 254 (1986)—U.S.S.R. (European Russia).
×austriacus *J. Herborg, Variabilität u. Sippenabgrenz. Senecio nemorensis-Gruppe europ. Teil. (Diss. Bot. 107)*: 195 (1987)—Austria. (S. cacaliaster × hercynicus).
chrysanthemifolius
subsp. aethnensis *(Jan ex DC.) J. Lambinon in Soc. Ech. Pl. Vasc. Eur. Bassin Médit. Bull.*, 21: 63 (1986): S. aethnensis.
×decipiens *J. Herborg, Variabilität u. Sippenabgrenz. Senecio nemorensis-Gruppe europ. Teil. (Diss. Bot. 107)*: 197 (1987)—Germany, Poland, Switzerland, Austria, Italy, Greece. (S. hercynicus × ovatus).
dolichodoryius *J. Cuatrecasas in Proc. Biol. Soc. Wash.*, 98(3): 623 (1985)—Peru.
germanicus
subsp. glabratus *J. Herborg, Variabilität u. Sippenabgrenz. Senecio nemorensis-Gruppe europ. Teil. (Diss. Bot. 107)*: 119 (1987)—Austria, Germany, Italy, Jugoslavia.
var. karaulensis *(Form.) J. Herborg, Variabilität u. Sippenabgrenz. Senecio nemorensis-Gruppe europ. Teil. (Diss. Bot. 107)*: 115 (1987): S. fuchsii var. karaulensis.
gesnerifolius *B.L. Turner in Phytologia*, 62(1): 75 (1987)—Mexico.
hercynicus *J. Herborg, Variabilität u. Sippenabgrenz. Senecio nemorensis-Gruppe europ. Teil. (Diss. Bot. 107)*: 160 (1987)—Germany, Belgium, France, Poland, Czechoslovakia, Hungary, Switzerland, Austria, Italy, Jugoslavia, Bulgaria, Greece.
subsp. durmitorensis *J. Herborg, Variabilität u. Sippenabgrenz. Senecio nemorensis-Gruppe europ. Teil. (Diss. Bot. 107)*: 177 (1987)—Jugoslavia.

SENECIO :-
var. expansus *(Boiss. & Heldr.) J. Herborg, Variabilität u. Sippenabgrenz. Senecio nemorensis-Gruppe europ. Teil. (Diss. Bot. 107):* 174 (1987): S. jacquinianus var. expansus.
kingbishopii *J. Cuatrecasas in Proc. Biol. Soc. Wash.*, 98(3): 625 (1985)—Peru.
ovatus
subsp. alpestris *(Gaudin) J. Herborg, Variabilität u. Sippenabgrenz. Senecio nemorensis-Gruppe europ. Teil. (Diss. Bot. 107):* 143 (1987): S. alpestris.
spartioides
var. multicapitatus *(Greenm. ex Rydb.) S.L. Welsh in Great Basin Nat.*, 43(2): 337 (1983): S. multicapitatus.

SENNA (Legumin.).
acuta *(Meyen ex Vogel) O. Zoellner & C. San Martín in An. Mus. Hist. Nat. Valparaíso,* 16: 25 (1983–85 publ. 1986): Cassia acuta.
alcaparra *(Phil.) O. Zoellner & C. San Martín in An. Mus. Hist. Nat. Valparaíso,* 16: 25 (1983–85 publ. 1986): Cassia alcaparra.
coquimbensis *(Vogel) O. Zoellner & C. San Martín in An. Mus. Hist. Nat. Valparaíso,* 16: 26 (1983–85 publ. 1986): Cassia coquimbensis.
foliosa *(Phil.) O. Zoellner & C. San Martín in An. Mus. Hist. Nat. Valparaíso,* 16: 21 (1983–85 publ. 1986): Cassia foliosa.
glanduligera *(St. John) A.C. Smith in Fl. Vitiensis Nova,* 3: 108 (1985): Cassia glanduligera.
hebecarpa
var. longipila *(E.L. Braun) C.F. Reed in Phytologia,* 63(5): 411 (1987): Cassia hebecarpa var. longipila.
huidobriana *(Phil.) O. Zoellner & C. San Martín in An. Mus. Hist. Nat. Valparaíso,* 16: 23 (1983–85 publ. 1986): Cassia huidobriana.
paposana *(Phil.) O. Zoellner & C. San Martín in An. Mus. Hist. Nat. Valparaíso,* 16: 23 (1983–85 publ. 1986): Cassia paposana.

SERAPIAS (Orchidac.).
nurrica
f. chlorantha *A. Scrugli & M.P. Grasso in Inform. Bot. Ital.,* 17(1–3): 110 (1985 publ. 1986)—Sardinia.

SERENOA (Palmae).
repens *(Bartram) J.K. Small in J. New York Bot. Gard.,* 27(321): 193, 197 (1926): Corypha repens.

SERJANIA (Sapindac.).
fluminensis *P. Acevedo-Rodríguez in Brittonia,* 39(3): 348 (1987)—Brazil.
unidentata *P. Acevedo-Rodríguez in Brittonia,* 39(3): 350 (1987)—Brazil.

SESBANIA (Legumin.).
coccinea
subsp. atollensis *(St. John) M.-H. Sachet in Bull. Mus. Nation. Hist. Nat., B, Adansonia,* Ser. 4, 9(1): 22 (1987): S. atollensis.
var. atollensis *(St. John) M.-H. Sachet in Bull. Mus. Nation. Hist. Nat., B, Adansonia,* Ser. 4, 9(1): 22 (1987): S. atollensis.
var. marchionica *(F. Brown) M.-H. Sachet in Bull. Mus. Nation. Hist. Nat., B, Adansonia,* Ser. 4, 9(1): 23 (1987): S. marchionica.
var. parkinsonii *M.-H. Sachet in Bull. Mus. Nation. Hist. Nat., B, Adansonia,* Ser. 4, 9(1): 24 (1987)—Society Is.
var. quaylei *M.-H. Sachet in Bull. Mus. Nation. Hist. Nat., B, Adansonia,* Ser. 4, 9(1): 23 (1987)—Marquesas Is.
var. tuamotensis *(F. Brown) M.-H. Sachet in Bull. Mus. Nation. Hist. Nat., B, Adansonia,* Ser. 4, 9(1): 24 (1987): S. speciosa var. tuamotensis.
mannii *M.-H. Sachet in Bull. Mus. Nation. Hist. Nat., B, Adansonia,* Ser. 4, 9(1): 13 (1987)—Hawaiian Is.

SESELI (Umbellif.).
libanotis
f. glabrum *(Duvigneaud) C. Benedí & J. Molero in Bol. Soc. Brot.,* 59: 61 (1986): Libanotis pyrenaica f. glabra.
ramosissimum *P. Hartvig & A. Strid in Bot. Jahrb.,* 108(2–3): 305 (1987)—Turkey.

SESLERIA (Gramin.).
pedemontana *(Reuter) Steudel, Syn. Pl. Glum.,* 1: 296 (1854): Oreochloa pedemontana.

SETARIA (Gramin.).
decipiens *C. Schimper ex P. Ascherson, Fl. Brandenburg,* 1: 809 (1864)—Locality not stated.
hunzikeri *A.M. Anton in Kurtziana,* 17: 139 (1984)—Argentina.
nana *(Dumort.) B.D. Jackson in Index Kewensis,* 2: 893 (1753–1885 publ. 1895): S. viridis * (subsp.) nana.
paraguayensis *J.F. Pensiero in Candollea,* 41(2): 469 (1986)—Paraguay.
parviflora *(Poiret) M. Kerguélen in Lejeunia,* 120: 161 (1987): Cenchrus parviflorus.

SIAMOSIA K. Larsen & T. Myndel Pedersen in Nordic J. Bot., 7(3): 271 (1987). *AMARANTHACEAE.*
thailandica *K. Larsen & T. Myndel Pedersen in Nordic J. Bot.,* 7(3): 271 (1987)—Thailand.

SIBBALDIA (Rosac.).
purpurea
var. pentaphylla *(J. Krause) B.K. Dixit in Bull. Bot. Soc. Bengal,* 38(1–2): 39 (1984 publ. 1987): S. pentaphylla.

SICYOS (Cucurbitac.).
erostratus *H. St. John in Phytologia,* 63(3): 187 (1987)—Hawaiian Is.
var. herbstii *H. St. John in Phytologia,* 63(3): 187 (1987)—Hawaiian Is.
var. kealiaensis *H. St. John in Phytologia,* 63(3): 187 (1987)—Hawaiian Is.
var. paucipilosus *H. St. John in Phytologia,* 63(3): 187 (1987)—Hawaiian Is.
hillebrandii
var. anunu *H. St. John in Phytologia,* 63(3): 187 (1987)—Hawaiian Is.
var. douglasii *H. St. John in Phytologia,* 63(3): 187 (1987)—Hawaiian Is.
var. lanaiensis *H. St. John in Phytologia,* 63(3): 188 (1987)—Hawaiian Is.
microcarpus
var. ewaensis *H. St. John in Phytologia,* 63(3): 190 (1987)—Hawaiian Is.
var. forbesii *H. St. John in Phytologia,* 63(3): 188 (1987)—Hawaiian Is.
var. haleakalaensis *H. St. John in Phytologia,* 63(3): 191 (1987)—Hawaiian Is.
var. kalaupapaensis *H. St. John in Phytologia,* 63(3): 188 (1987)—Hawaiian Is.
var. kamoaensis *H. St. John in Phytologia,* 63(3): 189 (1987)—Hawaiian Is.
subsp. kanepuuensis *H. St. John in Phytologia,* 63(3): 189 (1987)—Hawaiian Is.
var. kauensis *H. St. John in Phytologia,* 63(3): 191 (1987)—Hawaiian Is.
var. kealiaensis *H. St. John in Phytologia,* 63(3): 189 (1987)—Hawaiian Is.
var. kilaueaensis *H. St. John in Phytologia,* 63(3): 188 (1987)—Hawaiian Is.
var. koeleensis *H. St. John in Phytologia,* 63(3): 190 (1987)—Hawaiian Is.
var. koloaensis *H. St. John in Phytologia,* 63(3): 191 (1987)—Hawaiian Is.
subsp. lanaiensis *H. St. John in Phytologia,* 63(3): 189 (1987)—Hawaiian Is.
var. lanaiensis *H. St. John in Phytologia,* 63(3): 189 (1987)—Hawaiian Is.
subsp. laysanensis *(St. John) H. St. John in Phytologia,* 63(3): 190 (1987): S. laysanensis.
var. laysanensis *(St. John) H. St. John in Phytologia,* 63(3): 190 (1987): S. laysanensis.
var. meeboldii *H. St. John in Phytologia,* 63(3): 191 (1987)—Hawaiian Is.
var. obovoideis *H. St. John in Phytologia,* 63(3): 191 (1987)—Hawaiian Is.
var. oniniensis *H. St. John in Phytologia,* 63(3): 190 (1987)—Hawaiian Is.
var. paneensis *H. St. John in Phytologia,* 63(3): 188 (1987)—Hawaiian Is.
var. pohakeaensis *H. St. John in Phytologia,* 63(3): 188 (1987)—Hawaiian Is.
subsp. puberulus *H. St. John in Phytologia,* 63(3): 191 (1987)—Hawaiian Is.
var. trilobata *H. St. John in Phytologia,* 63(3): 190 (1987)—Hawaiian Is.

SICYOS :–
var. woolfordii *H. St. John in Phytologia,* 63(3): 189 (1987)—Hawaiian Is.
niger *H. St. John in Phytologia,* 63(3): 192 (1987)—Hawaiian Is.
waimanaloensis *H. St. John in Phytologia,* 63(3): 192 (1987)—Hawaiian Is.

SIDA (Malvac.).
anodifolia *P.A. Fryxell in Sida,* 12(1): 14 (1987)—Mexico.
fastuosa *P.A. Fryxell & S.D. Koch in Aliso,* 11(4): 554 (1987)—Mexico.
michoacana *P.A. Fryxell in Sida,* 12(1): 16 (1987)—Mexico.
monticola *P.A. Fryxell in Sida,* 12(1): 18 (1987)—Mexico.
prolifica *P.A. Fryxell & S.D. Koch in Aliso,* 11(4): 556 (1987)—Mexico.
rohlenae
var. mutica *(Bentham) P.A. Fryxell in Sida,* 12(1): 26 (1987): S. cordifolia var. mutica.

SIDERITIS (Labiat.).
flavovirens *(Rouy) F. Alcaraz, M. Peinado, J.M. Martínez-Parra, J.S. Carrión & P. Sánchez-Gómez in Acta Bot. Malacitana,* 12: 245 (1987): S. leucantha var. flavovirens.
granatensis *(Pau) F. Alcaraz, M. Peinado, J.M. Martínez-Parra, J.S. Carrión & P. Sánchez-Gómez in Acta Bot. Malacitana,* 12: 246 (1987): S. hirsuta var. granatensis.
leucantha
subsp. bourgeana *(Boissier) F. Alcaraz, M. Peinado, J.M. Martínez-Parra, J.S. Carrión & P. Sánchez-Gómez in Acta Bot. Malacitana,* 12: 246 (1987): S. bourgeana.
subsp. incana *(Willk.) F. Alcaraz, M. Peinado, J.M. Martínez-Parra, J.S. Carrión & P. Sánchez-Gómez in Acta Bot. Malacitana,* 12: 246 (1987): S. linearifolia var. incana.
osteoxylla *(Pau) F. Alcaraz, M. Peinado, J.M. Martínez-Parra, J.S. Carrión & P. Sánchez-Gómez in Acta Bot. Malacitana,* 12: 246 (1987): S. almeriensis var. osteoxylla.
romana
var. lazae *O. Socorro, I. Tárrega & M.L. Zafra in Lagascalia,* 14(1): 98 (1986)—Spain.

SIGESBECKIA (Compos.).
bogotensis *D.L. Schulz in Haussknechtia, Mitt. Thüring. Bot. Ges.,* 3: 61 (1987)—Colombia, Venezuela.

SILENE (Caryophyllac.).
subsect. Echinatae *C.L. Tang in Acta Bot. Yunnanica,* 9(1): 32 (1987).
subsect. Platypetalae *C.L. Tang in Acta Bot. Yunnanica,* 9(1): 31 (1987).
antri-jovis *W. Greuter & Burdet in Willdenowia,* 16(2): 440 (1987), nom. nov.: S. fruticulosa *Otth.*

SILENE :–
asclepiadea
var. dumicola *(W.W. Smith) C.L. Tang in Acta Bot. Yunnanica,* 9(1): 28 (1987): S. dumicola.
baldaccii *(J. Rohlena) A.W. Hill in Index Kewensis,* Supp. 6: 194 (1916–1920 publ. 1926): S. otites var. baldaccii.
boullui *(Rouy & Fouc.) M. Kerguélen in Lejeunia,* 120: 163 (1987): S. mirabilis subsp. boullui.
brevicalyx *P. Hartvig & A. Strid in Bot. Jahrb.,* 108(2–3): 323 (1987)—Turkey.
dioica
var. glandulosa *(Brügger) M. Kerguélen in Lejeunia,* 120: 163 (1987): Lychnis diurna var. glandulosa.
fabaria
subsp. thebana *(Orph. ex Boiss.) V. Melzheimer in Pl. Syst. Evol.,* 155(1–4): 252 (1987): S. thebana.
fissicalyx *G. Bocquet & A.O. Chater in Enum. Fl. Pl. Nepal,* 2: 55 (1979)—Nepal.
holosteifolia *G. Bocquet & A.O. Chater in Enum. Fl. Pl. Nepal,* 2: 55 (1979)—Nepal.
indica
var. cashmeriana *(Royle ex Benth.) Y. Nasir in Fl. Pakistan,* 175: 71 (1986): Lychnis cashmeriana.
var. edgeworthii *(Bocquet) Y. Nasir in Fl. Pakistan,* 175: 70 (1986): Silene edgeworthii.
jugorum *(F.N. Williams) G. Bocquet & A.O. Chater in Enum. Fl. Pl. Nepal,* 2: 56 (1979), isonym; see N.C. Majumdar *in Bull. Bot. Surv. India,* 15(1–2): 39 (1973 publ. 1976): Melandrium jugorum.
pseudocashmeriana *G. Bocquet & A.O. Chater in Enum. Fl. Pl. Nepal,* 2: 56 (1979)—Nepal.
scabriflora
subsp. tuberculata *(Ball) S. Talavera in Lagascalia,* 14(1): 131 (1986): S. hirsuta var. tuberculata.
stellariifolia *G. Bocquet & A.O. Chater in Enum. Fl. Pl. Nepal,* 2: 56 (1979)—Nepal.
uniflora
var. montana *(Arrondeau) M. Kerguélen in Soc. Ech. Pl. Vasc. Eur. Bassin Médit. Bull.,* 21: 51 (1986): S. montana.

SIMABA (Simaroubac.).
polyphylla *(Cavalcante) W.W. Thomas in Acta Amazonica,* 15(1–2) Suppl.: 76 (1985 publ. 1987): S. guianensis subsp. polyphylla.

SINARUNDINARIA (Gramin.).
longiuscula *J.R. Xue & Y.Y. Dai in J. Bamboo Res.,* 6(2): 19 (1987)—China.
yunnanensis *(Hsueh & Yi) J.R. Xue & D.Z. Li in J. Bamboo Res.,* 6(2): 21 (1987): Fargesia yunnanensis.

SINNINGIA (Gesneriac.).
reitzii *(Hoehne) L.E. Skog in Gloxinian,* 37(1): 35 (1987): Rechsteineria reitzii.

SINOBACOPA D.Y. Hong in Acta Phytotax. Sin., 25(5): 393 (1987). *SCROPHULARIACEAE.*
aquatica *D.Y. Hong in Acta Phytotax. Sin.,* 25(5): 395 (1987)—Hainan, China.

SINOBAMBUSA (Gramin.).
dushanensis *(C.D. Chu & J.G. Zhang) T.H. Wen in J. Bamboo Res.,* 6(3): 33 (1987): Arundinaria dushanensis.
nandanensis *T.H. Wen in J. Bamboo Res.,* 6(3): 29 (1987)—China.
parvifolia *T.H. Wen & S.Y. Chen in J. Bamboo Res.,* 6(3): 31 (1987), as 'parvofolia'—China.

SINODIELSIA (Umbellif.).
digitata *E.V. Kljuykov in Feddes Repert.,* 97(11–12): 756 (1986)—Bhutan.
microloba *E.V. Kljuykov in Feddes Repert.,* 97(11–12): 757 (1986)—Xizang.

SIPANEA (Rubiac.).
setacea *J.A. Steyermark in Ann. Missouri Bot. Gard.,* 74(1): 114 (1987)—Venezuela.

SISYRINCHIUM (Iridac.).
graminifolium
subsp. luteum *(Bertero ex Steudel) R. Rodriguez Rios in Mitt. Bot. Staatssamml. München,* 22: 141 (1986): S. luteum.
junceum
subsp. colchaguense *(Philippi) R. Rodriguez Rios in Mitt. Bot. Staatssamml. München,* 22: 145 (1986): S. colchaguense.
subsp. depauperatum *(Philippi) R. Rodriguez Rios in Mitt. Bot. Staatssamml. München,* 22: 146 (1986): S. depauperatum.
philippii
subsp. illapelinum *(Philippi) R. Rodriguez Rios in Mitt. Bot. Staatssamml. München,* 22: 158 (1986): S. illapelinum.
scirpoideum
subsp. leucanthum *(Colla) R. Rodriguez Rios in Mitt. Bot. Staatssamml. München,* 22: 164 (1986): S. leucanthum.
subsp. luridum *(Ravenna) R. Rodriguez Rios in Mitt. Bot. Staatssamml. München,* 22: 167 (1986): S. luridum.
subsp. scirpeum *(Philippi) R. Rodriguez Rios in Mitt. Bot. Staatssamml. München,* 22: 167 (1986): S. scirpeum.

SIUM (Umbellif.).
suave
var. floridanum *(Small) C.F. Reed in Phytologia,* 63(5): 411 (1987): S. floridanum.

SKIMMIA (Rutac.).
arborescens
subsp. nitida *N.P. Taylor & H.K. Airy Shaw in Kew Mag.,* 4(4): 185 (1987)—China, India, Bhutan, Burma, Thailand, Laos, Vietnam.

SKIMMIA :-
×confusa *N.P. Taylor in Kew Mag.*, 4(4): 193 (1987)—Cultivation. (S. anquetilia × japonica subsp. japonica).
japonica
var. euphlebia *(Merrill) N.P. Taylor in Kew Mag.*, 4(4): 184 (1987): S. euphlebia.
var. kwangsiensis *(C.C. Huang) N.P. Taylor in Kew Mag.*, 4(4): 183 (1987): S. kwangsiensis.
subsp. reevesiana *(Fortune) N.P. Taylor & H.K. Airy Shaw in Kew Mag.*, 4(4): 182 (1987): Ilex reevesiana.
var. reevesiana *(Fortune) N.P. Taylor in Kew Mag.*, 4(4): 183 (1987): Ilex reevesiana.
laureola
subsp. multinervia *(C.C. Huang) N.P. Taylor & H.K. Airy Shaw in Kew Mag.*, 4(4): 189 (1987): S. multinervia.

SMILACINA (Convallariac.).
subsect. Dioica *H. Hara in J. Fac. Sci. Univ. Tokyo, Bot.*, 14(2): 139 (1987).
sect. Tubifera *H. Hara in J. Fac. Sci. Univ. Tokyo, Bot.*, 14(2): 139 (1987).
purpurea
f. albiflora *(Wall.) H. Hara in J. Fac. Sci. Univ. Tokyo, Bot.*, 14(2): 147 (1987): S. purpurea albiflora.
racemosa
subsp. amplexicaulis *(Nutt.) J. McNeill in Nat. Canad.*, 108(3): 240 (1981): S. amplexicaulis.
tatsienensis
f. stenoloba *(Franch.) H. Hara in J. Fac. Sci. Univ. Tokyo, Bot.*, 14(2): 155 (1987): Tovaria stenoloba.

SOLANUM (Solanac.).
astleyi *J.G. Hawkes & J.P. Hjerting in Kew Bull.*, 42(2): 436 (1987)—Bolivia. †
badilloi *C.E. Benítez de Rojas in Ernstia*, 44: 1 (1987)—Venezuela.
cheesmaniae *(Riley) F.R. Fosberg in Phytologia*, 62(3): 181 (1987): Lycopersicum cheesmanii.
circaeifolium
subsp. quimense *J.G. Hawkes & J.P. Hjerting in Kew Bull.*, 42(2): 436 (1987)—Bolivia. †
darienense *S. Knapp in Ann. Missouri Bot. Gard.*, 73(4): 738 (1986 publ. 1987)—Panama.
dolosum *Morton ex S. Knapp in Ann. Missouri Bot. Gard.*, 73(4): 742 (1986 publ. 1987)—Colombia.
extremiorientale *M.S. Ignatov in Byull. Mosk. Obshch. Ispyt. Prir., Biol.*, 91(3): 113 (1986)—U.S.S.R (Soviet Far East).
flavescens *M.F. Dunal in J.L.M. Poiret, Encycl. Méthod., Bot.*, Supp. 3: 778 (1814)—Trinidad. †
glutinosum *M.F. Dunal in J.L.M. Poiret, Encycl. Méthod., Bot.*, Supp. 3: 769 (1814)—Locality not stated. †

SOLANUM :-
gourlayi
subsp. saltense *A.M. Clausen & K.A. Okada in Phytologia*, 62(3): 166 (1987)—Argentina.
herthae *(Werderm.) W.G. D'Arcy in Solanaceae Newsl.*, 2(4): 31 (1986): Pheliandra herthae.
hondelmannii *J.G. Hawkes & J.P. Hjerting in Kew Bull.*, 42(2): 436 (1987)—Bolivia. †
lasiocarpum
var. domesticum *C.B. Heiser in W.G. D'Arcy, Solanaceae: biol. & syst.*: 413 (1986)—Cultivation ex Thailand.
longiusculus *C. Ochoa in Phytologia*, 63(5): 329 (1987)—Peru.
macranthum *M.F. Dunal in J.L.M. Poiret, Encycl. Méthod., Bot.*, Supp. 3: 775 (1814)—Brazil. †
richardii *M.F. Dunal in J.L.M. Poiret, Encycl. Méthod., Bot.*, Supp. 3: 775 (1814)—Locality not stated. †
sessiliflorum *M.F. Dunal in J.L.M. Poiret, Encycl. Méthod., Bot.*, Supp. 3: 775 (1814)—Brazil. †
triflorum
f. malvinum *J.E. De Langhe & R. D'hose in Bull. Soc. Roy. Bot. Belg.*, 119(2): 160 (1986 publ. 1987)—Belgium.
unifoliatum *S. Knapp in Ann. Missouri Bot. Gard.*, 73(4): 739 (1986 publ. 1987)—Colombia.
vaillantii *M.F. Dunal in J.L.M. Poiret, Encycl. Méthod., Bot.*, Supp. 3: 772 (1814)—Locality not stated. †

SOLDANELLA (Primulac.).
pelia *T. Raus in Willdenowia*, 16(2): 337 (1987)—Greece.

SOLENA (Cucurbitac.).
delavayi *(Cogn.) C.Y. Wu in Fl. Xizangica*, 4: 553 (1985): Melothria delavayi.

SOLENANTHUS (Boraginac.).
hupehensis *R.R. Mill in Notes Roy. Bot. Gard. Edinburgh*, 44(2): 271 (1987)—China.

SOLENIDIOPSIS K. Senghas in Orchidee, 37(6): 274 (1986). *ORCHIDACEAE.*
tigroides *(C. Schweinf.) K. Senghas in Orchidee*, 37(6): 274 (1986): Odontoglossum tigroides.

SOLIDAGO (Compos.).
glutinosa
var. chlorolepis *(Fern.) G.S. Ringius in Taxon*, 36(1): 157 (1987): S. chlorolepis.
var. ontarioensis *G.S. Ringius in Phytologia*, 62(5): 432 (1987)—Canada (Ontario).
nemoralis
subsp. haleana *(Fern.) G.W. Douglas in Canad. J. Bot.*, 64(11): 2726 (1986): S. nemoralis var. haleana.
subsp. longipetiolata *(Mack. & Bush) G.W. Douglas in Canad. J. Bot.*, 64(11): 2726 (1986): S. longipetiolata.

SONERILA (Melastomatac.).
cannanorensis *G.S. Giri & M.P. Nayar in Bull. Bot. Surv. India,* 27(1–4): 86 (1985 publ. 1987)—India.
elegans
var. beddomei *G.S. Giri & M.P. Nayar in Bull. Bot. Surv. India,* 27(1–4): 72 (1985 publ. 1987)—India.
sadasivanii
var. kanniyakumariensis *G.S. Giri & M.P. Nayar in Bull. Bot. Surv. India,* 26(3–4): 189 (1984 publ. 1986)—India.
sahyadrica *G.S. Giri & M.P. Nayar in Bull. Bot. Surv. India,* 26(3–4): 175 (1984 publ. 1986)—India.
talbotii *G.S. Giri & M.P. Nayar in J. Jap. Bot.,* 61(11): 344 (1986)—India.

SOPHRONITIS (Orchidac.).
mantiqueirae
var. pygmaea *(Pabst) J.A. Fowlie in Orchid Dig.,* 51(1): 34 (1987): S. coccinea subsp. pygmaea.

SPARGANIUM (Sparganiac.).
emersum
subsp. acaule *(Beeby ex Macoun) C.D.K. Cook & M.S. Nicholls in Bot. Helvetica,* 96(2): 257 (1986): S. simplex var. acaule.
erectum
subsp. stoloniferum *(F. Hamilton ex Graebner) C.D.K. Cook & M.S. Nicholls in Bot. Helvetica,* 97(1): 21 (1987): S. ramosum subsp. stoloniferum.
eurycarpum
subsp. coreanum *(Léveillé) C.D.K. Cook & M.S. Nicholls in Bot. Helvetica,* 97(1): 28 (1987): S. coreanum.

SPATHIPHYLLUM (Arac.).
liesneri *T.B. Croat in Aroideana,* 9(1–4): 171 (1986)—Venezuela.

SPERMACOCE (Rubiac.).
dispersa *(Hook. f.) F.R. Fosberg in Phytologia,* 62(3): 182 (1987): Borreria dispersa.
ericaefolia *(Hook. f.) F.R. Fosberg in Phytologia,* 62(3): 182 (1987): Borreria ericaefolia.
linearifolia *(Hook. f.) F.R. Fosberg in Phytologia,* 62(3): 182 (1987): Borreria linearifolia.
malabarica *(Sivar. & Mani.) V.V. Sivarajan, R. Vasudevan Nair & T.U. Ahmed Kunju in Proc. Indian Acad. Sci., Pl. Sci.,* 97(4): 355 (1987): Borreria malabarica.
perpusilla *(Hook. f.) F.R. Fosberg in Phytologia,* 62(3): 182 (1987): Borreria perpusilla.
rotundifolia *(Anderss.) F.R. Fosberg in Phytologia,* 62(3): 183 (1987): Borreria rotundifolia.
suberecta *(Hook. f.) F.R. Fosberg in Phytologia,* 62(3): 183 (1987): Borreria suberecta.

SPHAEROSACME Wall. ex Royle, Ill. Bot. Himal. Mts.: 142 (1835); cf. D.J. Mabberley in Blumea, 31(1): 135 (1985). *MELIACEAE.* †

SPHYROSPERMUM (Ericac.).
campanulatum *J.L. Luteyn in Op. Bot.,* 92: 126 (1987)—Colombia.
dissimile *(Blake) J.L. Luteyn in Op. Bot.,* 92: 126 (1987), without basionym ref.: Vaccinium dissimile.
distichum *J.L. Luteyn in Op. Bot.,* 92: 124 (1987)—Colombia.
munchiqueense *J.L. Luteyn in Op. Bot.,* 92: 124 (1987)—Colombia.

SPIRADICLIS (Rubiac.).
guangdongensis *H.S. Lo in Acta Bot. Yunnanica,* 9(3): 299 (1987)—China.
hainanensis *H.S. Lo in Acta Bot. Yunnanica,* 9(3): 301 (1987)—Hainan.

SPIRAEA (Rosac.).
hemicryptophyta *A.J.C. Grierson in Notes Roy. Bot. Gard. Edinburgh,* 44(2): 262 (1987)—Sikkim, Nepal, Burma, Xizang, China.
schlothauerae *V.N. Voroshilov & M.S. Ignatov in Byull. Mosk. Obshch. Ispȳt. Prir., Biol.,* 92(1): 132 (1987)—U.S.S.R. (Soviet Far East).
tomentulosa *(Yu) Y.Z. Zhao in Acta Sci. Nat. Univ. Intramongolicae,* 18(2): 289 (1987): S. mongolica var. tomentulosa.

SPONGIOCARPELLA G.P. Yakovlev & N. Ulziĭkhumag in Bot. Zhurn., 72(2): 250 (1987). *LEGUMINOSAE.*
grubovii *(Ulzij.) G.P. Yakovlev in Bot. Zhurn.,* 72(2): 258 (1987): Oxytropis grubovii.
intermedia *G.P. Yakovlev in Bot. Zhurn.,* 72(2): 254 (1987)—Xizang.
nubigena *(D. Don) G.P. Yakovlev in Bot. Zhurn.,* 72(2): 252 (1987): Astragalus nubigenus.
subsp. kumaoensis *G.P. Yakovlev in Bot. Zhurn.,* 72(2): 252 (1987)—Nepal, India.
paucifoliolata *G.P. Yakovlev in Bot. Zhurn.,* 72(2): 254 (1987)—China.
polystichoides *(Hand.-Mazz.) G.P. Yakovlev in Bot. Zhurn.,* 72(2): 258 (1987): Calophaca polystichoides.
potaninii *G.P. Yakovlev in Bot. Zhurn.,* 72(2): 259 (1987)—Xinjiang.
purpurea *(P.C. Li) G.P. Yakovlev in Bot. Zhurn.,* 72(2): 253 (1987): Chesneya purpurea.
spinosa *(P.C. Li) G.P. Yakovlev in Bot. Zhurn.,* 72(2): 259 (1987): Chesneya spinosa.
yunnanensis *G.P. Yakovlev in Bot. Zhurn.,* 72(2): 256 (1987)—China.

SPOROBOLUS (Gramin.).
exilis *S.A. Renvoize in Kew Bull.,* 42(4): 923 (1987)—Brazil.

STACHYS (Labiat.).
ser. Lanatae *Knorr. ex Sh.I. Kutateladze in Fl. Gruzii, ed. 2,* 11: 238 (1987).
ser. Maritimae *Knorr. ex Sh.I. Kutateladze in Fl. Gruzii, ed. 2,* 11: 239 (1987).
ser. Palustres *Knorr. ex Sh.I. Kutateladze in Fl. Gruzii, ed. 2,* 11: 238 (1987).
ser. Persicae *Knorr. ex Sh.I. Kutateladze in Fl. Gruzii, ed. 2,* 11: 238 (1987).
ser. Rectae *Knorr. ex Sh.I. Kutateladze in Fl. Gruzii, ed. 2,* 11: 238 (1987).
ser. Setiferae *Knorr. ex Sh.I. Kutateladze in Fl. Gruzii, ed. 2,* 11: 238 (1987).
ser. Spectabiles *Knorr. ex Sh.I. Kutateladze in Fl. Gruzii, ed. 2,* 11: 238 (1987).
ser. Sylvaticae *Knorr. ex Sh.I. Kutateladze in Fl. Gruzii, ed. 2,* 11: 238 (1987).
choruhensis *Kit Tan & F. Sorger in Pl. Syst. Evol.,* 155(1–4): 99 (1987)—Turkey.
officinalis
subsp. peredae *M. Laínz in An. Jard. Bot. Madrid,* 43(2): 459 (1986 publ. 1987)—Spain.
tundjeliensis *Kit Tan & F. Sorger in Pl. Syst. Evol.,* 155(1–4): 99 (1987)—Turkey.

STAUNTONIA (Lardizabalac.).
conspicua *R.H. Chang in Acta Phytotax. Sin.,* 25(3): 235 (1987)—China.

STEFANOFFIA (Umbellif.).
insoluta *E.V. Klyuikov in Novosti Sist. Vyssh. Rast.,* 23: 90 (1986)—Turkey.

STEGANOTROPIS J.G.C. Lehmann in Sem. Hort. Bot. Hamb.: 18 (1826). *LEGUMINOSAE.* †
conjugata *J.G.C. Lehmann in Sem. Hort. Bot. Hamb.:* 18 (1826)—Locality not stated. †

STEGIA (Malvac.).
trimestris *(L.) T. Luque & J.A. Devesa in Lagascalia,* 14(2): 237 (1986): Lavatera trimestris.
var. brachypoda *(Pérez Lara) T. Luque & J.A. Devesa in Lagascalia,* 14(2): 238 (1986): Lavatera trimestris var. brachypoda.

STEGNOSPERMA (Phytolaccac.).
sanchezii *F. González-Medrano & R. Medina Lemos in Bol. Soc. Bot. Méx.,* 46: 37 (1984 publ. 1986)—Mexico.

STEINCHISMA (Gramin.).
spathellosa *(Doell) S.A. Renvoize in Kew Bull.,* 42(4): 921 (1987): Panicum spathellosum.

STELIS (Orchidac.).
choriantha *D.D. Dod in Moscosoa,* 4: 160 (1986)—Hispaniola (Haiti).
glacensis *D.D. Dod in Moscosoa,* 4: 166 (1986)—Hispaniola (Haiti).
minipetala *D.D. Dod in Moscosoa,* 4: 162 (1986)—Hispaniola (Haiti).

STELIS :–
mucrouncata *D.D. Dod in Moscosoa,* 4: 164 (1986)—Hispaniola (Dominican Republic).
triapiculata *D.D. Dod in Moscosoa,* 4: 168 (1986)—Hispaniola (Haiti).

STELLARIA (Caryophyllac.).
antoniana *C.R. Volponi in Kurtziana,* 18: 95 (1986)—Argentina.
kotschyana
subsp. glabra *(Bornm.) S.A. Ghazanfar in Fl. Pakistan,* 175: 30 (1986): S. kotschyana var. glabra.
montioides *(Edgew. & Hook.f.) S.A. Ghazanfar in Fl. Pakistan,* 175: 25 (1986): S. graminea var. montioides.
patens
var. ovatifolia *H. Hara in Enum. Fl. Pl. Nepal,* 2: 58 (1979)—Nepal.
pedersenii *C.R. Volponi in Kurtziana,* 18: 97 (1986)—Argentina.

STENOCEPHALUM (Compos.).
hystrix *(Chodat) H. Robinson in Proc. Biol. Soc. Wash.,* 100(3): 583 (1987): Vernonia hystrix.
jucundum *(Gleason) H. Robinson in Proc. Biol. Soc. Wash.,* 100(3): 583 (1987): Vernonia jucunda.

STENOGYNE (Labiat.).
acuta *H. St. John in Phytologia,* 63(3): 159 (1987)—Hawaiian Is.
adpressa *H. St. John in Phytologia,* 63(3): 159 (1987)—Hawaiian Is.
adscendens *H. St. John in Phytologia,* 63(3): 159 (1987)—Hawaiian Is.
alakaiensis *H. St. John in Phytologia,* 63(3): 159 (1987)—Hawaiian Is.
alba *H. St. John in Phytologia,* 63(3): 159 (1987)—Hawaiian Is.
albimontis *H. St. John in Phytologia,* 63(3): 159 (1987)—Hawaiian Is.
alternans *H. St. John in Phytologia,* 63(3): 159 (1987)—Hawaiian Is.
ambiobtuta *H. St. John in Phytologia,* 63(3): 160 (1987)—Hawaiian Is.
angularis *H. St. John in Phytologia,* 63(3): 160 (1987)—Hawaiian Is.
bracteosa *H. St. John in Phytologia,* 63(3): 160 (1987)—Hawaiian Is.
calaminthoides
var. kulaniensis *H. St. John in Phytologia,* 63(3): 160 (1987)—Hawaiian Is.
calvata *H. St. John in Phytologia,* 63(3): 160 (1987)—Hawaiian Is.
ciliata *H. St. John in Phytologia,* 63(3): 160 (1987)—Hawaiian Is.
conscandens *H. St. John in Phytologia,* 63(3): 160 (1987)—Hawaiian Is.
coriacea *H. St. John in Phytologia,* 63(3): 160 (1987)—Hawaiian Is.
davisiae *H. St. John in Phytologia,* 63(3): 160 (1987), as 'davisae'—Hawaiian Is.
decumbens *H. St. John in Phytologia,* 63(3): 161 (1987)—Hawaiian Is.

STENOGYNE :-
deltoidea *H. St. John in Phytologia,* 63(3): 161 (1987)—Hawaiian Is.
diantha *H. St. John in Phytologia,* 63(3): 161 (1987)—Hawaiian Is.
douglasii *H. St. John in Phytologia,* 63(3): 161 (1987)—Hawaiian Is.
earina *H. St. John in Phytologia,* 63(3): 161 (1987)—Hawaiian Is.
elliptica *H. St. John in Phytologia,* 63(3): 161 (1987)—Hawaiian Is.
elliptilobata *H. St. John in Phytologia,* 63(3): 161 (1987)—Hawaiian Is.
flava *H. St. John in Phytologia,* 63(3): 161 (1987)—Hawaiian Is.
forbesii *(Sherff) H. St. John in Phytologia,* 63(3): 162 (1987): S. purpurea var. forbesii.
fruticosa *H. St. John in Phytologia,* 63(3): 162 (1987)—Hawaiian Is.
fusca *H. St. John in Phytologia,* 63(3): 162 (1987)—Hawaiian Is.
gracilis *(Sherff) H. St. John in Phytologia,* 63(3): 162 (1987): S. macrantha var. gracilis.
hamakuaensis *H. St. John in Phytologia,* 63(3): 162 (1987)—Hawaiian Is.
hawaiiensis *H. St. John in Phytologia,* 63(3): 162 (1987)—Hawaiian Is.
hemidasys *H. St. John in Phytologia,* 63(3): 162 (1987)—Hawaiian Is.
hiloensis *H. St. John in Phytologia,* 63(3): 162 (1987)—Hawaiian Is.
hosakae *H. St. John in Phytologia,* 63(3): 162 (1987)—Hawaiian Is.
huluhuluensis *H. St. John in Phytologia,* 63(3): 163 (1987)—Hawaiian Is.
humululaensis *H. St. John in Phytologia,* 63(3): 163 (1987)—Hawaiian Is.
integra *H. St. John in Phytologia,* 63(3): 163 (1987)—Hawaiian Is.
kaalae
var. obatae *H. St. John in Phytologia,* 63(3): 163 (1987)—Hawaiian Is.
var. pearsallii *H. St. John in Phytologia,* 63(3): 163 (1987)—Hawaiian Is.
kamehamehae
f. albiflora *(Sherff) H. St. John in Phytologia,* 63(3): 163 (1987): S. kamehamehae var. albiflora.
kauaiensis *H. St. John in Phytologia,* 63(3): 163 (1987)—Hawaiian Is.
kilaueae *H. St. John in Phytologia,* 63(3): 163 (1987)—Hawaiian Is.
kipahuluensis *H. St. John in Phytologia,* 63(3): 163 (1987)—Hawaiian Is.
konaensis *H. St. John in Phytologia,* 63(3): 163 (1987)—Hawaiian Is.
kukuiensis *H. St. John in Phytologia,* 63(3): 164 (1987)—Hawaiian Is.
kulaensis *H. St. John in Phytologia,* 63(3): 164 (1987)—Hawaiian Is.
lactea *H. St. John in Phytologia,* 63(3): 164 (1987)—Hawaiian Is.
laculata *H. St. John in Phytologia,* 63(3): 164 (1987)—Hawaiian Is.

STENOGYNE :-
laevis *H. St. John in Phytologia,* 63(3): 164 (1987)—Hawaiian Is.
lanceolata *H. St. John in Phytologia,* 63(3): 164 (1987)—Hawaiian Is.
latisepala *(Sherff) H. St. John in Phytologia,* 63(3): 164 (1987): S. kaalae var. latisepala.
lehuensis *H. St. John in Phytologia,* 63(3): 164 (1987)—Hawaiian Is.
leptophylla *(Sherff) H. St. John in Phytologia,* 63(3): 164 (1987): S. purpurea var. leptophylla.
var. retrorsa *H. St. John in Phytologia,* 63(3): 164 (1987)—Hawaiian Is.
linearis *H. St. John in Phytologia,* 63(3): 165 (1987)—Hawaiian Is.
longiloba *H. St. John in Phytologia,* 63(3): 165 (1987)—Hawaiian Is.
lutea *H. St. John in Phytologia,* 63(3): 165 (1987)—Hawaiian Is.
mamanecola *H. St. John in Phytologia,* 63(3): 165 (1987)—Hawaiian Is.
meeboldii *(Sherff) H. St. John in Phytologia,* 63(3): 165 (1987): S. angustifolia var. meeboldii.
mollissima *(Skottsb.) H. St. John in Phytologia,* 63(3): 165 (1987): S. rugosa f. mollissima.
montana *H. St. John in Phytologia,* 63(3): 165 (1987)—Hawaiian Is.
montideae *H. St. John in Phytologia,* 63(3): 165 (1987)—Hawaiian Is.
multiflora *H. St. John in Phytologia,* 63(3): 165 (1987)—Hawaiian Is.
nahukuensis *H. St. John in Phytologia,* 63(3): 167 (1987)—Hawaiian Is.
nigra *H. St. John in Phytologia,* 63(3): 167 (1987)—Hawaiian Is.
novalimontis *H. St. John in Phytologia,* 63(3): 167 (1987)—Hawaiian Is.
oblonga *(Sherff) H. St. John in Phytologia,* 63(3): 167 (1987): S. rotundifolia var. oblonga.
ohiascandens *H. St. John in Phytologia,* 63(3): 167 (1987)—Hawaiian Is.
olaaensis *H. St. John in Phytologia,* 63(3): 167 (1987)—Hawaiian Is.
olowaluensis *H. St. John in Phytologia,* 63(3): 167 (1987)—Hawaiian Is.
oonoides *H. St. John in Phytologia,* 63(3): 167 (1987)—Hawaiian Is.
ovata *H. St. John in Phytologia,* 63(3): 167 (1987)—Hawaiian Is.
oxyodonta *(Sherff) H. St. John in Phytologia,* 63(3): 166 (1987): S. calaminthoides var. oxyodonta.
pedunculata *H. St. John in Phytologia,* 63(3): 166 (1987)—Hawaiian Is.
pilosa *H. St. John in Phytologia,* 63(3): 166 (1987)—Hawaiian Is.
pubicostata *H. St. John in Phytologia,* 63(3): 166 (1987)—Hawaiian Is.
recta *H. St. John in Phytologia,* 63(3): 166 (1987)—Hawaiian Is.
remyi *(Sherff) H. St. John in Phytologia,* 63(3): 166 (1987): S. scrophularioides var. remyi.

STENOGYNE :-
repens *H. St. John in Phytologia,* 63(3): 166 (1987)—Hawaiian Is.
retrorsa *H. St. John in Phytologia,* 63(3): 166 (1987)—Hawaiian Is.
rhuakos *H. St. John in Phytologia,* 63(3): 166 (1987)—Hawaiian Is.
rockii *H. St. John in Phytologia,* 63(3): 168 (1987)—Hawaiian Is.
rotundifolia
var. salebrosa *H. St. John in Phytologia,* 63(3): 168 (1987)—Hawaiian Is.
rubra *H. St. John in Phytologia,* 63(3): 168 (1987)—Hawaiian Is.
rugosa
var. imberbis *H. St. John in Phytologia,* 63(3): 168 (1987)—Hawaiian Is.
var. monticola *H. St. John in Phytologia,* 63(3): 168 (1987)—Hawaiian Is.
var. pubens *H. St. John in Phytologia,* 63(3): 168 (1987)—Hawaiian Is.
var. pubinervis *H. St. John in Phytologia,* 63(3): 168 (1987)—Hawaiian Is.
var. punaensis *H. St. John in Phytologia,* 63(3): 169 (1987)—Hawaiian Is.
var. remota *H. St. John in Phytologia,* 63(3): 169 (1987)—Hawaiian Is.
semipilosa *H. St. John in Phytologia,* 63(3): 169 (1987)—Hawaiian Is.
septentrionalis *H. St. John in Phytologia,* 63(3): 169 (1987)—Hawaiian Is.
serrata *H. St. John in Phytologia,* 63(3): 169 (1987)—Hawaiian Is.
serrulata *H. St. John in Phytologia,* 63(3): 169 (1987)—Hawaiian Is.
sessilis
var. hanaulaensis *H. St. John in Phytologia,* 63(3): 169 (1987)—Hawaiian Is.
var. lanceolata *H. St. John in Phytologia,* 63(3): 169 (1987)—Hawaiian Is.
var. pilosa *H. St. John in Phytologia,* 63(3): 169 (1987)—Hawaiian Is.
var. wailukuensis *H. St. John in Phytologia,* 63(3): 170 (1987)—Hawaiian Is.
subsessilis *H. St. John in Phytologia,* 63(3): 170 (1987)—Hawaiian Is.
teres *H. St. John in Phytologia,* 63(3): 170 (1987)—Hawaiian Is.
triangularis *H. St. John in Phytologia,* 63(3): 170 (1987)—Hawaiian Is.
villosa *H. St. John in Phytologia,* 63(3): 170 (1987)—Hawaiian Is.
viridialba *H. St. John in Phytologia,* 63(3): 170 (1987)—Hawaiian Is.
volanica *H. St. John in Phytologia,* 63(3): 170 (1987)—Hawaiian Is.
waimeana *(Sherff) H. St. John in Phytologia,* 63(3): 170 (1987): S. calaminthoides var. waimeana.
warshaueri *H. St. John in Phytologia,* 63(3): 170 (1987)—Hawaiian Is.
wilkesii *(Sherff) H. St. John in Phytologia,* 63(3): 171 (1987): S. sessilis var. wilkesii.

STEPHANOMERIA (Compos.).
tenuifolia
var. uintaensis *S. Goodrich & S.L. Welsh in Great Basin Nat.,* 43(2): 375 (1983)—U.S.A. (Utah).

STERCULIA (Sterculiac.).
caribaea *R. Brown, Pterocymbium:* 228 (4 June 1844)—Caribbean Is. †
ceramica *R. Brown, Pterocymbium:* 233 (4 June 1844)—Moluccas. †
cuneata *R. Brown, Pterocymbium:* 232 (4 June 1844)—Philippine Is. †
cuspidata *R. Brown, Pterocymbium:* 232 (4 June 1844)—Sumatra. †
ferruginea *R. Brown, Pterocymbium:* 232 (4 June 1844)—Philippine Is. †
grandifolia *R. Brown, Pterocymbium:* 232 (4 June 1844)—Java. †
insularis *R. Brown, Pterocymbium:* 232 (4 June 1844)—Moluccas. †
javanica *R. Brown, Pterocymbium:* 230 (4 June 1844)—Java. †
madagascariensis *R. Brown, Pterocymbium:* 229 (4 June 1844)—Madagascar. †
mexicana *R. Brown, Pterocymbium:* 227 (4 June 1844)—Mexico. †
oblongata *R. Brown, Pterocymbium:* 232 (4 June 1844)—Philippine Is. †
obovata *R. Brown, Pterocymbium:* 233 (4 June 1844)—Tropical Africa. †
polyphylla *R. Brown, Pterocymbium:* 227 (4 June 1844)—Sumatra. †
propinqua *R. Brown, Pterocymbium:* 229 (4 June 1844)—Surinam. †
quadrifida *R. Brown, Pterocymbium:* 233 (4 June 1844)—Australia. †
rugosa *R. Brown, Pterocymbium:* 229 (4 June 1844)—Guyana. †
spangleri *R. Brown, Pterocymbium:* 230 (4 June 1844)—Java. †
stipularis *R. Brown, Pterocymbium:* 232 (4 June 1844)—Philippine Is. †
surinamensis *R. Brown, Pterocymbium:* 229 (4 June 1844)—Surinam. †
tantraensis *P. Morat in Bull. Mus. Nation. Hist. Nat., B, Adansonia,* Ser. 4, 8(4): 362 (1986 publ. 1987), nom. nov.: Pterocymbium schumannianum *Lauterb.*
triphaca *R. Brown, Pterocymbium:* 228 (4 June 1844)—S.E. Africa. †

STEVENIA (Crucif.).
sergievskajae *(Krasnob.) R.V. Kamelin & I.A. Gubanov in Byull. Mosk. Obshch. Ispyt. Prir., Biol.,* 91(6): 91 (1986): Alyssum sergievskajae.

STEVIA (Compos.).
alpina
var. glutinosa *L. Ariza Espinar & M.M. Cerana in Bol. Acad. Nac. Cienc. (Cordoba),* 57(3–4): 393 (1986)—Argentina.

STEVIA :–
chamaedrys
var. glandulifera *L. Ariza Espinar & M.M. Cerana in Bol. Acad. Nac. Cienc. (Cordoba),* 57(3–4): 391 (1986)—Argentina.
gilliesii
var. glandulosa *L. Ariza Espinar & M.M. Cerana in Bol. Acad. Nac. Cienc. (Cordoba),* 57(3–4): 390 (1986)—Argentina.
vaga
var. effusa *(Hieron.) L. Ariza Espinar & M.M. Cerana in Bol. Acad. Nac. Cienc. (Cordoba),* 57(3–4): 386 (1986): S. effusa.
var. saltensis *(Hieron.) L. Ariza Espinar & M.M. Cerana in Bol. Acad. Nac. Cienc. (Cordoba),* 57(3–4): 386 (1986): S. effusa var. saltensis.

STEWARTIELLA (Umbellif.).
crucifolia *(Gilli) I.C. Hedge & J.M. Lamond in Fl. Iran.,* 162: 213 (1987): Ferula crucifolia.

STEYERBROMELIA L.B. Smith in Ann. Missouri Bot. Gard., 73(4): 699 (1986 publ. 1987). *BROMELIACEAE.*
diffusa *L.B. Smith, J.A. Steyermark & H. Robinson in Ann. Missouri Bot. Gard.,* 73(4): 700 (1986 publ. 1987)—Venezuela.

STIGMAPHYLLON (Malpighiac.).
laciniatum *(Ekman ex Nied.) C. Anderson in Contrib. Univ. Mich. Herb.,* 16: 29 (1987): S. angulosum f. laciniatum.
panamense *C. Anderson in Contrib. Univ. Mich. Herb.,* 16: 35 (1987)—Panama.
tarapotense *C. Anderson in Contrib. Univ. Mich. Herb.,* 16: 49 (1987)—Peru.
tonduzii *C. Anderson in Contrib. Univ. Mich. Herb.,* 16: 45 (1987)—Costa Rica.
yungasense *C. Anderson in Contrib. Univ. Mich. Herb.,* 16: 50 (1987)—Bolivia.

STIGMATODACTYLUS (Orchidac.).
serratus *(Deori) A. Nageswara Rao in J. Econ. Taxon. Bot.,* 9(2): 255 (1987): Pantlingia serrata.

STIPA (Gramin.).
ser. Barbatae *B. Moraldo in Webbia,* 40(2): 211 (1986).
ser. Capenses *B. Moraldo in Webbia,* 40(2): 212 (1986).
ser. Siculae *B. Moraldo in Webbia,* 40(2): 211 (1986).
aquarii *J.W. Vickery, S.W.L. Jacobs & J. Everett in Telopea,* 3(1): 27 (1986)—Australia (N. Terr.).
austroaltaica *Yu.A. Kotukhov in Bot. Zhurn.,* 72(9): 1254 (1987)—U.S.S.R. (Siberia).
subsp. appendiculata *(Čelak.) B. Moraldo in Webbia,* 40(2): 254 (1986): S. pennata var. appendiculata.
blakei *J.W. Vickery, S.W.L. Jacobs & J. Everett in Telopea,* 3(1): 34 (1986)—Australia (Queensland).

STIPA :–
breviseta *(Hitchc.) C. Marticorena & M. Quezada in Gayana, Bot.,* 42(1–2): 152 (1985): Ortachne breviseta.
centralis *J.W. Vickery, S.W.L. Jacobs & J. Everett in Telopea,* 3(1): 39 (1986)—Australia (N. Terr.).
dongicola *J.W. Vickery, S.W.L. Jacobs & J. Everett in Telopea,* 3(1): 46 (1986)—Australia (W. Australia).
echinata *J.W. Vickery, S.W.L. Jacobs & J. Everett in Telopea,* 3(1): 50 (1986)—Australia (S. Australia).
epilosa
subsp. montana *B. Moraldo in Webbia,* 40(2): 258 (1986)—Italy.
etrusca *B. Moraldo in Webbia,* 40(2): 236 (1986)—Italy.
feresetacea *J.W. Vickery, S.W.L. Jacobs & J. Everett in Telopea,* 3(1): 58 (1986)—Australia (N. Terr.).
gussonei *B. Moraldo in Webbia,* 40(2): 221 (1986)—Sicily, Italy.
lanata *J.W. Vickery, S.W.L. Jacobs & J. Everett in Telopea,* 3(1): 68 (1986)—Australia (W. Australia, S. Australia).
nicorae *F.A. Roig in Giorn. Bot. Ital.,* 121(1–2): 41 (1987)—Argentina.
nullarborensis *J.W. Vickery, S.W.L. Jacobs & J. Everett in Telopea,* 3(1): 88 (1986)—Australia (W. Australia).
oligotricha *B. Moraldo in Webbia,* 40(2): 248 (1986)—Italy.
subsp. kiemii *(Martinovský) B. Moraldo in Webbia,* 40(2): 250 (1986): S. pennata subsp. kiemii.
pilata *S.W.L. Jacobs & J. Everett in Telopea,* 3(1): 92 (1986)—Australia (S. Australia).
scabra
subsp. falcata *(Hughes) J.W. Vickery, S.W.L. Jacobs & J. Everett in Telopea,* 3(1): 110 (1986): S. falcata.
speciosa
var. ciliata *F.A. Roig in Giorn. Bot. Ital.,* 121(1–2): 43 (1987)—Argentina.
velutina *J.W. Vickery, S.W.L Jacobs & J. Everett in Telopea,* 3(1): 126 (1986)—Australia (S. Australia, W. Australia).
veneta *B. Moraldo in Webbia,* 40(2): 238 (1986)—Italy.
wakoolica *J.W. Vickery, S.W.L. Jacobs & J. Everett in Telopea,* 3(1): 129 (1986)—Australia (New South Wales).
yatesensis *C. Marticorena & M. Quezada in Gayana, Bot.,* 43(1–4): 149 (1986), nom. nov.: Ortachne breviseta *Hitchc.*

STIPAGROSTIS (Gramin.).
szovitsiana *(Trin. & Rupr.) S.G. Musaev in Bot. Zhurn.,* 72(1): 94 (1987): Aristida plumosa var. szovitsiana.

STRAMENTOPAPPUS H. Robinson & V.A. Funk in Bot. Jahrb., 108(2–3): 227 (1987). *COMPOSITAE.*

STRAMENTOPAPPUS :–
pooleae *(B. Turner) H. Robinson & V.A. Funk in Bot. Jahrb.*, 108(2–3): 227 (1987): Vernonia pooleae.

STROMANTHE (Marantac.).
macrochlamys *(Woodson & Standley) H. Kennedy & D.H. Nicolson in Brittonia*, 39(1): 139 (1987): Calathea macrochlamys.

STRUTHANTHUS (Loranthac.).
aequatoris *J. Kuijt in Fl. Ecuador*, 24(32C): 149 (1986)—Ecuador.

STRYCHNOS (Loganiac.).
sect. Angustiflorae *Trân Công Khánh in Feddes Repert.*, 98(1–2): 88 (1987).
subsect. Nitidae *Trân Công Khánh in Feddes Repert.*, 98(1–2): 83 (1987).
subsect. Wallichianae *Trân Công Khánh in Feddes Repert.*, 98(1–2): 84 (1987).
daclacensis *Trâñ Công Khánh in Feddes Repert.*, 97(11–12): 748 (1986)—Vietnam.
sonlaensis *Trâñ Công Khánh in Feddes Repert.*, 97(11–12): 749 (1986)—Vietnam.

SUAEDA (Chenopodiac.).
pelagica *G. Bartolo, S. Brullo & P. Pavone in Willdenowia*, 16(2): 391 (1987)—Italy (Lampedusa Is.).

SULCOREBUTIA (Cactac.).
fischeriana *K. Augustin in Kakt. And. Sukk.*, 38(9): 216 (1987)—Bolivia.

SWERTIA (Gentianac.).
ser. Ciliatae *Sunita Garg née Agrawal, Gentianac. N.W. Himalayas (Internat. Biosci. Monog. 17):* 167 (1987).
ser. Racemosae *Sunita Garg née Agrawal, Gentianac. N.W. Himalayas (Internat. Biosci. Monog. 17):* 185 (1987).
alpina *U.C. Bhattacharyya & Sunita Agarwal in Bull. Bot. Soc. Bengal*, 36(1–2): 47 (1982 publ. 1986)—India.
alpina *Sunita Garg née Agrawal, Gentianac. N.W. Himalayas (Internat. Biosci. Monog. 17):* 181 (1987)—India.
bifolia
var. wardii *(Marq.) T.N. He in Fl. Xizangica*, 3: 985 (1986): S. wardii.
hispidicalyx
var. subglabra *(Marq.) T.N. He in Fl. Xizangica*, 3: 992 (1986): S. hispidicalyx f. subglabra.
jiendeensis *Y.Y. Fang in Bull. Bot. Res. North-East. Forest. Inst.*, 7(1): 90 (1987)—China.

SYMPHYTUM (Boraginac.).
carpaticum *Yu.M. Frolov in Bot. Zhurn.*, 72(10): 1396 (1987)—U.S.S.R. (Ukraine).

SYMPLOCOS (Symplocac.).
ascidiiformis *Y.F. Wu in Fl. Xizangica*, 3: 860 (1986), nom. illegit., type different from S. ascidiformis Y.F. Wu (1982)—Xizang.

SYMPLOCOS :–
hottae *H. Nagamasu in Acta Phytotax. Geobot.*, 37(4–6): 149 (1986)—Sumatra.

SYNADENIUM (Euphorbiac.).
calycinum *S. Carter in Kew Bull.*, 42(3): 670 (1987)—Tanzania.
compactum
var. rubrum *S. Carter in Kew Bull.*, 42(3): 669 (1987)—Kenya.
glabratum *S. Carter in Kew Bull.*, 42(3): 667 (1987)—Zambia, Tanzania.

SYNOTIS (Compos.).
borii *(Raizada) R. Mathur in J. Econ. Taxon. Bot.*, 8(2): 482 (1986): Senecio borii.
jowaiensis *(Balak.) R. Mathur in J. Econ. Taxon. Bot.*, 8(2): 482 (1986): Senecio jowaiensis.

SYZYGIUM (Myrtac.).
cordemoyi *J. Bosser & T. Cadet in Bull. Mus. Nation. Hist. Nat., B, Adansonia*, Ser. 4, 9(1): 30 (1987)—Réunion.
coriaceum *J. Bosser & J. Guého in Bull. Mus. Nation. Hist. Nat., B, Adansonia*, Ser. 4, 9(1): 34 (1987)—Mauritius.
dubium *(Perry) A.C. Smith in Fl. Vitiensis Nova*, 3: 324 (1985): S. brackenridgei var. dubium.
mamillatum *J. Bosser & J. Guého in Bull. Mus. Nation. Hist. Nat., B, Adansonia*, Ser. 4, 9(1): 36 (1987)—Mauritius.
minus *A.C. Smith in Fl. Vitiensis Nova*, 3: 336 (1985)—Fiji.
petrinense *J. Bosser & J. Guého in Bull. Mus. Nation. Hist. Nat., B, Adansonia*, Ser. 4, 9(1): 36 (1987)—Mauritius.
purpureum *(Perry) A.C. Smith in Fl. Vitiensis Nova*, 3: 332 (1985): S. diffusum var. purpureum.
xizangense *H.T. Chang & Miau in Fl. Xizangica*, 3: 343 (1986)—Xizang.

T

TACHIADENUS (Gentianac.).
boivinii *Humbert ex J. Klackenberg in Bull. Mus. Nation. Hist. Nat., B, Adansonia*, Ser. 4, 9(1): 68 (1987)—Madagascar.
pervillei *Humbert ex J. Klackenberg in Bull. Mus. Nation. Hist. Nat., B, Adansonia*, Ser. 4, 9(1): 68 (1987)—Madagascar.
platypterus
subsp. angustialatus *(Humbert) J. Klackenberg in Bull. Mus. Nation. Hist. Nat., B, Adansonia*, Ser. 4, 9(1): 65 (1987): T. platypterus var. angustialatus.
umbellatus *J. Klackenberg in Bull. Mus. Nation. Hist. Nat., B, Adansonia*, Ser. 4, 9(1): 73 (1987)—Madagascar.

TAINIA (Orchidac.).
taiwaniana *S.S. Ying in Quart. J. Chinese Forest.*, 20(2): 55 (1987)—Taiwan.

TANACETOPSIS (Compos.).
korovinii *S.S. Kovalevskaya in Novosti Sist. Vÿssh. Rast.*, 23: 242 (1986)—U.S.S.R. (Middle Asia).
popovii *R. Kam. & S.S. Kovalevskaya in Novosti Sist. Vÿssh. Rast.*, 23: 244 (1986)—U.S.S.R. (Middle Asia).

TANACETUM (Compos.).
douglasii *A.P. de Candolle, Prodromus,* 6: 128 (1837 publ. 1838)—North America. †
mindshelkense *S.S. Kovalevskaya in Novosti Sist. Vÿssh. Rast.*, 23: 240 (1986)—U.S.S.R. (Middle Asia).
zangezuricum *N.S. Khandzhyan in Biol. Zhurn. Arm.*, 35(1): 72 (1982), nom. nov.: Pyrethrum komarovii *Sosn.*

TAPIPHYLLUM (Rubiac.).
cinerascens
var. inaequale *(Robyns) B. Verdcourt in Kew Bull.*, 42(1): 144 (1987): T. inaequale.
var. laetum *(Robyns) B. Verdcourt in Kew Bull.*, 42(1): 145 (1987): T. laetum.
var. laevius *(K. Schum.) B. Verdcourt in Kew Bull.*, 42(1): 144 (1987): Vangueria velutina var. laevior.
var. richardsii *(Robyns) B. Verdcourt in Kew Bull.*, 42(1): 145 (1987): T. richardsii.

TARAXACUM (Compos.).
sect. Ruderalia *J. Kirschner, H. øllgaard & J. Štěpánek in Taxon,* 36(3): 615 (1987).
sect. Sonchidium *(DC.) J. Kirschner & J. Štěpánek in Taxon,* 36(3): 617 (1987): Dubyaea sect. Sonchidium.
bachczisaraicum *N.N. Tsvelev in Novosti Sist. Vÿssh. Rast.*, 23: 258 (1986)—U.S.S.R. (Crimea).
badzhalense *V.N. Voroshilov & S.D. Shlotgauér in Byull. Mosk. Obshch. Ispÿt. Prir., Biol.*, 91(3): 115 (1986)—U.S.S.R. (Soviet Far East).
gaditanum *S. Talavera in Lagascalia,* 14(1): 161 (1986)—Spain.
hypanicum *N.N. Tsvelev in Novosti Sist. Vÿssh. Rast.*, 23: 256 (1986)—U.S.S.R. (Ukraine).
indonesicum *C.E. Sonck in Ann. Bot. Fenn.*, 24(3): 307 (1987)—Java.
neosivaschicum *N.N. Tsvelev in Novosti Sist. Vÿssh. Rast.*, 23: 259 (1986)—U.S.S.R. (Ukraine).
ohritense *C.E. Sonck in Ann. Bot. Fenn.*, 24(2): 225 (1987)—Albania.
oppidanum *C.E. Sonck in Ann. Bot. Fenn.*, 24(2): 226 (1987)—Albania.
pseudomurbeckianum *N.N. Tsvelev in Novosti Sist. Vÿssh. Rast.*, 23: 257 (1986)—U.S.S.R. (Crimea).
skanderbegii *C.E. Sonck in Ann. Bot. Fenn.*, 24(2): 229 (1987)—Albania.

TARENNA (Rubiac.).
acutisepala *W.C. Chen in Acta Phytotax. Sin.*, 24(6): 477 (1986)—China.
brevicymigera *W.C. Chen in Acta Bot. Austro Sinica,* 3: 26 (1986)—Vietnam.
luhomeroensis *D.M. Bridson in Kew Bull.*, 42(1): 257 (1987)—Tanzania.
roseicosta *D.M. Bridson in Kew Bull.*, 42(1): 259 (1987)—Tanzania.
uzungwaensis *D.M. Bridson in Kew Bull.*, 42(1): 260 (1987)—Tanzania.

TAUSCHIA (Umbellif.).
hintoniorum *L. Constance & J. Affolter in Syst. Bot.*, 12(2): 286 (1987)—Mexico.
infernicola *L. Constance & J. Affolter in Syst. Bot.*, 12(2): 289 (1987)—Mexico.
moorei *L. Constance & J. Affolter in Syst. Bot.*, 12(2): 291 (1987)—Mexico.

TEPHROCACTUS (Cactac.).
weberi
var. deminutus *W. Rausch in Succulenta,* 65(12): 251 (1986)—Argentina.

TEPHROSERIS (Compos.).
helenitis
var. discoidea *(DC.) M. Kerguélen in Lejeunia,* 120: 173 (1987): Senecio brachychaetus var. discoideus.

TEPHROSIA (Legumin.).
albissima
subsp. zuluensis *(H.M.L. Forbes) B.D. Schrire in Bothalia,* 17(1): 13 (1987): T. zuluensis.
brummittii *B.D. Schrire in Bothalia,* 17(1): 12 (1987)—South Africa (Natal), Swaziland, Mozambique.
glomeruliflora
subsp. meisneri *(Hutch. & Burtt Davy) B.D. Schrire in Bothalia,* 17(1): 14 (1987): T. meisneri.
macropoda
var. diffusa *(E. Mey.) B.D. Schrire in Bothalia,* 17(1): 12 (1987): Apodynomene diffusa.
natalensis
subsp. pseudocapitata *(H.M.L. Forbes) B.D. Schrire in Bothalia,* 17(1): 11 (1987): T. pseudocapitata.
quartiniana *Cuf. ex W. Greuter & Burdet in Willdenowia,* 16(2): 446 (1987), nom. nov.: T. vicioides *A. Rich.*

TERMINALIA (Combretac.).
psilantha *A.C. Smith in Fl. Vitiensis Nova,* 3: 428 (1985)—Fiji.

TETRADIA R Brown, Pterocymbium: 233 (4 June 1844). *STERCULIACEAE.* †
horsfieldii *R. Brown, Pterocymbium:* 233 (4 June 1844)—Java. †

TETRAMOLOPIUM (Compos.).
sect. Alpinum *T.K. Lowrey in Allertonia,* 4(4): 241 (1986).
sect. Sandwicense *T.K. Lowrey in Allertonia,* 4(4): 254 (1986).
arenarium
subsp. laxum *T.K. Lowrey in Allertonia,* 4(4): 259 (1986)—Hawaiian Is.
consanguineum
var. kauense *T.K. Lowrey in Allertonia,* 4(4): 262 (1986)—Hawaiian Is.
subsp. leptophyllum *(Sherff) T.K. Lowrey in Allertonia,* 4(4): 260 (1986): T. consanguineum var. leptophyllum.
filiforme
var. polyphyllum *(Sherff) T.K. Lowrey in Allertonia,* 4(4): 247 (1986): T. polyphyllum.
humile
subsp. haleakalae *T.K. Lowrey in Allertonia,* 4(4): 243 (1986)—Hawaiian Is.
lepidotum
subsp. arbusculum *(A. Gray) T.K. Lowrey in Allertonia,* 4(4): 256 (1986): Vittadinia chamissonis var. arbuscula.
rockii
var. calcisabulorum *(St. John) T.K. Lowrey in Allertonia,* 4(4): 250 (1986): Luteidiscus calcisabulorum.
sylvae *T.K. Lowrey in Allertonia,* 4(4): 252 (1986)—Hawaiian Is., Cook Is.

TETRAPLANDRA (Euphorbiac.).
amazonica *M. Emmerich in Arq. Mus. Nac. (Rio de Janeiro),* 56: 95 (1981)—Brazil.
bahiensis *M. Emmerich in Arq. Mus. Nac. (Rio de Janeiro),* 56: 95 (1981)—Brazil.
dimitrii *M. Emmerich in Arq. Mus. Nac. (Rio de Janeiro),* 56: 96 (1981)—Brazil.
kuhlmannii *M. Emmerich in Arq. Mus. Nac. (Rio de Janeiro),* 56: 96 (1981)—Brazil.

TETRAPOGON (Gramin.).
ferrugineus *(Renvoize) S.M. Phillips in Kew Bull.,* 42(2): 478 (1987): Chloris ferruginea.

TETRAPTERYS (Malpighiac.).
cordifolia *W.R. Anderson in Contrib. Univ. Mich. Herb.,* 16: 107 (1987)—Brazil.

TETRASTIGMA (Vitac.).
gamblei *B.V. Shetty & P. Singh in Kew Bull.,* 42(4): 933 (1987)—India.

TETRATAENIUM (Umbellif.).
bivittatum *(Boissieu) I.P. Mandenova in Zam. Sist. Geogr. Rast.,* 41: 45 (1986): Heracleum bivittatum.
burmanicum *(Kurz) I.P. Mandenova in Zam. Sist. Geogr. Rast.,* 41: 46 (1986): Heracleum burmanicum.
candicans *(Wall. ex DC.) I.P. Mandenova in Zam. Sist. Geogr. Rast.,* 41: 44 (1986): Heracleum candicans.

TETRATAENIUM :-
cardiocarpum *(Rech. f. & H. Riedl) I. Mandenova in Fl. Iran.,* 162: 505 (1987): Heracleum cardiocarpum.
ceylanicum *(Gardn. ex C.B. Clarke) I.P. Mandenova in Zam. Sist. Geogr. Rast.,* 41: 46 (1986): Heracleum ceylanicum.
lasiopetalum *(Boiss.) I. Mandenova in Fl. Iran.,* 162: 504 (1987): Heracleum lasiopetalum.
leucocarpum *(Aitch. & Hemsl.) I. Mandenova in Fl. Iran.,* 162: 506 (1987): Heracleum leucocarpum.
nephrophyllum *(Leute) I. Mandenova in Fl. Iran.,* 162: 505 (1987): Heracleum nephrophyllum.
sublineare *(C.B. Clarke) I.P. Mandenova in Zam. Sist. Geogr. Rast.,* 41: 47 (1986): Heracleum sublineare.
wallichii *(DC.) I.P. Mandenova in Zam. Sist. Geogr. Rast.,* 41: 47 (1986): Heracleum wallichii.

THALICTRUM (Ranunculac.).
minus
var. arenicola *V. Osvačilová in Biológia, A (Czechoslovakia),* 38(5): 456 (1983)—Czechoslovakia.
subsp. carpaticum *(Kotula) V. Osvačilová in Fl. Slovenska,* 3: 237 (1982): T. minus var. carpaticum.
subsp. elatum *(Jacq.) M. Kerguélen in Lejeunia,* 120: 175 (1987): T. elatum.
f. jacquinianum *(Koch) V. Osvačilová in Fl. Slovenska,* 3: 231 (1982): T. jacquinianum.
f. macrophyllum *(Schur) V. Osvačilová in Fl. Slovenska,* 3: 230 (1982): T. minus b. macrophyllum.
var. petrosum *(Schur) V. Osvačilová in Fl. Slovenska,* 3: 230 (1982): T. petrosum.
subsp. saxicola *V. Osvačilová in Biológia, A (Czechoslovakia),* 38(5): 456 (1983)—Czechoslovakia.

THELESPERMA (Compos.).
pubescens *R.D. Dorn in Great Basin Nat.,* 43(4): 749 (1983)—U.S.A. (Wyoming).
subnudum
var. alpinum *S.L. Welsh in Great Basin Nat.,* 43(2): 369 (1983)—U.S.A. (Utah).

THELOCACTUS (Cactac.).
bicolor
var. schwarzii *(Backeb.) E.F. Anderson in Bradleya,* 5: 61 (1987): T. schwarzii.
durangensis *(Ruenge) G.D. Rowley in Excelsa,* 12: 31 (1986): Echinocactus durangensis.
setispinus *(Engelm.) E.F. Anderson in Bradleya,* 5: 59 (1987): Echinocactus setispinus.
tulensis
var. buekii *(Klein) E.F. Anderson in Bradleya,* 5: 65 (1987): Echinocactus buekii.

THELOCACTUS :–
var. matudae *(Sanchez-Mej. & A. Lau) E.F. Anderson in Bradleya,* 5: 66 (1987): T. matudae.
unguispinus *(Eng.) G.D. Rowley in Excelsa,* 12: 30 (1986): Echinocactus unguispinus.

THIBAUDIA (Ericac.).
foreroi *J.L. Luteyn in Op. Bot.,* 92: 127 (1987)—Colombia.
harlingii *J.L. Luteyn in Op. Bot.,* 92: 127 (1987)—Ecuador.

THINOPYRUM (Gramin.).
ponticum *(Podp.) M.E. Barkworth & D.R. Dewey in Great Basin Nat.,* 43(4): 570 (1983), without basionym ref.: Basionym not stated.

THISMIA (Burmanniac.).
melanomitra *P.J.M. Maas & H. Maas in Op. Bot.,* 92: 141 (1987)—Ecuador.
saulensis *H. Maas & P.J.M. Maas in Brittonia,* 39(3): 376 (1987)—French Guiana.

THLADIANTHA (Cucurbitac.).
pustulata *(Lévl.) C. Jeffrey ex A.M. Lu & Z.Y. Zhang in Fl. Reipubl. Popul. Sin.,* 73(1): 141 (1986): Melothria pustulata.

THLASPI (Crucif.).
alpestre
subsp. sylvium *(Gaudin) M. Kerguélen in Lejeunia,* 120: 175 (1987).: T. sylvium.

THRYPTOMENE (Myrtac.).
biseriata *J.W. Green in Fl. S. Australia,* 2: 948 (1986)—Australia (W. Australia, S. Australia).
longifolia *J.W. Green in Fl. S. Australia,* 2: 950 (1986)—Australia (S. Australia).

THYMELAEA (Thymelaeac.).
dioica
subsp. glauca *S. Talavera & J. Muñoz in Lagascalia,* 14(1): 149 (1986)—Spain.
tartonraira
subsp. austroiberica *J. Lambinon in Soc. Ech. Pl. Vasc. Eur. Bassin Médit. Bull.,* 21: 21 (1986), nom. nov.: Passerina tartonraira var. angustifolia *Boiss.*

THYMOPHYLLA (Compos.).
acerosa *(DC.) J.L. Strother in Sida,* 11(4): 376 (1986): Dyssodia acerosa.
aurea
var. polychaeta *(A. Gray) J.L. Strother in Sida,* 11(4): 376 (1986): Hymenatherum polychaetum.
concinna *(A. Gray) J.L. Strother in Sida,* 11(4): 376 (1986): Hymenatherum concinnum.
gentryi *(M. Johnston) J.L. Strother in Sida,* 11(4): 376 (1986): Dyssodia gentryi.
gypsophila *(B. Turner) J.L. Strother in Sida,* 11(4): 377 (1986): Dyssodia gypsophila.

THYMOPHYLLA :–
micropoides *(DC.) J.L. Strother in Sida,* 11(4): 377 (1986): Gnaphalopsis micropoides.
mutica *(M. Johnston) J.L. Strother in Sida,* 11(4): 377 (1986): Dyssodia mutica.
pentachaeta
var. belenidium *(DC.) J.L. Strother in Sida,* 11(4): 377 (1986): Hymenatherum belenidium.
var. hartwegii *(A. Gray) J.L. Strother in Sida,* 11(4): 377 (1986): Hymenatherum hartwegii.
var. puberula *(Rydb.) J.L. Strother in Sida,* 11(4): 377 (1986): T. puberula.
setifolia
var. radiata *(A. Gray) J.L. Strother in Sida,* 11(4): 378 (1986): T. greggii var. radiata.
tenuiloba
var. texana *(Cory) J.L. Strother in Sida,* 11(4): 378 (1986): Dyssodia texana.
var. treculii *(A. Gray) J.L. Strother in Sida,* 11(4): 378 (1986): Hymenatherum treculii.
var. wrightii *(A. Gray) J.L. Strother in Sida,* 11(4): 378 (1986): Hymenatherum wrightii.
tephroleuca *(S.F. Blake) J.L. Strother in Sida,* 11(4): 378 (1986): Dyssodia tephroleuca.

THYMUS (Labiat.).
×brachychaetus
nothosubsp. toletanus *(Pau) R. Morales Valverde in An. Jard. Bot. Madrid,* 43(1): 39 (1986): T. × mixtus var. toletanus.
×bractichina *R. Morales Valverde in An. Jard. Bot. Madrid,* 43(1): 39 (1986), nom. nov.: T. × pectinatus *R. Morales.*
funkii
var. sabulicola *(Cosson) R. Morales Valverde in An. Jard. Bot. Madrid,* 43(1): 36 (1986): T. sabulicola.
×hyeronymi
nothosubsp. hurtadoi *(O. Socorro et al.) R. Morales Valverde in An. Jard. Bot. Madrid,* 43(1): 39 (1986): T. × hurtadoi.
×josephi-angeli *J. Mansanet & A. Aguilella in Mediterranea,* 8: 84 (1985)—Spain. (T. piperella × vulgaris).
kondratjukii *V.M. Ostapko in Ukr. Bot. Zhurn.,* 44(2): 47 (1987)—U.S.S.R. (Ukraine).
leucospermus *P. Hartvig in Pl. Syst. Evol.,* 155(1–4): 206 (1987)—Greece.
millefloris *D. Rivera Núñez, J. Flores & J. Laencina in An. Jard. Bot. Madrid,* 44(1): 47 (1987)—Spain.
polytrichus
subsp. britannicus *(Ronniger) M. Kerguélen in Lejeunia,* 120: 177 (1987): T. britannicus.
subsp. vallicola *(H. Braun) M. Debray ex M. Kerguélen in Lejeunia,* 120: 177 (1987): T. trachselianus var. vallicola.
rechingeri *P. Hartvig in Pl. Syst. Evol.,* 155(1–4): 205 (1987)—Greece.
subsp. macrocalyx *P. Hartvig in Pl. Syst. Evol.,* 155(1–4): 206 (1987)—Greece.

THYMUS :–
rigidulus *M. Kerguélen in Lejeunia,* 120: 178 (1987), nom. nov.: T. augustifolius var. rigidus *Wimm. & Grab.*
teucrioides
subsp. alpinus *P. Hartvig in Pl. Syst. Evol.,* 155(1–4): 204 (1987)—Greece.
subsp. candilicus *(Beauverd) P. Hartvig in Pl. Syst. Evol.,* 155(1–4): 204 (1987): T. teucrioides var. candilicus.
×viciosoi *(Pau) R. Morales Valverde in An. Jard. Bot. Madrid,* 43(1): 41 (1986): T. bracteatus f. viciosoi.

THYRSOSALACIA (Celastrac.).
pararacemosa *N. Hallé in Fl. Gabon,* 29: 24 (1986)—Cameroun.
racemosa *(Loesener ex Harms) N. Hallé in Fl. Gabon,* 29: 26 (1986): Salacia racemosa.

THYSANOTUS (Anthericac.).
newbeyi *N.H. Brittan in Fl. Australia,* 45: 495, 323 (1987)—Australia (W. Australia).

TIGRIDIA (Iridac.).
martinezii *G. Calderón de Rzedowski in Bol. Soc. Bot. Méx.,* 47: 3 (1987)—Mexico.

TILIA (Tiliac.).
angustibracteata *H.T. Chang in Acta Sci. Nat. Univ. Sunyatseni,* 1987(2): 41 (1987)—China.
fulvosa *H.T. Chang in Acta Sci. Nat. Univ. Sunyatseni,* 1987(2): 42 (1987)—China.
gracilis *H.T. Chang in Acta Sci. Nat. Univ. Sunyatseni,* 1987(2): 42 (1987)—China.

TILLANDSIA (Bromeliac.).
abysmophila *L.B. Smith & J.A. Steyermark in Ann. Missouri Bot. Gard.,* 73(4): 714 (1986 publ. 1987)—Venezuela.
almeriae *W. Rauh in Abh. Akad. Wiss. Lit. Mainz, Math.-Naturwiss. Klasse, Trop. Subtrop. Pflanzenwelt,* 58: 19 (1986)—Colombia.
alvareziae *W. Rauh in Abh. Akad. Wiss. Lit. Mainz, Math.-Naturwiss. Klasse, Trop. Subtrop. Pflanzenwelt,* 58: 22 (1986), as 'alvarezii'—Mexico.
chapeuensis *W. Rauh in Abh. Akad. Wiss. Lit. Mainz, Math.-Naturwiss. Klasse, Trop. Subtrop. Pflanzenwelt,* 58: 24 (1986)—Brazil.
nolleriana *R. Ehlers ex W. Rauh in Abh. Akad. Wiss. Lit. Mainz, Math.-Naturwiss. Klasse, Trop. Subtrop. Pflanzenwelt,* 58: 29 (1986)—Mexico.
peiranoi
var. alba *J. Rutschmann ex W. Rauh in Abh. Akad. Wiss. Lit. Mainz, Math.-Naturwiss. Klasse, Trop. Subtrop. Pflanzenwelt,* 58: 32 (1986)—Argentina.
pseudosetacea *R. Ehlers & W. Rauh in Abh. Akad. Wiss. Lit. Mainz, Math.-Naturwiss. Klasse, Trop. Subtrop. Pflanzenwelt,* 58: 35 (1986)—Mexico.

TILLANDSIA :–
reclinata *Anon. in Bradea,* 4(33): errata slip (1986), non Pereira & Martinelli (1982)—Brazil.
schusteri *W. Rauh in Abh. Akad. Wiss. Lit. Mainz, Math.-Naturwiss. Klasse, Trop. Subtrop. Pflanzenwelt,* 58: 37 (1986)—Mexico.
weberi *L. Hromadnik & P. Schneider in J. Bromeliad Soc.,* 37(3): 105 (1987)—Mexico.

TIPULARIA (Orchidac.).
harae *(Maekawa) S.C. Chen in Acta Phytotax. Sin.,* 25(5): 335 (1987): T. japonica var. harae.

TORULINIUM (Cyperac.).
macrocephalum
var. davidsei *C.D. Adams in Phytologia,* 63(4): 300 (1987)—Trinidad.

TOUCHARDIA (Urticac.).
christensenii *H. St. John in Phytologia,* 63(3): 183 (1987)—Hawaiian Is.
glabra *H. St. John in Phytologia,* 63(3): 183 (1987)—Hawaiian Is.
haupuensis *H. St. John in Phytologia,* 63(3): 183 (1987)—Hawaiian Is.
iaoensis *H. St. John in Phytologia,* 63(3): 183 (1987)—Hawaiian Is.
konaensis *H. St. John in Phytologia,* 63(3): 183 (1987)—Hawaiian Is.
lanaiensis *H. St. John in Phytologia,* 63(3): 183 (1987)—Hawaiian Is.
molokaiensis *H. St. John in Phytologia,* 63(3): 184 (1987)—Hawaiian Is.
nana *H. St. John in Phytologia,* 63(3): 184 (1987)—Hawaiian Is.
napaliensis *H. St. John in Phytologia,* 63(3): 184 (1987)—Hawaiian Is.
occidentalis *H. St. John in Phytologia,* 63(3): 184 (1987)—Hawaiian Is.
wailauensis *H. St. John in Phytologia,* 63(3): 184 (1987)—Hawaiian Is.
wainihaensis *H. St. John in Phytologia,* 63(3): 184 (1987)—Hawaiian Is.

TOVOMITOPSIS (Guttif.).
silvicola *B.E. Hammel in Selbyana,* 9(1): 113 (1986)—Costa Rica.

TOWNSENDIA (Compos.).
jonesii
var. lutea *S.L. Welsh in Great Basin Nat.,* 43(2): 369 (1983)—U.S.A. (Utah).
montana
var. caelilinensis *S.L. Welsh in Great Basin Nat.,* 43(2): 370 (1983)—U.S.A. (Utah).

TOXICODENDRON (Anacardiac.).
toxicarum
f. elobatum *(Fern.) C.F. Reed in Phytologia,* 63(5): 411 (1987): Rhus toxicodendron f. elobatum.
f. leiocarpum *(Fern.) C.F. Reed in Phytologia,* 63(5): 411 (1987): Rhus toxicodendron f. leiocarpa.

TRACHYDIUM (Umbellif.).
pauciradiatum *(Boiss. & Hohen.) K.H. Rechinger in Fl. Iran.,* 162: 183 (1987): Prangos pauciradiata.

TRACHYSPERMUM (Umbellif.).
confusum *I.C. Hedge, J.M. Lamond & K.H. Rechinger in Fl. Iran.,* 162: 340 (1987)—Afghanistan.
gedrosiacum *(Bornm.) I.C. Hedge, J.M. Lamond & K.H. Rechinger in Fl. Iran.,* 162: 340 (1987): Pimpinella gedrosiaca.
halophilum *I.C. Hedge, J.M. Lamond & K.H. Rechinger in Fl. Iran.,* 162: 339 (1987)—Afghanistan.
microcarpum *I.C. Hedge, J.M. Lamond & K.H. Rechinger in Fl. Iran.,* 162: 342 (1987)—Afghanistan.
paktianum *I.C. Hedge, J.M. Lamond & K.H. Rechinger in Fl. Iran.,* 162: 343 (1987)—Afghanistan.
papillare *(Boiss.) I.C. Hedge, J.M. Lamond & K.H. Rechinger in Fl. Iran.,* 162: 342 (1987): Reutera papillaris.
podlechii *(Leute) I.C. Hedge, J.M. Lamond & K.H. Rechinger in Fl. Iran.,* 162: 339 (1987): Seseli podlechii.
stewartii *(Dunn) I.C. Hedge, J.M. Lamond & K.H. Rechinger in Fl. Iran.,* 162: 338 (1987): Pituranthos stewartii.

TRAGIA (Euphorbiac.).
giardelliae *M.M. Gutiérrez de Sanguinetti & M.E. Múlgura de Romero in Darwiniana,* 27(1–4): 491 (1986), as 'giardellii'—Argentina.
glabrata
var. hispida *A. Radcliffe-Smith in Kew Bull.,* 42(2): 396 (1987)—Mozambique.
mazoensis *A. Radcliffe-Smith in Kew Bull.,* 42(2): 396 (1987)—Zimbabwe.
prionoides *A. Radcliffe-Smith in Kew Bull.,* 42(2): 397 (1987)—Zimbabwe, South Africa (Transvaal).
prostrata *A. Radcliffe-Smith in Kew Bull.,* 42(2): 398 (1987)—Zambia.
shirensis
var. glabriuscula *A. Radcliffe-Smith in Kew Bull.,* 42(2): 399 (1987)—Mozambique.

TRAGOPOGON (Compos.).
kashmirianus *G. Singh in G. Singh & P. Kachroo, Forest. Fl. Srinagar:* 123 (1987)—India.
oligolepis *P. Hartvig & A. Strid in Bot. Jahrb.,* 108(2–3): 312 (1987)—Turkey.

TRAPA (Trapac.).
litwinowii
var. chihuensis *S.F. Guan & Q. Lang in Bull. Bot. Res. North-East. Forest. Inst.,* 7(1): 77 (1987)—China.

TRETORHIZA (Gentianac.).
cruciata
subsp. phlogifolia *(Schott & Kotschy) Á. Löve & D. Löve in Taxon,* 35(4): 899 (1986): Gentiana phlogifolia.
decumbens *(L.f.) Á. Löve & D. Löve in Taxon,* 35(4): 899 (1986): Gentiana decumbens.

TRIBONANTHES (Haemodorac.).
purpurea *T.D. Macfarlane & S.D. Hopper in Fl. Australia,* 45: 465, 134 (1987)—Australia (W. Australia).

TRICALYSIA (Rubiac.).
sect. Rosea *(Klotzsch) E. Robbrecht in Bull. Jard. Bot. Nation. Belg.,* 57(1–2): 180 (1987): gen. Rosea.
amplexicaulis *E. Robbrecht in Bull. Jard. Bot. Nation. Belg.,* 57(1–2): 156 (1987)—Cameroun.
bagshawei
subsp. malaissei *E. Robbrecht in Bull. Jard. Bot. Nation. Belg.,* 57(1–2): 140 (1987)—Zaïre, Zambia.
capensis
var. galpinii *(Schinz) E. Robbrecht in Bull. Jard. Bot. Nation. Belg.,* 57(1–2): 187 (1987): T. galpinii.
var. transvaalensis *E. Robbrecht in Bull. Jard. Bot. Nation. Belg.,* 57(1–2): 186 (1987)—South Africa (Transvaal), Swaziland.
coriacea
subsp. angustifolia *(Garcia) E. Robbrecht in Bull. Jard. Bot. Nation. Belg.,* 57(1–2): 87 (1987): T. nyassae var. angustifolia.
fangana *(N. Hallé) E. Robbrecht in Bull. Jard. Bot. Nation. Belg.,* 57(1–2): 126 (1987): T. biafrana var. fangana.
griseiflora
var. barotseana *E. Robbrecht in Bull. Jard. Bot. Nation. Belg.,* 57(1–2): 169 (1987)—Zambia, Zaïre, Angola.
var. benguellensis *(Welw. ex Hiern) E. Robbrecht in Bull. Jard. Bot. Nation. Belg.,* 57(1–2): 168 (1987): T. benguellensis.
jasminiflora
var. kirkii *(Hiern) E. Robbrecht in Bull. Jard. Bot. Nation. Belg.,* 57(1–2): 202 (1987): T. kirkii.
lasiodelphys
subsp. anomalura *(N. Hallé) E. Robbrecht in Bull. Jard. Bot. Nation. Belg.,* 57(1–2): 75 (1987): T. anomalura.
niamniamensis
var. djurensis *(Schweinf. ex Hiern) E. Robbrecht in Bull. Jard. Bot. Nation. Belg.,* 57(1–2): 173 (1987): T. djurensis.
var. nodosa *E. Robbrecht in Bull. Jard. Bot. Nation. Belg.,* 57(1–2): 174 (1987)—Zimbabwe, Malawi, Zambia, Uganda, Zaïre.

TRICALYSIA :–
obanensis
subsp. kwangoensis *E. Robbrecht in Bull. Jard. Bot. Nation. Belg.*, 57(1–2): 98 (1987)—Zaïre, Angola.
pedicellata *E. Robbrecht in Bull. Jard. Bot. Nation. Belg.*, 57(1–2): 194 (1987)—Tanzania.
pedunculosa *(N. Hallé) E. Robbrecht in Bull. Jard. Bot. Nation. Belg.*, 57(1–2): 105 (1987): T. soyauxii var. pedunculosa.
var. pilosula *(N. Hallé) E. Robbrecht in Bull. Jard. Bot. Nation. Belg.*, 57(1–2): 106 (1987): T. soyauxii var. pilosula.
var. walkeriana *(N. Hallé) E. Robbrecht in Bull. Jard. Bot. Nation. Belg.*, 57(1–2): 106 (1987): T. soyauxii var. walkeriana.
repens *E. Robbrecht in Bull. Jard. Bot. Nation. Belg.*, 57(1–2): 152 (1987)—Zambia, Zaïre, Angola.
schliebenii *E. Robbrecht in Bull. Jard. Bot. Nation. Belg.*, 57(1–2): 196 (1987)—Tanzania.
sylvae *E. Robbrecht in Bull. Jard. Bot. Nation. Belg.*, 57(1–2): 154 (1987)—Cameroun, Gabon.
vadensis *E. Robbrecht in Bull. Jard. Bot. Nation. Belg.*, 57(1–2): 78 (1987)—Cameroun.
velutina *E. Robbrecht in Bull. Jard. Bot. Nation. Belg.*, 57(1–2): 179 (1987)—Tanzania.
yangambiensis *(N. Hallé) E. Robbrecht in Bull. Jard. Bot. Nation. Belg.*, 57(1–2): 110 (1987): T. soyauxii var. yangambiensis.
zambesiaca *E. Robbrecht in Bull. Jard. Bot. Nation. Belg.*, 57(1–2): 141 (1987)—Zambia.

TRICHOLAENA (Gramin.).
capensis
subsp. arenaria *(Nees) G. Zizka, Revis. Melinideae Hitchcock:* 75 (1987): T. arenaria.
teneriffae
subsp. eichingeri *(Mez) G. Zizka, Revis. Melinideae Hitchcock:* 69 (1987): Melinis eichingeri.

TRICHOLASER (Umbellif.).
cachemiricum *(C.B. Clarke) R. Alava in Fl. Iran.*, 162: 491 (1987): Heracleum cachemiricum.

TRICHOSALPINX (Orchidac.).
drosoides *G. Carnevali F.C. & I. Ramírez de Carnevali in Ernstia,* 42: 1 (1987)—Venezuela.
greenwoodiana *M.A. Soto Arenas in Orquídea (México),* 10(2): 257 (1987)—Mexico, Salvador.
nageliana *M.A. Soto Arenas in Orquídea (México),* 10(2): 259 (1987)—Mexico.
spathuliglossa *(Hoehne) C.A. Luer, Ic. Pleurothallid. III Syst. Pleurothallis (Monogr. Syst. Bot. Missouri Bot. Gard. 20):* 16 (1986): Pleurothallis spathuliglossa.

TRICHOSALPINX :–
tamayoana *M.A. Soto Arenas in Orquídea (México),* 10(2): 264 (1987)—Mexico.

TRICHOSANTHES (Cucurbitac.).
rosthornii
var. multicirrata *(C.Y. Cheng & Yueh) S.K. Chen in Fl. Reipubl. Popul. Sin.*, 73(1): 244 (1986): T. multicirrata.
var. scabrella *(Yueh & D.F. Gao) S.K. Chen in Fl. Reipubl. Popul. Sin.*, 73(1): 244 (1986): T. scabrella.

TRICORYNE (Anthericac.).
anceps
subsp. pterocaulon *(Baker) A. Thongpukdee in Fl. Australia,* 45: 496 (1987): T. pterocaulon.

TRIDENS (Gramin.).
nicorae *A.M. Anton in Kurtziana,* 10: 51 (1977)—Argentina.

TRIFOLIUM (Legumin.).
egrissicum *A.D. Mikheev & A.Yu. Magulaev in Bot. Zhurn.*, 72(9): 1257 (1987)—U.S.S.R. (Caucasus).
gordeievii *(Kom.) E. Small in Taxon,* 36(3): 580 (1987): Medicago gordeievi.
pannonicum
var. moldavicum *A.T. Szabó in Not. Bot. Horti Agrobot. Cluj-Napoca,* 17: 61 (1987)—U.S.S.R. (Moldavia).
piligerum *(J. Rohlena) A.W. Hill in Index Kewensis,* Supp. 6: 212 (1916–1920 publ. 1926): T. pignantii f. piligerum.
xanthinoides *(J. Rohlena) A.W. Hill in Index Kewensis,* Supp. 6: 212 (1916–1920 publ. 1926): T. stellatum f. xanthinoides.

TRIGONELLA (Legumin.).
sect. Lopholobus *(Boiss.) P. Lassen in Willdenowia,* 16(2): 447 (1987): Melilotus sect. Lopholobus.
bicolor *(Boiss. & Balansa) P. Lassen in Willdenowia,* 16(2): 446 (1987): Melilotus bicolor.
glabra
subsp. uncata *(Boiss. & Noë) P. Lassen in Willdenowia,* 16(2): 447 (1987): T. uncata.

TRIGONOSCIADIUM (Umbellif.).
brachytaenium *(Boiss.) R. Alava in Fl. Iran.*, 162: 513 (1987): Malabaila brachytaenia.
lasiocarpum *(Boiss.) R. Alava in Fl. Iran.*, 162: 514 (1987): Malabaila lasiocarpa.

TRIKERAIA (Gramin.).
oreophila *T.A. Cope in Kew Bull.*, 42(2): 350 (1987)—Bhutan, Xizang.

TRILLIUM (Trilliac.).
recurvatum
f. esepalum *J.D. Freeman & T.E. Heineke in Rhodora,* 89(857): 17 (1987)—U.S.A. (Tennessee).

TRIPLEUROSPERMUM (Compos.).
perforatum *(Mérat) G. Wagenitz in G. Hegi, Illustr. Fl. Mitteleur.*, ed. 2, 6(4): 1356 (1987): Matricaria perforata.

TRISETUM (Gramin.).
kochianum *I. Hernández Torres in Phytologia,* 61(7): 454 (1987), nom. nov.: Deschampsia pringlei *Scribn.*

×TRITICOSECALE Wittmack ex A. Camus in Bull. Mus. Nation. Hist. Nat. Paris, 33: 539 (1927); cf. C.A. Stace in Taxon, 36: 447 (1987). *GRAMINEAE.* (Secale × Triticum).

TRITICUM (Gramin.).
phoenicoides *(L.) Brotero, Fl. Lusit.*, 1: 121 (1804): Festuca phoenicoides.
×teres *H.R. Jiang & X.X. Kong in Acta Bot. Bor.-Occid. Sin.*, 6(3): 206 (1986)—Cultivation. (T. tauschii × urartu).
vavilovii
var. lorenze *S.Kh. Galstyan-Avanesyan in Biol. Zhurn. Arm.*, 39(6): 522 (1986)—Cultivation.
var. ramocoeruleum *S.Kh. Galstyan-Avanesyan in Biol. Zhurn. Arm.*, 39(6): 522 (1986)—Cultivation.
var. ramomuticum *S.Kh. Galstyan-Avanesyan in Biol. Zhurn. Arm.*, 39(6): 521 (1986)—Cultivation.
var. sisianicum *S.Kh. Galstyan-Avanesyan in Biol. Zhurn. Arm.*, 39(6): 521 (1986)—Cultivation.

TRIUNIA (Proteac.).
erythrocarpa *D.B. Foreman in Muelleria,* 6(5): 302 (1987)—Australia (Queensland).

TRUKIA (Rubiac.).
dryadum *(S. Moore) F.R. Fosberg in Phytologia,* 62(3): 175 (1987): Gardenia dryadum.
fitzalanii *(F. Muell.) F.R. Fosberg in Phytologia,* 62(3): 175 (1987): Gardenia fitzalanii.
macarthurii *(F. Muell.) F.R. Fosberg in Phytologia,* 62(3): 176 (1987): Randia macarthurii.
tahitensis *(Nadeaud) F.R. Fosberg in Phytologia,* 62(3): 174 (1987): Randia tahitensis.

TUPISTRA (Convallariac.).
heensis *Y. Wan & X.H. Lu in Bull. Bot. Res. North-East. Forest. Inst.*, 7(1): 81 (1987)—China.
verruculosa *Q.H. Chen in Acta Phytotax. Sin.*, 25(1): 69 (1987)—China.

TURBINICARPUS (Cactac.).
swobodae *L. Diers in Kakt. And. Sukk.*, 38(4): 90 (1987)—Mexico.

TURRAEA (Meliac.).
sect. Nurmonia *(Harms) F. White in Bothalia,* 16(2): 151 (1986), without exact basionym page: gen. Nurmonia.
streyi *F. White & B.T. Styles in Bothalia,* 16(2): 153 (1986)—South Africa (Natal).

TURRILLIA A.C. Smith in Fl. Vitiensis Nova, 3: 753 (1985). *PROTEACEAE.*
bleasdalei *(F. v. Muell.) A.C. Smith in Fl. Vitiensis Nova,* 3: 754 (1985): Grevillea bleasdalei.
ferruginea *(A.C. Sm.) A.C. Smith in Fl. Vitiensis Nova,* 3: 757 (1985): Kermadecia ferruginea.
lutea *(Guillaumin) A.C. Smith in Fl. Vitiensis Nova,* 3: 754 (1985): Kermadecia lutea.
papuana *(Diels) A.C. Smith in Fl. Vitiensis Nova,* 3: 754 (1985): Euplassa papuana.
vitiensis *(Turrill) A.C. Smith in Fl. Vitiensis Nova,* 3: 755 (1985): Kermadecia vitiensis.

TWEEDIA (Asclepiadac.).
aucaensis *G.H. Rua in Darwiniana,* 27(1–4): 503 (1986)—Argentina.

TYLERIA (Ochnac.).
apiculata *C. Sastre in Phytologia,* 59(5): 313 (21 February 1986); *et in Bull. Mus. Nation. Hist. Nat., B, Adansonia, Ser. 4,* 8(1): 14 (8 July 1986); *see Bull. Mus. Nation. Hist. Nat., B, Adansonia, Ser. 4,* 9(2): 124 (1987)— Venezuela. †

U

ULEX (Legumin.).
baeticus
subsp. scaber *(G. Kunze) P. Cubas in Lagascalia,* 14(1): 140 (1986): U. scaber.
borgiae *S. Rivas Martínez in Lagascalia,* 14(1): 140 (1986)—Spain.

ULMUS (Ulmac.).
harbinensis *S.Q. Nie & G.Q. Huang in Bull. Bot. Res. North-East. Forest. Inst.*, 7(1): 151 (1987)—China.
macrocarpa
var. glabra *S.Q. Nie & G.Q. Huang in Bull. Bot. Res. North-East. Forest. Inst.*, 7(1): 153 (1987)—China.

URERA (Urticac.).
nitida *(Vellozo) P. Brack in Napaea, Rev. Bot.*, 1: 7 (1987): Urtica nitida.

URERA :–
trinervis *(Hochst.) I. Friis & K. Immelman in Nordic J. Bot.*, 7(2): 126 (1987): Elatostema trinerve.

UROCHLOA (Gramin.).
advena *(Vick.) R.D. Webster, Australian Paniceae (Poac.):* 231 (1987): Brachiaria advena.
atrisola *R.D. Webster, Australian Paniceae (Poac.):* 232 (1987)—Australia (N. Terr., Queensland).
brizantha *(A. Rich.) R.D. Webster, Australian Paniceae (Poac.):* 233 (1987): Panicum brizanthum.
decumbens *(Stapf) R.D. Webster, Australian Paniceae (Poac.):* 234 (1987): Brachiaria decumbens.
fasciculata
var. reticulata *(Vick.) R.D. Webster, Australian Paniceae (Poac.):* 235 (1987), specific name not validly publ.: Brachiaria fasciculata var. reticulata.
foliosa *(R. Br.) R.D. Webster, Australian Paniceae (Poac.):* 236 (1987): Panicum foliosum.
gilesii
subsp. occidentalis *(C.A. Gardn. & C.E. Hubb.) R.D. Webster, Australian Paniceae (Poac.):* 238 (1987): Brachiaria occidentalis.
holosericea *(R. Br.) R.D. Webster, Australian Paniceae (Poac.):* 239 (1987): Panicum holosericeum.
subsp. velutina *R.D. Webster, Australian Paniceae (Poac.):* 240 (1987)—Australia (N. Terr., W. Australia, Queensland).
kurzii *(Hook.f.) R.D. Webster, Australian Paniceae (Poac.):* 241 (1987: Panicum kurzii.
maxima *(Jacq.) R.D. Webster, Australian Paniceae (Poac.):* 241 (1987): Panicum maximum.
var. trichoglume *(Robyns) R.D. Webster, Australian Paniceae (Poac.):* 242 (1987): Panicum maximum var. trichoglume.
piligera *(F. Muell. ex Benth.) R.D. Webster, Australian Paniceae (Poac.):* 246 (1987): Panicum piligerum.
polyphylla *(R. Br.) R.D. Webster, Australian Paniceae (Poac.):* 248 (1987): Panicum polyphyllum.
pubigera *(R. & S.) R.D. Webster, Australian Paniceae (Poac.):* 250 (1987): Panicum pubigerum.
ramosa *(L.) R.D. Webster, Australian Paniceae (Poac.):* 251 (1987): Panicum ramosum.
subquadripara *(Trin.) R.D. Webster, Australian Paniceae (Poac.):* 252 (1987): Panicum subquadriparum.
texana *(Buckl.) R.D. Webster, Australian Paniceae (Poac.):* 253 (1987): Panicum texanum.
whiteana *(Domin) R.D. Webster, Australian Paniceae (Poac.):* 254 (1987): Panicum whiteanum.

V

VACCINIUM (Ericac.).
bocatorense *R.L. Wilbur in Sida,* 11(4): 441 (1986), as 'bocatorensis'—Panama.
elvirae *J.L. Luteyn in Op. Bot.,* 92: 129 (1987)—Peru.
iteophyllum
var. glandulosum *C.Y. Wu & R.C. Fang in Fl. Xizangica,* 3: 729 (1986)—Xizang.
nummularia
var. oblongifolium *C.Y. Wu & R.C. Fang in Fl. Xizangica,* 3: 724 (1986)—Xizang.
ovalifolium
var. alpinum *(Tatewaki) T. Yamazaki in J. Jap. Bot.,* 62(4): 128 (1987): V. chamissonis var. alpinum.
var. sachalinense *T. Yamazaki in J. Jap. .Bot.,* 62(4): 128 (1987)—Sakhalin Is., Japan.
oxycoccos
subsp. microphyllum *(Lange) J. Feilberg in Medd. Grønland, Biosci.,* 15: 12 (1984): Oxycoccus palustris f. microphylla.
scalarinervium *C.Y. Wu & R.C. Fang in Fl. Xizangica,* 3: 730 (1986)—Xizang.

VALERIANA (Valerianac.).
kamelinii *B.A. Sharipova in Izv. Akad. Nauk Tadzh. SSR, Biol. Nauk,* 1(98): 16 (1985)—U.S.S.R. (Middle Asia).
subsp. albiflora *B.A. Sharipova in Izv. Akad. Nauk Tadzh. SSR, Biol. Nauk,* 1(98): 18 (1985)—U.S.S.R. (Middle Asia).
muliensis *S.K. Wu in Acta Phytotax. Sin.,* 25(3): 239 (1987)—China.
schachristanica *R. Kam. & B.A. Sharipova in Izv. Akad. Nauk Tadzh. SSR, Biol. Nauk,* 1(98): 18 (1985)—U.S.S.R. (Middle Asia).

VALERIANELLA (Valerianac.).
ovczinnikovii *B.A. Sharipova in Izv. Akad. Nauk Tadzh. SSR, Biol. Nauk,* 3(92): 95 (1983)—U.S.S.R. (Middle Asia).

VALERIOANTHUS (Myrsinac.).
hirsutissimus *(Lundell) C.L. Lundell in Phytologia,* 63(2): 78 (1987): Ardisia hirsutissima.

VANDA (Orchidac.).
flabellata *(Rolfe ex Downie) E.A. Christenson in Indian Orchid J.,* 1(4): 156 (1985); *et in Selbyana,* 9(1): 169 (1986): Aerides flabellata.

VANGUERIA (Rubiac.).
subgen. Itigi *B. Verdcourt in Kew Bull.,* 42(1): 186 (1987).

VANGUERIELLA B. Verdcourt in Kew Bull., 42(1): 189 (1987), nom. nov.: Vangueriopsis subgen. Brachyanthus Robyns (Rubiac.).
sect. Stenosepalae *(Robyns) B. Verdcourt in Kew Bull.,* 42(1): 194 (1987): Vangueriopsis sect. Stenosepalae.

VANGUERIELLA :–
campylacantha *(Mildbr.) B. Verdcourt in Kew Bull.,* 42(1): 196 (1987): Plectronia campylacantha.
chlorantha *(K. Schum.) B. Verdcourt in Kew Bull.,* 42(1): 193 (1987): Plectronia chlorantha.
discolor *(Benth.) B. Verdcourt in Kew Bull.,* 42(1): 193 (1987): Canthium discolor.
georgesii *B. Verdcourt in Kew Bull.,* 42(1): 195 (1987)—Gabon.
glabrescens *(Robyns) B. Verdcourt in Kew Bull.,* 42(1): 197 (1987): Vangueriopsis glabrescens.
laxiflora *(K. Schum.) B. Verdcourt in Kew Bull.,* 42(1): 192 (1987): Plectronia laxiflora.
letestui *B. Verdcourt in Kew Bull.,* 42(1): 194 (1987)—Gabon.
nigerica *(Robyns) B. Verdcourt in Kew Bull.,* 42(1): 193 (1987): Vangueriopsis nigerica.
nigricans *(Robyns) B. Verdcourt in Kew Bull.,* 42(1): 193 (1987): Vangueriopsis nigricans.
olacifolia *(Robyns) B. Verdcourt in Kew Bull.,* 42(1): 193 (1987): Vangueriopsis olacifolia.
orthacantha *(Mildbr.) D.M. Bridson & B. Verdcourt in Kew Bull.,* 42(1): 198 (1987): Plectronia orthacantha.
rhamnoides *(Hiern) B. Verdcourt in Kew Bull.,* 42(1): 197 (1987): Canthium rhamnoides.
rufa *(Robyns) B. Verdcourt in Kew Bull.,* 42(1): 195 (1987): Vangueriopsis rufa.
sapinii *(De Wild.) B. Verdcourt in Kew Bull.,* 42(1): 196 (1987): Vangueria sapini.
soyauxii *(K. Schum.) B. Verdcourt in Kew Bull.,* 42(1): 196 (1987): Plectronia soyauxii.
spinosa *(Schumach. & Thonn.) B. Verdcourt in Kew Bull.,* 42(1): 194 (1987): Phallaria spinosa.
vanguerioides *(Hiern) B. Verdcourt in Kew Bull.,* 42(1): 193 (1987): Canthium vanguerioides.
zenkeri *B. Verdcourt in Kew Bull.,* 42(1): 192 (1987), nom. nov.: Plectronia macrocarpa *K. Schum.*

VANGUERIOPSIS (Rubiac.).
subgen. Guranivea *B. Verdcourt in Kew Bull.,* 42(1): 188 (1987).

VANILLA (Orchidac.).
taiwaniana *S.S. Ying in Quart. J. Chinese Forest.,* 20(2): 55 (1987)—Taiwan.

VASSOBIA (Solanac.).
iochromoides *A.T. Hunziker in Lorentzia,* 6: 3 (1986)—Argentina.

VAUQUELINIA (Rosac.).
californica
subsp. pauciflora *(Standl.) W.J. Hess & J. Henrickson in Sida,* 12(1): 135 (1987): V. pauciflora.
subsp. retherfordii *(I.M. Johnston) W.J. Hess & J. Henrickson in Sida,* 12(1): 132 (1987): V. retherfordii.

VAUQUELINIA :–
subsp. sonorensis *W.J. Hess & J. Henrickson in Sida,* 12(1): 130 (1987)—U.S.A. (Arizona), Mexico.
corymbosa
subsp. angustifolia *(Rydb.) W.J. Hess & J. Henrickson in Sida,* 12(1): 148 (1987): V. angustifolia.
subsp. heterodon *(I.M. Johnston) W.J. Hess & J. Henrickson in Sida,* 12(1): 145 (1987): V. heterodon.
subsp. karwinskyi *(Maxim.) W.J. Hess & J. Henrickson in Sida,* 12(1): 152 (1987): V. karwinskyi.
subsp. latifolia *(Rydb. ex Standl.) W.J. Hess & J. Henrickson in Sida,* 12(1): 159 (1987): V. latifolia.
subsp. saltilloensis *W.J. Hess & J. Henrickson in Sida,* 12(1): 156 (1987)—Mexico.

VEITCHIA (Palmae).
canterburyana *(C. Moore & F. Mueller) H.A. Wendl ex Anon. in Gard. Chron.,* 1872: 372 (1872): Kentia canterburyana. †

×**VEJVARUTARA** hort. in Orchid Rev., 95(1130): centre page pull-out p. 9 (1987). *ORCHIDACEAE.* (Broughtonia × Cattleya × Cattleyopsis).

VERBASCUM (Scrophulariac.).
aqranse *A.K. Al-Bermani & A.H.E. Al-Musawi in Feddes Repert.,* 97(7–8): 393 (1986)—Iraq.
ballii *(Batt.) M. Qaiser in Fl. Libya,* 88: 26 (1982): Celsia ballii.
giganteum
subsp. martinezii *B. Valdés in Lagascalia,* 14(1): 89 (1986)—Spain.
simplex
var. compositum *(Mariz) C. Benedí & J. Molero in Bol. Soc. Brot.,* 59: 62 (1986): V. linckianum var. compositum.

VERBENA (Verbenac.).
parvula
f. albiflora *H.N. Moldenke in Phytologia,* 61(7): 444 (1987)—Bolivia.

VERBESINA (Compos.).
guerreroana *B.L. Turner in Phytologia,* 63(1): 7 (1987)—Mexico.
macvaughii *B.L. Turner in Phytologia,* 63(1): 8 (1987)—Mexico.
mexiae *B.L. Turner in Phytologia,* 63(1): 9 (1987)—Mexico.
nayaritensis *B.L. Turner in Phytologia,* 63(1): 10 (1987)—Mexico.
neotenoriensis *B.L. Turner in Phytologia,* 63(1): 11 (1987)—Mexico.
virgata
var. oreopola *(Rob. & Greenm.) B.L. Turner in Phytologia,* 63(1): 12 (1987): V. oreopola.

VERNONIA (Compos.).
chiriquiensis *S.C. Keeley in Brittonia,* 39(1): 45 (1987)—Panama.
cocleana *S.C. Keeley in Brittonia,* 39(1): 44 (1987)—Panama.
duvigneaudii *K. Kalanda in Bull. Jard. Bot. Nation. Belg.*, 57(1–2): 209 (1987)—Zaïre.
evrardiana *K. Kalanda in Bull. Jard. Bot. Nation. Belg.*, 56(3–4): 388 (1986)—Zaïre.
kapolowensis *K. Kalanda in Bull. Jard. Bot. Nation. Belg.*, 57(1–2): 210 (1987)—Zaïre.
timpermaniana *K. Kalanda in Bull. Jard. Bot. Nation. Belg.*, 57(1–2): 211 (1987)—Zaïre.

VERONICA (Scrophulariac.).
telephiifolia
subsp. glareosa *(Somm. & Levier) M.A. Fischer in Zam. Sist. Geogr. Rast.*, 41: 97 (1986): V. glareosa.

VERONICASTRUM (Scrophulariac.).
serpyllifolium
subsp. humifusum *(Dicks.) W.A. Weber in Phytologia,* 62(5): 437 (1987): Veronica humifusa.

VEXIBIA (Legumin.).
alopecuroides *(L.) W.A. Weber in Phytologia,* 62(5): 438 (1987): Sophora alopecuroides.
nuttalliana *(B.L. Turner) W.A. Weber in Phytologia,* 62(5): 437 (1987): Sophora nuttalliana.
pachycarpa *(C.A. Mey.) W.A. Weber in Phytologia,* 62(5): 438 (1987): Sophora pachycarpa.
prodanii *(E. Anderson) W.A. Weber in Phytologia,* 62(5): 438 (1987): Sophora prodanii.
stenophylla *(A. Gray) W.A. Weber in Phytologia,* 62(5): 438 (1987): Sophora stenophylla.

VIBURNUM (Caprifoliac.).
thibeticum *C.Y. Wu & Y.F. Huang in Fl. Xizangica,* 4: 480 (1985)—Xizang.

VICIA (Legumin.).
linearifolia *Y.N. Lee in Korean J. Bot.*, 25(4): 176 (1982)—Korea.
lutea
subsp. cavanillesii *(Martínez) C. Romero Zarco in Lagascalia,* 14(1): 141 (1986): V. cavanillesii.
ochroleuca
subsp. atlantica *(Pomel) Quézel & Santa ex W. Greuter & Burdet in Willdenowia,* 16(2): 447 (1987): V. atlantica.
subsp. baborensis *(Batt. & Trabut) Quézel & Santa ex W. Greuter & Burdet in Willdenowia,* 16(2): 447 (1987): V. baborensis.

VICIA :–
sativa
subsp. terana *(Losa) C. Benedí & J. Molero in Bol. Soc. Brot.*, 59: 64 (1986): Vicia terana.
tenuifolia
subsp. atroviolacea *(Bornm.) W. Greuter & Burdet in Willdenowia,* 16(2): 447 (1987): V. tenuifolia var. atroviolacea.
villosa
subsp. ambigua *(Guss.) M. Kerguélen in Lejeunia,* 120: 183 (1987): V. ambigua.
subsp. elegantissima *(Shuttlew. ex Rouy) G. Bosc & M. Kerguélen in Lejeunia,* 120: 183 (1987): V. elegantissima.

VIGNA (Legumin.).
hainiana *C.R. Babu, Gopinathan & S.K. Sharma in Bull. Bot. Surv. India,* 27(1–4): 15 (1985 publ. 1987), nom. nov.: Phaseolus wightii *W. & A.*
minima
f. dimorphaphylla *T.L. Wu in Acta Bot. Austro Sinica,* 3: 20 (1986), nom. nov.: Phaseolus heterophyllus *Hayata.*
prainiana *C.R. Babu & S.K. Sharma in Bull. Bot. Surv. India,* 27(1–4): 19 (1985 publ. 1987), nom. nov.: Phaseolus calcaratus var. gracilis *Prain.*
ricciardiana *(Tenore) C.R. Babu & S.K. Sharma in Bull. Bot. Surv. India,* 27(1–4): 21 (1985 publ. 1987): Phaseolus ricciardianus.
sublobata *(Roxb.) C.R. Babu & S.K. Sharma in Bull. Bot. Surv. India,* 27(1–4): 21 (1985 publ. 1987): Phaseolus sublobatus.
vexillata
var. sepiaria *(Dalz.) C.R. Babu & S.K. Sharma in Bull. Bot. Surv. India,* 27(1–4): 27 (1985 publ. 1987): Phaseolus sepiarius.
var. wightii *(Benth. ex Baker) C.R. Babu & S.K. Sharma in Bull. Bot. Surv. India,* 27(1–4): 28 (1985 publ. 1987): V. wightii.

VINCETOXICUM (Asclepiadac.).
daniloi *(J. Rohlena) A.W. Hill in Index Kewensis,* Supp. 6: 218 (1916–1920 publ. 1926): V. hirundinaria var. daniloi.
hirundinaria
f. puberulum *(Beck) K. Sutorý in Čas. Morav. Muz., Vědy Přír.*, 68: 211 (1983): V. officinale var. puberulum.
var. timbal-lagravei *K. Sutorý in Čas. Morav. Muz., Vědy Přír.*, 68: 212 (1983), nom. nov.: V. puberulum *Timb.-Lagr.*

VIOLA (Violac.).
acrocerauniensis *M. Erben in Mitt. Bot. Staatssamml. München,* 22: 493 (1986)—Albania.
arborescens
f. latifolia *(DC.) M.A. Fernández Casado in Lagascalia,* 14(1): 136 (1986): V. arborescens var. latifolia.

VIOLA :–
f. willkommii *M.A. Fernández Casado in Lagascalia,* 14(1): 136 (1986), without type, as nom. nov. but contrary to Art. 33.3 ICBN 1983—*Spain.*
bezdelevae *V.N. Voroshilov in Byull. Mosk. Obshch. Ispyt. Prir., Biol.,* 92(1): 134 (1987)—Kuril Is.
binayensis *M. Okamoto & K. Ueda in M. Kato, Taxon. Stud. Pl. Seram Is.:* 143 (1986), without latin descr. or type; *et in Acta Phytotax. Geobot.,* 37(1–3): 1 (1986), with latin descr. & type— Moluccas.
boissieuana
var. pseudoselkirki *(Nakai) T. Yahara in J. Fac. Sci. Univ. Tokyo, Bot.,* 14(2): 96 (1987): V. pseudoselkirki.
cleistogamoides *(L. Adams) R.D. Seppelt in Fl. S. Australia,* 2: 870 (1986): V. hederacea subsp. cleistogamoides.
etrusca *M. Erben in Mitt. Bot. Staatssamml. München,* 22: 499 (1986)—Italy.
nudicaulis *(W. Beck.) S.Y. Chen in Fl. Xizangica,* 3: 297 (1986), without basionym ref.: V. biflora var. nudicaulis.

VITEX (Verbenac.).
velutinifolia *A.A. Munir in J. Adelaide Bot. Gard.,* 10(1): 59 (1987)—Australia (W. Australia).

VITIS (Vitac.).
cinerea
var. baileyana *(Munson) B.L. Comeaux in Castanea,* 52(3): 212 (1987), basionym ref. date '1883' in error for '1893': V. baileyana.

VRIESEA (Bromeliac.).
dressleri *W. Rauh in Abh. Akad. Wiss. Lit. Mainz, Math.-Naturwiss. Klasse, Trop. Subtrop. Pflanzenwelt,* 58: 41 (1986)—Panama.
ochracea *W. Rauh & E. Gross in J. Bromeliad Soc.,* 37(5): 196 (1987)—Peru.
parvula *W. Rauh in Abh. Akad. Wiss. Lit. Mainz, Math.-Naturwiss. Klasse, Trop. Subtrop. Pflanzenwelt,* 58: 44 (1986)—Brazil.
punctulata *E. Pereira & I. de A. Penna in Bradea,* 4(34): 272 (1986)—Brazil.
weberi *E. Pereira & I. de A. Penna in Bradea,* 4(34): 273 (1986)—Brazil.

VULPIA (Gramin.).
hispanica
subsp. montana *(Boiss. & Reuter) J.A. Devesa in Lagascalia,* 14(1): 165 (1986): Nardurus montanus.
incrassata *Steudel, Nomencl. Bot.,* ed. 2, 1: 229 (1840), nom. nov., illegit.: Bromus tenuis *Tineo.*

W

WAHLENBERGIA (Campanulac.).
aridicola *P.J. Smith in Fl. S. Australia,* 3: 1377 (1986)—Australia (S. Australia, New South Wales).
litticola *P.J. Smith in Fl. S. Australia,* 3: 1380 (1986)—Australia (Victoria, New South Wales, W. Australia, S. Australia), Tasmania, New Zealand.
luteola *P.J. Smith in Fl. S. Australia,* 3: 1380 (1986)—Australia (New South Wales, Victoria, S. Australia).
malaissei *M. Thulin in Nordic J. Bot.,* 7(3): 265 (1987)—Zaïre.
marunguensis *M. Thulin in Nordic J. Bot.,* 7(3): 264 (1987)—Zaïre.
petraea *M. Thulin in Nordic J. Bot.,* 7(3): 262 (1987)—Burundi.
polyphylla *M. Thulin in Nordic J. Bot.,* 7(3): 264 (1987)—Zaïre.
queenslandica *R.C. Carolin ex P.J. Smith in Fl. S. Australia,* 3: 1381 (1986)—Australia (Queensland, New South Wales, N. Terr., W. Australia, S. Australia).
tetramera *M. Thulin in Nordic J. Bot.,* 7(3): 261 (1987)—South Africa (Natal).
tumidifructa *P.J. Smith in Fl. S. Australia,* 3: 1383 (1986)—Australia (New South Wales, Victoria, Queensland, N. Terr., W. Australia, S. Australia).

WALLENIA (Myrsinac.).
hughsonii *Alain H. Liogier in Phytologia,* 63(1): 65 (1987)—Hispaniola (Dominican Republic).

WEDELIA (Compos.).
subalpestris *(Baker) J.U.M. dos Santos in Bol. Mus. Paraense Emílio Goeldi, Bot.,* 2(1): 12 (1985): Aspilia subalpestris.

WEINGARTIA (Cactac.).
columnaris *F.H. Brandt in Kakt. Orchid. Rundschau,* 11(2–3): 18 (1986)—Bolivia.
minima *(Rausch) F.H. Brandt in Kakt. Orchid. Rundschau,* 11(1): 2 (1986): Sulcorebutia taratensis var. minima.

WENDLANDIA (Rubiac.).
sericea *W.C. Chen in Acta Bot. Austro Sinica,* 2: 41 (1986)—Laos.

WESTRINGIA (Labiat.).
davidii *B.J. Conn in Muelleria,* 6(5): 321 (1987)—Australia (New South Wales).
saxatilis *B.J. Conn in Muelleria,* 6(5): 325 (1987)—Australia (New South Wales).

WIKSTROEMIA (Thymelaeac.).
linearifolia *H.F. Zhou ex C.Y. Chang in Guihaia,* 6(4): 270 (1986)—China.

WILLIAMODENDRON K. Kubitzki & H.G. Richter in Bot. Jahrb., 109(1): 50 (1987). *LAURACEAE.*

WILLIAMODENDRON :–
glaucophyllum *(van der Werff) K. Kubitzki & H.G. Richter in Bot. Jahrb.*, 109(1): 58 (1987): Mezilaurus glaucophylla.
quadrilocellatum *(van der Werff) K. Kubitzki & H.G. Richter in Bot. Jahrb.*, 109(1): 58 (1987): Mezilaurus quadrilocellata.
spectabile *K. Kubitzki & H.G. Richter in Bot. Jahrb.*, 109(1): 50 (1987)—Brazil, Colombia.

WITHERINGIA (Solanac.).
laxissima *(Standl.) W.G. D'Arcy, Fl. Panama Checklist & Index, 2 (Monogr. Syst. Bot. Missouri Bot. Gard. 18):* 663, 362 (1987): Lycianthes laxissima.

WOLFFIA (Lemnac.).
sect. Elongatae *E. Landolt in Veröff. Geobot. Inst. Rübel,* 71: 452 (1986).
sect. Pigmentatae *E. Landolt in Veröff. Geobot. Inst. Rübel,* 71: 452 (1986).
sect. Pseudorrhizae *E. Landolt in Veröff. Geobot. Inst. Rübel,* 71: 451 (1986).

WOLFFIELLA (Lemnac.).
sect. Rotundae *E. Landolt in Veröff. Geobot. Inst. Rübel,* 71: 449 (1986).

X

XANTHOSOMA (Arac.).
akkermansii *(Bunting) T.B. Croat in Aroideana,* 9(1–4): 176 (1986): Caladium akkermansii.

XEROPHYTA (Velloziac.).
rippsteinii *L.B. Smith, J.-P. Lebrun & A.L. Stork in Bull. Mus. Nation. Hist. Nat., B, Adansonia,* Ser. 4, 8(3): 297 (1986)—Ethiopia.

XEROSPHAERA J. Soják in Čas. Nár. Muz. (Prague), 154(1): 35 (1985 publ. 1986). *LEGUMINOSAE.*
bullata *(Boiss. & Hausskn.) J. Soják in Čas. Nár. Muz. (Prague),* 154(1): 35 (1985 publ. 1986): Trifolium bullatum.
clusii *(Godr. & Gren.) J. Soják in Čas. Nár. Muz. (Prague),* 154(1): 35 (1985 publ. 1986): Trifolium clusii.
fragifera *(L.) J. Soják in Čas. Nár. Muz. (Prague),* 154(1): 35 (1985 publ. 1986): Trifolium fragiferum.
subsp. bonannii *(Presl) J. Soják in Čas. Nár. Muz. (Prague),* 154(1): 35 (1985 publ. 1986): Trifolium bonannii.
physodes *(Stev. ex Marschall-Bieberstein) J. Soják in Čas. Nár. Muz. (Prague),* 154(1): 35 (1985 publ. 1986): Trifolium physodes.

XEROSPHAERA :–
raddeana *(Trautv.) J. Soják in Čas. Nár. Muz. (Prague),* 154(1): 35 (1985 publ. 1986): Trifolium raddeanum.
resupinata *(L.) J. Soják in Čas. Nár. Muz. (Prague),* 154(1): 35 (1985 publ. 1986): Trifolium resupinatum.
tomentosa *(L.) J. Soják in Čas. Nár. Muz. (Prague),* 154(1): 35 (1985 publ. 1986): Trifolium tomentosum.
tumens *(Stev. ex Marschall-Bieberstein) J. Soják in Čas. Nár. Muz. (Prague),* 154(1): 35 (1985 publ. 1986): Trifolium tumens.

XYLINABARIOPSIS (Apocynac.).
xylinabariopsoides *(Tsiang) Trân Đình Lý in Feddes Repert.,* 97(9–10): 679 (1986): Chunechites xylinabariopsoides.

XYLOCALYX (Scrophulariac.).
carterae *M. Thulin in Nordic J. Bot.,* 7(3): 267 (1987), as 'carteriae'—Somali Republic.

XYLOMELUM (Proteac.).
cunninghamianum *D.B. Foreman in Muelleria,* 6(5): 299 (1987)—Australia (New South Wales, Queensland).

XYRIS (Xyridac.).
coutensis *M. das Graças L. Wanderley & T.M. Cerati in Brittonia,* 39(2): 300 (1987)—Brazil.
jolyi *M. das Graças L. Wanderley & T.M. Cerati in Brittonia,* 39(2): 298 (1987)—Brazil.

Y

YAKIRRA (Gramin.).
australiensis
var. intermedia *R.D. Webster, Australian Paniceae (Poac.):* 266 (1987)—Australia (W. Australia, N. Terr., Queensland).
foliolosa *(Hook. f.) W.D. Clayton in Kew Bull.,* 42(2): 403 (1987): Ichnanthus foliolosus.

YINSHANIA (Crucif.).
ser. Henryanae *Y.H. Zhang in Acta Phytotax. Sin.,* 25(3): 213 (1987).
sect. Microcarpa *Y.H. Zhang in Acta Phytotax. Sin.,* 25(3): 210 (1987).
acutangula *(O.E. Schulz) Y.H. Zhang in Acta Phytotax. Sin.,* 25(3): 217 (1987): Cochlearia acutangula.
var. albiflora *(Ma & Y.Z. Zhao) Y.H. Zhang in Acta Phytotax. Sin.,* 25(3): 217 (1987): Y. albiflora.
furcatopilosa *(Kuan) Y.H. Zhang in Acta Phytotax. Sin.,* 25(3): 214 (1987): Cochlearia furcatopilosa.

YINSHANIA :–
ganluoensis *Y.H. Zhang in Acta Phytotax. Sin.*, 25(3): 211 (1987)—China.
henryi *(Oliv.) Y.H. Zhang in Acta Phytotax. Sin.*, 25(3): 213 (1987): Nasturtium henryi.
microcarpa *(Kuan) Y.H. Zhang in Acta Phytotax. Sin.*, 25(3): 211 (1987): Cochlearia microcarpa.
qianningensis *Y.H. Zhang in Acta Phytotax. Sin.*, 25(3): 212 (1987)—China.
var. brachybotrys *Y.H. Zhang in Acta Phytotax. Sin.*, 25(3): 213 (1987)—China.
wenxianensis *Y.H. Zhang in Acta Phytotax. Sin.*, 25(3): 215 (1987)—China.
var. songpanensis *Y.H. Zhang in Acta Phytotax. Sin.*, 25(3): 216 (1987)—China.
zayuensis *Y.H. Zhang in Acta Phytotax. Sin.*, 25(3): 214 (1987)—Xizang, China.

YPSILANDRA (Melanthiac.).
kansuensis *R.N. Zhao & Z.X. Peng in Acta Bot. Bor.-Occid. Sin.*, 7(1): 57 (1987)—China.

YUSHANIA (Gramin.).
megalothyrsa *(Hand.-Mazz.) T.H. Wen in J. Bamboo Res.*, 6(3): 34 (1987): Arundinaria megalothyrsa.
yunnanensis *(Hsueh & Yi) P.C. Keng & Wen in J. Bamboo Res.*, 6(4): 16 (1987): Fargesia yunnanensis.

Z

ZAMIA (Zamiac.).
amblyphyllidia *D.W. Stevenson in Bull. Fairchild Trop. Gard.*, 42(3): 26 (1987)—Puerto Rico.

ZANNICHELLIA (Zannichelliac.).
obtusifolia *S. Talavera, P. García Murillo & H. Smit in Lagascalia*, 14(2): 249 (1986)—Spain, France, Sardinia, Portugal, Morocco.

ZANTHOXYLUM (Rutac.).
armatum
var. ferrugineum *(Rehd. & Wils.) C.C. Huang in Guihaia*, 7(1): 1 (1987): Z. alatum f. ferrugineum.
arnoldii *C. Beurton in Feddes Repert.*, 98(1–2): 65 (1987)—Cuba.
bissei *C. Beurton in Rev. Jard. Bot. Nacion. Univ. Habana*, 6(2): 3 (1985 publ. 1986)—Cuba.
ignoratum *C. Beurton in Feddes Repert.*, 98(1–2): 61 (1987)—Cuba.
liboense *C.C. Huang in Guihaia*, 7(1): 6 (1987)—China.
nitidum
var. tomentosum *C.C. Huang in Guihaia*, 7(1): 5 (1987)—China.

ZANTHOXYLUM :–
ovalifolium
var. multifoliolatum *(Huang) C.C. Huang in Guihaia*, 7(1): 4 (1987): Z. dimorphophyllum var. multifoliolatum.
var. spinifolium *(Rehd. & Wils.) C.C. Huang in Guihaia*, 7(1): 4 (1987): Z. dimorphophyllum var. spinifolium.
pseudodumosum *C. Beurton in Feddes Repert.*, 98(1–2): 57 (1987)—Cuba.
stipitatum *C.C. Huang in Guihaia*, 7(1): 2 (1987)—China.
yakumontanum *(Sugimoto) H. Nagamasu in Acta Phytotax. Geobot.*, 37(4–6): 137 (1986): Fagara ailanthoides var. yakumontana.

ZAPOTECA H.M. Hernández in Ann. Missouri Bot. Gard., 73(4): 757 (1986 publ. 1987). *LEGUMINOSAE.*
caracasana *(Jacq.) H.M. Hernández in Ann. Missouri Bot. Gard.*, 73(4): 757 (1986 publ. 1987): Mimosa caracasana.
formosa *(Kunth) H.M. Hernández in Ann. Missouri Bot. Gard.*, 73(4): 757 (1986 publ. 1987): Acacia formosa.
lambertiana *(G. Don) H.M. Hernández in Ann. Missouri Bot. Gard.*, 73(4): 757 (1986 publ. 1987): Acacia lambertiana.
media *(M. Martens & Galeotti) H.M. Hernández in Ann. Missouri Bot. Gard.*, 73(4): 757 (1986 publ. 1987): Acacia media.
mollis *(Standley) H.M. Hernández in Ann. Missouri Bot. Gard.*, 73(4): 757 (1986 publ. 1987): Calliandra mollis.
portoricensis *(Jacq.) H.M. Hernández in Ann. Missouri Bot. Gard.*, 73(4): 758 (1986 publ. 1987): Mimosa portoricensis.
tetragona *(Willd.) H.M. Hernández in Ann. Missouri Bot. Gard.*, 73(4): 757 (1986 publ. 1987): Acacia tetragona.

ZERAVSCHANIA (Umbellif.).
aucheri *(Boiss.) M.G. Pimenov in Fl. Iran.*, 162: 459 (1987): Peucedanum aucheri.
ferulaefolia *(Gilli) M.G. Pimenov in Fl. Iran.*, 162: 460 (1987): Peucedanum ferulaefolium.
membranacea *(Boiss.) M.G. Pimenov in Fl. Iran.*, 162: 458 (1987): Peucedanum membranaceum.
minjanensis *(Rech. f.) K.H. Rechinger in Fl. Iran.*, 162: 461 (1987): Eleutherospermum minjanense.
pauciradiata *(Tamamsch.) M.G. Pimenov in Fl. Iran.*, 162: 459 (1987): Peucedanum pauciradiatum.

ZEUXINE (Orchidac.).
gracilis
var. sakagutii *(Tuyama) T. Hashimoto in Ann. Tsukuba Bot. Gard.*, 5: 28 (1986): Z. sakagutii.
var. tenuifolia *(Tuyama) T. Hashimoto in Ann. Tsukuba Bot. Gard.*, 5: 28 (1986): Z. tenuifolia.
seidenfadenii *S. Deva & H.B. Naithani, Orchid Fl. N.W. Himalaya:* 95 (1986)—India.

strateumatica
f. rupicola *(Fukuyama) T. Hashimoto in Ann. Tsukuba Bot. Gard.*, 5: 27 (1986): Z. rupicola.
taiwaniana *S.S. Ying in Quart. J. Chinese Forest.*, 20(2): 57 (1987)—Taiwan.

ZINGIBER (Zingiberac.).
flavomaculosum *S.Q. Tong in Acta Phytotax. Sin.*, 25(2): 144 (1987), as 'flavo-maculosum'—China.
longiligulatum *S.Q. Tong in Acta Phytotax. Sin.*, 25(2): 140 (1987)—China.
menghaiense *S.Q. Tong in Acta Phytotax. Sin.*, 25(2): 145 (1987)—China.
stipitatum *S.Q. Tong in Acta Phytotax. Sin.*, 25(2): 146 (1987)—China.
truncatum *S.Q. Tong in Acta Phytotax. Sin.*, 25(2): 147 (1987)—China.
wandingense *S.Q. Tong in Acta Phytotax. Sin.*, 25(2): 143 (1987)—China.
yingjiangense *S.Q. Tong in Acta Phytotax. Sin.*, 25(2): 142 (1987)—China.

ZINOWIEWIA (Celastrac.).
revoluta *C.L. Lundell in Phytologia,* 63(2): 73 (1987)—Costa Rica.

×ZYGODISANTHUS hort. in Orchid Rev., 95(1125): centre page pull-out p. 9 (1987). *ORCHIDACEAE.* (Paradisanthus × Zygopetalum).

ZYGOPHYLLUM (Zygophyllac.).
macrocarpon *E. Retief in Bothalia,* 17(2): 189 (1987)—South Africa (Cape Province), Namibia.

APPENDIX — FERNS & FERN ALLIES

NAMES OF ALL RANKS ABOVE GENUS UP TO AND INCLUDING FAMILY

PTERIDACEAE
subfam. **Adiantoideae** (Presl) R. Tryon in Amer. Fern J., 76(4): 184 (1986): tribus Adiantaceae Presl.
subfam. **Ceratopteridoideae** (J. Smith) R. Tryon in Amer. Fern J., 76(4): 184 (1986): tribus Ceratopterideae J. Smith.
subfam. **Cheilanthoideae** (J. Smith) R. Tryon in Amer. Fern J., 76(4): 184 (1986): tribus Cheilantheae J. Smith.

PTERIDACEAE :-
subfam. **Taenitidoideae** (Presl) R. Tryon in Amer. Fern J., 76(4): 184 (1986): tribus Taenitideae Presl.

TECTARIACEAE
fam. **Tectariaceae** G. Panigrahi in J. Orissa Bot. Soc., 8(1): 41 (1986).

NAMES AT GENERIC RANK AND BELOW

A

ADIANTUM
hispidulum
var. whitei *(Bailey) P.D. Bostock in Austrobaileya,* 2(4): 362 (1987): Adiantium whitei Bailey
×spurium *A.C. Jermy & T.G. Walker in Fern Gaz. (U.K.),* 13(3): 188 (1987), nom. nov.: Adiantum × villosolucidum *A.C. Jermy & T.G. Walker*

ASPLENIUM
aethiopicum
subsp. dodecaploideum *A.F. Braithwaite in Bot. J. Linn. Soc.,* 93(4): 360 (1986)—South Africa (Transvaal, Cape Province), Lesotho.
subsp. filare *(Forsk.) A.F. Braithwaite in Bot. J. Linn. Soc.,* 93(4): 362 (1986): Acrostichum filare Forsk.
subsp. tripinnatum *(Bak.) A.F. Braithwaite in Bot. J. Linn. Soc.,* 93(4): 355 (1986): Asplenium furcatum var. tripinnatum Bak.
×badense *(D.E. Meyer) A.C. Jermy in Sommerfeltia,* 6: viii (1987): × Asplenoceterach badense D.E. Meyer
· ceterach
nothosubsp. mantoniae *(Varoczky & Vida) A.C. Jermy in Sommerfeltia,* 6: viii (1987): Asplenium × mantoniae Varoczky & Vida

ASPLENIUM :-
cuspidatum
var. triculum *D.B. Lellinger in Proc. Biol. Soc. Wash.,* 98(2): 367 (1985)—Costa Rica, Honduras, Panama, Venezuela, Peru.

excelsum *D.B. Lellinger in Proc. Biol. Soc. Wash.,* 98(2): 369 (1985)—Panama, Nicaragua, Costa Rica.

gomezianum *D.B. Lellinger in Proc. Biol. Soc. Wash.,* 98(2): 371 (1985)—Costa Rica, Panama.

latibasis *(Ching & Shing) T. Nakaike in Bull. Nation. Sci. Mus., B (Tokyo),* 12(2): 45 (1986): Ceterachopsis latibasis Ching & Shing

maxonii *D.B. Lellinger in Proc. Biol. Soc. Wash.,* 98(2): 372 (1985)—Panama, Costa Rica, Colombia, Ecuador, Peru, Bolivia.

qiujiangense *(Ching & Fu) T. Nakaike in Bull. Nation. Sci. Mus., B (Tokyo),* 12(2): 45 (1986): Ceterachopsis qiujiangensis Ching & Fu

×reichsteinii *H.W. Bennert, H. Rasbach & K. Rasbach in Fern Gaz. (U.K.),* 13(3): 134 (1987)—Balearic Is. (A. fontanum × majoricum).

tadei *C.R. Fraser-Jenkins & J.J. Schneller in Fern Gaz. (U.K.),* 13(3): 166 (1987)—Turkey.

ATHYRIUM

acutidentatum *R.C. Ching in Acta Bot. Bor.-Occid. Sin.,* 6(3): 150 (1986), as 'acuti-dentatum'—China.

acutiserratum *R.C. Ching in Acta Bot. Bor.-Occid. Sin.,* 6(3): 149 (1986)—China.

austrooccidentale *R.C. Ching in Acta Bot. Bor.-Occid. Sin.,* 6(3): 152 (1986), as 'austro-occidentale'—China.

baishanzuense *R.C. Ching & Y.T. Hsieh in Acta Bot. Bor.-Occid. Sin.,* 6(3): 157 (1986)—China.

bijiangense *Y.T. Hsieh & W.M. Chu in Acta Bot. Bor.-Occid. Sin.,* 7(1): 52 (1987)—China.

confertipinnum *R.C. Ching in Acta Bot. Bor.-Occid. Sin.,* 6(3): 155 (1986)—China.

dentatum *R.C. Ching in Acta Bot. Bor.-Occid. Sin.,* 6(3): 150 (1986)—China.

erythrocaulon *R.C. Ching in Acta Bot. Bor.-Occid. Sin.,* 6(3): 154 (1986)—China.

glabrescens *R.C. Ching in Acta Bot. Bor.-Occid. Sin.,* 6(3): 156 (1987), as 'glabrescense'—China.

gongshanense *R.C. Ching in Acta Bot. Bor.-Occid. Sin.,* 6(3): 153 (1986)—China.

gracillimum *R.C. Ching in Acta Bot. Bor.-Occid. Sin.,* 6(3): 149 (1986)—China.

habaense *R.C. Ching in Acta Bot. Bor.-Occid. Sin.,* 6(3): 154 (1986)—China.

hebeiense *R.C. Ching in Acta Bot. Bor.-Occid. Sin.,* 6(3): 158 (1986)—China.

jieguishanense *R.C. Ching in Acta Bot. Bor.-Occid. Sin.,* 6(3): 157 (1986)—China.

leiboense *R.C. Ching & Y.T. Hsieh in Acta Bot. Bor.-Occid. Sin.,* 6(3): 158 (1986)—China.

mabianense *R.C. Ching & Y.T. Hsieh in Acta Bot. Bor.-Occid. Sin.,* 6(3): 157 (1986)—China.

maoshanense *R.C. Ching & P.S. Chiu in Acta Bot. Bor.-Occid. Sin.,* 6(3): 157 (1986)—China.

minimum *R.C. Ching in Acta Bot. Bor.-Occid. Sin.,* 6(3): 151 (1986)—Taiwan.

multipinnum *Y.T. Hsieh & Z.B. Wang in Acta Bot. Bor.-Occid. Sin.,* 7(1): 55 (1987)—China.

nanyueense *R.C. Ching in Acta Bot. Bor.-Occid. Sin.,* 6(3): 152 (1986)—China.

neowardii *R.C. Ching in Acta Bot. Bor.-Occid. Sin.,* 6(3): 155 (1986)—China.

nudifrons *R.C. Ching in Acta Bot. Bor.-Occid. Sin.,* 6(3): 158 (1986)—China.

pseudoepirachis *R.C. Ching in Acta Bot. Bor.-Occid. Sin.,* 6(3): 156 (1986)—China.

remotisorum *R.C. Ching & H.S. Kung in Acta Bot. Bor.-Occid. Sin.,* 6(3): 153 (1986)—China.

remotum *R.C. Ching in Acta Bot. Bor.-Occid. Sin.,* 6(3): 149 (1986)—China.

×saitoanum *(Sugimoto) S. Serizawa in J. Phytogeogr. Taxon.,* 34(2): 68 (1986): Athyrium sheareri f. saitoanum Sugimoto (A. niponicum × sheareri).

ATHYRIUM :-

sessile *R.C. Ching in Acta Bot. Bor.-Occid. Sin.,* 6(3): 154 (1986)—China.

shaaxiense *R.C. Ching & Y.T. Hsieh in Acta Bot. Bor.-Occid. Sin.,* 6(3): 150 (1986)—China.

spinosissimum *R.C. Ching in Acta Bot. Bor.-Occid. Sin.,* 6(3): 155 (1986)—China.

subcoriaceum *R.C. Ching in Acta Bot. Bor.-Occid. Sin.,* 6(3): 151 (1986)—China.

supranigrescens *R.C. Ching in Acta Bot. Bor.-Occid. Sin.,* 6(3): 150 (1986)—China.

wugongshanense *R.C. Ching & Y.T. Hsieh in Acta Bot. Bor.-Occid. Sin.,* 6(3): 155 (1986)—China.

wuliangshanense *R.C. Ching in Acta Bot. Bor.-Occid. Sin.,* 6(3): 153 (1986)—China.

yindeense *R.C. Ching in Acta Bot. Bor.-Occid. Sin.,* 6(3): 152 (1986)—China.

yuanyangense *Y.T. Hsieh & W.M. Chu in Acta Bot. Bor.-Occid. Sin.,* 7(1): 54 (1987)—China.

zhenfengense *R.C. Ching in Acta Bot. Bor.-Occid. Sin.,* 6(3): 151 (1986)—China.

B

BLECHNUM

occidentale

f. pubirhachis *(Rosenst.) D.B. Lellinger in Proc. Biol. Soc. Wash.,* 98(2): 373 (1985): Blechnum occidentale var. pubirhachis Rosenst.

BOLBITIS

×nanjenensis *C.M. Kuo, Stud. Ferns Ken-Ting Nation. Park (Taiwan):* 53 (1986)—Taiwan. (B. appendiculata × heteroclita).

C

CEROSORA

microphylla *(Hook.) R. Tryon in Amer. Fern J.,* 76(4): 185 (1986): Gymnogramma microphylla Hook.

CHEILANTHES

smithii *(C. Chr.) R. Tryon in Amer. Fern J.,* 76(4): 185 (1986): Pellaea smithii C. Chr.

CHLAMYDOGRAMME R.E. Holttum in Gard. Bull. Singapore, 39(2): 157 (1986 publ. 1987).

hollrungii *(Kuhn) R.E. Holttum in Gard. Bull. Singapore,* 39(2): 157 (1986 publ. 1987): Gymnopteris hollrungii Kuhn

COLYSIS
loxogrammoides *(Copel.) M.G. Price in Contrib. Univ. Mich. Herb.*, 16: 193 (1987): Polypodium loxogrammoides Copel.

COVENIELLA M.D. Tindale in Gard. Bull. Singapore, 39(2): 169 (1986 publ. 1987).
poecilophlebia *(Hook.) M.D. Tindale in Gard. Bull. Singapore,* 39(2): 169 (1986 publ. 1987): Polypodium poecilophlebium Hook.

CTENITIS
bidecorata *D.B. Lellinger in Proc. Biol. Soc. Wash.*, 98(2): 373 (1985)—Costa Rica.
skutchii *D.B. Lellinger in Proc. Biol. Soc. Wash.*, 98(2): 375 (1985)—Costa Rica, Panama.

CYATHEA
assurgens *R. Tryon in Fl. Ecuador,* 27(13): 43 (1986)—Ecuador, Colombia.
balakrishnanii *R.D. Dixit & A.K. Tripathi in Bull. Bot. Surv. India,* 26(3–4): 170 (1984 publ. 1986)—India.
heliophila *R. Tryon in Fl. Ecuador,* 27(13): 43 (1986)—Ecuador.
nicobarica *N.P. Balakrishnan & R.D. Dixit in Indian J. Forest.*, 10(1): 44 (1987)—Nicobar Is.
pseudonanna *(Gómez) D.B. Lellinger in Proc. Biol. Soc. Wash.*, 98(2): 376 (1985): Trichopteris pseudonanna Gómez

D

DANAEA
oblanceolata *R.G. Stolze in Amer. Fern J.*, 77(1): 33 (1987)—Peru.

DENNSTAEDTIA
subgen. Mehraea *P. Kachroo in Indian Fern J.*, 3(1–2): 2 (1986).

DIPHASIASTRUM
complanatum
subsp. zeilleri *(Rouy) I. Kukkonen in Ann. Bot. Fenn.*, 23(3): 265 (1986): Lycopodium complanatum race zeilleri Rouy

DIPLAZIUM
allantoideum *M.G. Price in Contrib. Univ. Mich. Herb.*, 16: 195 (1987), nom. nov.: Allantodia sylvatica *Bl.*
beamanii *M.G. Price in Contrib. Univ. Mich. Herb.*, 16: 193 (1987)—Borneo (Sabah).
ferulaceum *(Moore ex Hook.) D.B. Lellinger in Proc. Biol. Soc. Wash.*, 98(2): 376 (1985): Asplenium ferulaceum Moore ex Hook.
mollifrons *(C. Chr.) M.G. Price in Contrib. Univ. Mich. Herb.*, 16: 195 (1987): Athyrium mollifrons C. Chr.

DIPLAZIUM :–
multigemmatum *D.B. Lellinger in Proc. Biol. Soc. Wash.*, 98(2): 376 (1985)—Costa Rica.
myriomerum *(Christ) D.B. Lellinger in Proc. Biol. Soc. Wash.*, 98(2): 376 (1985): Athyrium myriomerum Christ
navarrense *D.B. Lellinger in Proc. Biol. Soc. Wash.*, 98(2): 377 (1985)—Costa Rica.
ordinatum *(Christ) D.B. Lellinger in Proc. Biol. Soc. Wash.*, 98(2): 377 (1985): Athyrium ordinatum Christ
queenslandicum *M.D. Tindale in Gard. Bull. Singapore,* 39(2): 174 (1986 publ. 1987)—Australia (Queensland).
skutchii *D.B. Lellinger in Proc. Biol. Soc. Wash.*, 98(2): 377 (1985)—Costa Rica, Nicaragua, Panama.
solutum *(Christ) D.B. Lellinger in Proc. Biol. Soc. Wash.*, 98(2): 379 (1985): Athyrium solutum Christ

DRYOPTERIS
affinis
subsp. cambrensis *C.R. Fraser-Jenkins in Sommerfeltia,* 6: xi (1987)—Great Britain, Austria, Belgium, Bulgaria, Corsica, Czechoslovakia, Denmark, France, Germany, Ireland, Spain, Italy, Jugoslavia, Portugal, Norway, Poland, Sardinia, U.S.S.R. (Caucasus).
var. distans *(Fiori) C.R. Fraser-Jenkins in Sommerfeltia,* 6: xi (1987): Nephrodium filix-mas var. distans Fiori
var. setosa *(Christ) C.R. Fraser-Jenkins in Sommerfeltia,* 6: xi (1987): Aspidium filix-mas var. setosum Christ
×complexa *C.R. Fraser-Jenkins in Sommerfeltia,* 6: xi (1987)—Germany. (D. affinis × filix-mas).
nothosubsp. contorta *C.R. Fraser-Jenkins in Sommerfeltia,* 6: xii (1987)—Great Britain, ?Corsica. (D. affinis subsp. cambrensis × filix-mas).
nothosubsp. critica *C.R. Fraser-Jenkins in Sommerfeltia,* 6: xii (1987)—Germany, Austria, Great Britain, Bulgaria, France, Ireland, Switzerland, Hungary, Romania, Spain. (D. affinis subsp. borreri × filix-mas).

E

ELAPHOGLOSSUM
anceps *J.T. Mickel in Brittonia,* 39(3): 313 (1987)—Venezuela.
appressum *J.T. Mickel in Brittonia,* 39(3): 314 (1987)—Venezuela.
atrorubens *J.T. Mickel in Brittonia,* 39(3): 314 (1987)—Venezuela.

ELAPHOGLOSSUM :-
atrosquamatum *J.T. Mickel in Brittonia,* 39(3): 315 (1987)—Venezuela, Colombia.
chrysopogon *J.T. Mickel in Brittonia,* 39(3): 316 (1987)—Venezuela.
crispatum *J.T. Mickel in Brittonia,* 39(3): 316 (1987)—Venezuela.
var. beitelii *J.T. Mickel in Brittonia,* 39(3): 316 (1987)—Venezuela.
delicatulum *J.T. Mickel in Brittonia,* 39(3): 317 (1987)—Venezuela.
dolichopus *J.T. Mickel in Brittonia,* 39(3): 317 (1987)—Venezuela.
drewianum *J.T. Mickel in Brittonia,* 39(3): 318 (1987)—Venezuela.
eriopus *J.T. Mickel in Brittonia,* 39(3): 318 (1987)—Venezuela.
floccosum *J.T. Mickel in Brittonia,* 39(3): 319 (1987)—Venezuela.
grallator *J.T. Mickel in Brittonia,* 39(3): 319 (1987)—Venezuela.
hieracioides *J.T. Mickel in Brittonia,* 39(3): 320 (1987)—Venezuela.
incubus *J.T. Mickel in Brittonia,* 39(3): 320 (1987)—Venezuela.
luteynii *J.T. Mickel in Brittonia,* 39(3): 321 (1987)—Venezuela.
maguirei *J.T. Mickel in Brittonia,* 39(3): 321 (1987)—Venezuela.
nigrocostatum *J.T. Mickel in Brittonia,* 39(3): 322 (1987)—Venezuela.
obovatum *J.T. Mickel in Brittonia,* 39(3): 322 (1987)—Venezuela, Peru.
ornithoglossum *J.T. Mickel in Brittonia,* 39(3): 323 (1987)—Venezuela.
ortegae *J.T. Mickel in Brittonia,* 39(3): 323 (1987)—Venezuela, Colombia.
pilosius *J.T. Mickel in Brittonia,* 39(3): 324 (1987)—Venezuela.
praetermissum *J.T. Mickel in Brittonia,* 39(3): 324 (1987)—Guyana, Venezuela.
stenoglossum *J.T. Mickel in Brittonia,* 39(3): 325 (1987)—Colombia, Venezuela.
stergiosii *J.T. Mickel in Brittonia,* 39(3): 325 (1987)—Venezuela.
steyermarkii *J.T. Mickel in Brittonia,* 39(3): 326 (1987)—Venezuela.
styriacum *J.T. Mickel in Brittonia,* 39(3): 326 (1987)—Venezuela.
succubus *J.T. Mickel in Brittonia,* 39(3): 327 (1987)—Venezuela.
tachirense *J.T. Mickel in Brittonia,* 39(3): 327 (1987)—Venezuela.
tantalinum *J.T. Mickel in Brittonia,* 39(3): 328 (1987)—Venezuela.
urophyllum *J.T. Mickel in Brittonia,* 39(3): 328 (1987)—Venezuela.
vanderwerffii *J.T. Mickel in Brittonia,* 39(3): 329 (1987)—Venezuela.
vareschianum *J.T. Mickel in Brittonia,* 39(3): 329 (1987)—Venezuela.
variolatum *J.T. Mickel in Brittonia,* 39(3): 330 (1987)—Venezuela.

G

GRAMMITIS
barbensis *D.B. Lellinger in Proc. Biol. Soc. Wash.,* 98(2): 379 (1985)—Costa Rica.
cornuta *D.B. Lellinger in Proc. Biol. Soc. Wash.,* 98(2): 381 (1985)—Costa Rica, Panama.
micula *D.B. Lellinger in Proc. Biol. Soc. Wash.,* 98(2): 381 (1985)—Panama, Costa Rica.
palasaba *M.G. Price in Contrib. Univ. Mich. Herb.,* 16: 196 (1987), nom. nov.: Acrosorus nudicarpus *Zamora & Co*
pseudomitchellae *D.B. Lellinger in Proc. Biol. Soc. Wash.,* 98(2): 383 (1985)—Costa Rica, Panama.
rigescens *(Bory ex Willd.) D.B. Lellinger in Proc. Biol. Soc. Wash.,* 98(2): 383 (1985): Polypodium rigescens Bory ex Willd.
zeledoniana *D.B. Lellinger in Proc. Biol. Soc. Wash.,* 98(2): 383 (1985), nom. nov.: Polypodium taxifolium var. fragillimum *Christ*

H

HUPERZIA
binervia *(Herter) B. Øllgaard in Op. Bot.,* 92: 169 (1987): Lycopodium binervium Herter
catacachiensis *(Nessel) B. Øllgaard in Op. Bot.,* 92: 169 (1987): Urostachys catacachiensis Nessel
chamaeleon *(Herter) B. Øllgaard in Op. Bot.,* 92: 169 (1987): Urostachys chamaeleon Herter
deminuens *(Herter) B. Øllgaard in Op. Bot.,* 92: 169 (1987): Lycopodium deminuens Herter
dianae *(Herter) B. Øllgaard in Op. Bot.,* 92: 169 (1987): Urostachys dianae Herter
engleri *(Herter) B. Øllgaard in Op. Bot.,* 92: 169 (1987): Lycopodium engleri Herter
flexibilis *(Fée) B. Øllgaard in Op. Bot.,* 92: 169 (1987): Lycopodium flexibile Fée
hemleri *(Nessel & Hoehne) B. Øllgaard in Op. Bot.,* 92: 169 (1987): Urostachys hemleri Nessel & Hoehne
killippii *(Herter) B. Øllgaard in Op. Bot.,* 92: 169 (1987): Urostachys killippii Herter
kuesteri *(Nessel) B. Øllgaard in Op. Bot.,* 92: 169 (1987): Urostachys kuesteri Nessel
macbridei *(Herter) B. Øllgaard in Op. Bot.,* 92: 169 (1987): Urostachys macbridei Herter
minimadenta *J.F. Cheng in Bull. Bot. Res. North-East. Forest. Inst.,* 7(3): 113 (1987)—China.
polycarpos *(Kunze) B. Øllgaard in Op. Bot.,* 92: 169 (1987): Lycopodium polycarpos Kunze

HUPERZIA :–

pungentifolia *(Silveira) B. Øllgaard in Op. Bot.*, 92: 169 (1987): Lycopodium pungentifolium Silveira

riobambensis *(Nessel) B. Øllgaard in Op. Bot.*, 92: 169 (1987): Urostachys affinis var. riobambensis Nessel

schlimii *(Herter) B. Øllgaard in Op. Bot.*, 92: 169 (1987): Lycopodium schlimii Herter

sellowiana *(Herter) B. Øllgaard in Op. Bot.*, 92: 169 (1987): Lycopodium sellowianum Herter

tenuicaulis *(Underw. & Lloyd) B. Øllgaard in Op. Bot.*, 92: 169 (1987): Lycopodium tenuicaule Underw. & Lloyd

treitubensis *(Silveira) B. Øllgaard in Op. Bot.*, 92: 170 (1987): Lycopodium treitubense Silveira

wilsonii *(Underw. & Lloyd) B. Øllgaard in Op. Bot.*, 92: 170 (1987): Lycopodium wilsonii Underw. & Lloyd

HYMENOCYSTIS

fragilis *(Trev.) A.M. Askerov in Izv. Akad. Nauk Az. SSR, Biol. Nauk,* 1986(3): 52 (1986): Dicksonia fragilis Trev.

HYMENOPHYLLUM

mortonianum *D.B. Lellinger in Proc. Biol. Soc. Wash.*, 98(2): 385 (1985)—Colombia.

HYPOLEPIS

elegans

subsp. carolinensis *P.J. Brownsey in Blumea,* 32(2): 266 (1987)—Caroline Is.

glandulifera *P.J. Brownsey & R.J. Chinnock in J. Adelaide Bot. Gard.*, 10(1): 16 (1987)—Australia (Queensland, New South Wales, Victoria), India, Sri Lanka, New Guinea (Papua New Guinea), New Caledonia, Indonesia.

grandis *D.B. Lellinger in Proc. Biol. Soc. Wash.*, 98(2): 385 (1985)—Costa Rica, Panama.

hawaiiensis *P.J. Brownsey in Blumea,* 32(2): 240 (1987)—Hawaiian Is.

malesiana *P.J. Brownsey in Blumea,* 32(2): 248 (1987)—New Guinea (Papua New Guinea), Philippine Is., ?Sumatra.

scabristipes *P.J. Brownsey in Blumea,* 32(2): 238 (1987)—New Guinea (Papua New Guinea), Malaya, Sumatra.

I

ISOETES

bischlerae *H.P. Fuchs-Eckert in Proc. Kon. Nederl. Akad. Wetensch., C,* 85(1): 113 (1982)—Colombia.

boyacensis *H.P. Fuchs-Eckert in Proc. Kon. Nederl. Akad. Wetensch., C,* 85(1): 107 (1982)—Colombia.

brasiliensis *H.P. Fuchs-Eckert in Fl. Illustr. Catar.*, 1 (Isoetác.): 27 (1986)—Brazil.

ISOETES :–

fuscomarginata *H.P. Fuchs-Eckert in Fl. Illustr. Catar.*, 1 (Isoetác.): 12 (1986), as 'fusco-marginata'—Brazil.

muricata

subsp. asiatica *(T. Makino) H.P. Fuchs-Eckert in Proc. Kon. Nederl. Akad. Wetensch., C,* 85(2): 239 (1982): Isoetes echinospora var. asiatica T. Makino

subsp. truncata *(A.A. Eaton) H.P. Fuchs-Eckert in Proc. Kon. Nederl. Akad. Wetensch., C,* 85(2): 239 (1982): Isoetes echinospora var. truncata A.A. Eaton

rimbachiana *H.P. Fuchs-Eckert in Proc. Kon. Nederl. Akad. Wetensch., C,* 85(2): 226 (1982), nom. nov.: Isoetes lechleri var. anomala *T.C. Palmer*

sehnemii *H.P. Fuchs-Eckert in Fl. Illustr. Catar.*, 1 (Isoetác.): 38 (1986)—Brazil.

smithii *H.P. Fuchs-Eckert in Fl. Illustr. Catar.*, 1 (Isoetác.): 17 (1986)—Brazil.

spannagelii *H.P. Fuchs-Eckert in Fl. Illustr. Catar.*, 1 (Isoetác.): 33 (1986)—Brazil.

L

LASTREOPSIS

tinarooensis *M.D. Tindale in Gard. Bull. Singapore,* 39(2): 170 (1986 publ. 1987)—Australia (Queensland).

walleri *M.D. Tindale in Gard. Bull. Singapore,* 39(2): 172 (1986 publ. 1987)—Australia (Queensland).

LECANOPTERIS

celebica *E. Hennipman in Blumea,* 32(2): 316 (1987)—Celebes.

holttumii *E. Hennipman in Blumea,* 32(2): 317 (1987)—Celebes.

luzonensis *E. Hennipman in Blumea,* 32(2): 311 (1987)—Philippine Is.

LYCOPODIELLA

sect. Campylostachys *(K. Müller) B. Øllgaard in Op. Bot.*, 92: 175 (1987): Lycopodium sect. Campylostachys K. Müller

sect. Caroliniana *(Bruce) B. Øllgaard in Op. Bot.*, 92: 174 (1987): Lycopodium sect. Caroliniana Bruce

sect. Lateristachys *(Holub) B. Øllgaard in Op. Bot.*, 92: 175 (1987): gen. Lateristachys Holub

bradei *(Herter) B. Øllgaard in Op. Bot.*, 92: 176 (1987): Lycopodium bradei Herter

carnosa *(Silveira) B. Øllgaard in Op. Bot.*, 92: 176 (1987): Lycopodium carnosum Silveira

diffusa *(R. Br.) B. Øllgaard in Op. Bot.*, 92: 176 (1987): Lycopodium diffusum R. Br.

eichleri *(Fée) B. Øllgaard in Op. Bot.*, 92: 176 (1987): Lycopodium eichleri Fée

glaucescens *(Presl) B. Øllgaard in Op. Bot.*, 92: 176 (1987): Lycopodium glaucescens Presl

LYCOPODIELLA :–
iuliformis *(Underw. & Lloyd) B. Øllgaard in Op. Bot.*, 92: 176 (1987): Lycopodium iuliforme Underw. & Lloyd
lateralis *(R. Br.) B. Øllgaard in Op. Bot.*, 92: 176 (1987): Lycopodium laterale R. Br.
lehmannii *(Hieron.) B. Øllgaard in Op. Bot.*, 92: 176 (1987): Lycopodium lehmannii Hieron.
pendulina *(Hook.) B. Øllgaard in Op. Bot.*, 92: 176 (1987): Lycopodium pendulinum Hook.
ramulosa *(T. Kirk) B. Øllgaard in Op. Bot.*, 92: 176 (1987): Lycopodium ramulosum T. Kirk
riofrioi *(Sodiro) B. Øllgaard in Op. Bot.*, 92: 176 (1987): Lycopodium riofrioi Sodiro
serpentina *(Kunze) B. Øllgaard in Op. Bot.*, 92: 176 (1987): Lycopodium serpentinum Kunze
trianae *(Hieron.) B. Øllgaard in Op. Bot.*, 92: 176 (1987): Lycopodium trianae Hieron.

LYCOPODIUM
sect. Diphasium *(Presl ex Rothm.) B. Øllgaard in Op. Bot.*, 92: 172 (1987): gen. Diphasium Presl ex Rothm.
sect. Lycopodiastrum *(Holub) B. Øllgaard in Op. Bot.*, 92: 172 (1987): gen. Lycopodiastrum Holub
sect. Magellanica *B. Øllgaard in Op. Bot.*, 92: 172 (1987).
sect. Pseudodiphasium *(Holub) B. Øllgaard in Op. Bot.*, 92: 173 (1987): gen. Pseudodiphasium Holub
sect. Pseudolycopodium *(Holub) B. Øllgaard in Op. Bot.*, 92: 173 (1987): gen. Pseudolycopodium Holub

M

MEGALASTRUM R.E. Holttum in Gard. Bull. Singapore, 39(2): 161 (1986 publ. 1987).
canacae *(Holttum) R.E. Holttum in Gard. Bull. Singapore,* 39(2): 161 (1986 publ. 1987): Ctenitis canacae Holttum
exaggeratum *(Bak.) R.E. Holttum in Gard. Bull. Singapore,* 39(2): 161 (1986 publ. 1987): Nephrodium crinitum var. exaggeratum Bak.
lanatum *(Fée) R.E. Holttum in Gard. Bull. Singapore,* 39(2): 161 (1986 publ. 1987): Phegopteris lanata Fée
lanuginosum *(Kaulf.) R.E. Holttum in Gard. Bull. Singapore,* 39(2): 161 (1986 publ. 1987): Aspidium lanuginosum Kaulf.
magnum *(Bak.) R.E. Holttum in Gard. Bull. Singapore,* 39(2): 161 (1986 publ. 1987): Nephrodium magnum Bak.
villosum *(L.) R.E. Holttum in Gard. Bull. Singapore,* 39(2): 161 (1986 publ. 1987): Polypodium villosum L.

MONOMELANGIUM
dinghushanicum *Ching & S.H. Wu in Acta Bot. Austro Sinica,* 2: 5 (1986)—China.

P

PARACETERACH
bipinnata *(Christ) R. Tryon in Amer. Fern J.*, 76(4): 186 (1986): Gymnopteris bipinnata Christ
delavayi *(Baker) R. Tryon in Amer. Fern J.*, 76(4): 186 (1986): Gymnogramma delavayi Baker
marantae *(L.) R. Tryon in Amer. Fern J.*, 76(4): 186 (1986): Acrostichum marantae L.
sargentii *(Christ) R. Tryon in Amer. Fern J.*, 76(4): 186 (1986): Gymnopteris sargentii Christ
vestita *(Hook.) R. Tryon in Amer. Fern J.*, 76(4): 186 (1986): Gymnogramma vestita Hook.

PECLUMA
ptilodon
var. caespitosa *(Jenm.) D.B. Lellinger in Proc. Biol. Soc. Wash.*, 98(2): 387 (1985): Polypodium pectinatum var. caespitosum Jenm.

PHANEROPHLEBIOPSIS
coadunata *R.C. Ching in Bull. Bot. Res. North-East. Forest. Inst.*, 7(1): 73 (1987)—China.
falcata *R.C. Ching in Bull. Bot. Res. North-East. Forest. Inst.*, 7(1): 74 (1987)—China.
hunanensis *R.C. Ching in Bull. Bot. Res. North-East. Forest. Inst.*, 7(1): 74 (1987)—China.
intermedia *R.C. Ching in Bull. Bot. Res. North-East. Forest. Inst.*, 7(1): 72 (1987)—China.

PITYROGRAMMA
ebenea
var. aurata *(Moore) D.B. Lellinger in Proc. Biol. Soc. Wash.*, 98(2): 387 (1985): Gymnogramma tartarea var. aurata Moore

PLAGIOGYRIA
japonica
var. pseudojaponica *(Nakaike) K. Iwatsuki in J. Fac. Sci. Univ. Tokyo, Bot.*, 14(2): 71 (1987): Plagiogyria pseudo-japonica Nakaike

PLATYCERIUM
platylobum *A. Bidin & R. Jaman in Gard. Bull. Singapore,* 39(2): 149 (1986 publ. 1987)—Malaya.

PLEOPELTIS
fallax *(Schlecht. & Cham.) J.T. Mickel & J.M. Beitel in Amer. Fern J.*, 77(1): 21 (1987): Polypodium fallax Schlecht. & Cham.

PLEOPELTIS :–

interjecta *(Weath.) J.T. Mickel & J.M. Beitel in Amer. Fern J.*, 77(1): 21 (1987): Polypodium peltatum var. interjectum Weath.

×melanoneuron *J.T. Mickel & J.M. Beitel in Amer. Fern J.*, 77(1): 25 (1987)—Mexico. (P. crassinervata × fallax).

mexicana *(Fée) J.T. Mickel & J.M. Beitel in Amer. Fern J.*, 77(1): 21 (1987): Drynaria mexicana Fée

×sordidula *(Maxon & Weath.) J.T. Mickel & J.M. Beitel in Amer. Fern J.*, 77(1): 21 (1987): Polypodium sordidulum Maxon & Weath.

×PLEOPODIUM

bartlettii *(Weath.) J.T. Mickel & J.M. Beitel in Amer. Fern J.*, 77(1): 20 (1987): Polypodium bartlettii Weath.

fallacissimum *(Maxon) J.T. Mickel & J.M. Beitel in Amer. Fern J.*, 77(1): 20 (1987): Polypodium fallacissimum Maxon

leucosporum *(Klotzsch) J.T. Mickel & J.M. Beitel in Amer. Fern J.*, 77(1): 16 (1987): Polypodium leucosporum Klotzsch

tricholepis *J.T. Mickel & J.M. Beitel in Amer. Fern J.*, 77(1): 17 (1987)—Mexico. (Pleopeltis mexicana × Polypodium thyssanolepis).

POLYPODIUM

subviride *D.B. Lellinger in Proc. Biol. Soc. Wash.*, 98(2): 389 (1985)—Panama, Colombia.

POLYSTICHUM

mehrae *C.R. Fraser-Jenkins & S.P. Khullar in Indian Fern J.*, 3(1–2): 121 (1986)—India.

×midoriense *T. Oka & S. Ohtani in Sci. Rep. Yokosuka City Mus.*, 27: 57 (1980)—Japan. (P. fibrillosopaleaceum × ovatopaleaceum).

×yokohamense *T. Oka & S. Ohtani in Sci. Rep. Yokosuka City Mus.*, 27: 57 (1980)—Japan. (P. fibrillosopaleaceum × tagawanum).

PROSAPTIA

venulosa *(Bl.) M.G. Price in Contrib. Univ. Mich. Herb.*, 16: 197 (1987): Polypodium venulosum Bl.

PTERIDIUM

aquilinum

subsp. herediae *(Colmeiro ex Clemente) A.C. Jermy in Sommerfeltia*, 6: xiii (1987): Pteris herediae Colmeiro ex Clemente

falcatum *Ching ex S.H. Wu in Acta Bot. Austro Sinica*, 2: 2 (1986)—China.

lineare *Ching ex S.H. Wu in Acta Bot. Austro Sinica*, 2: 3 (1986)—China.

revolutum

var. muricatulum *Ching & S.H. Wu in Acta Bot. Austro Sinica*, 2: 2 (1986)—China.

yunnanense *Ching & S.H. Wu in Acta Bot. Austro Sinica*, 2: 4 (1986)—China.

PTERIS

×sintenensis *(Masamune) C.M. Kuo, Stud. Ferns Ken-Ting Nation. Park (Taiwan)*: 63 (1986): Pteris plumbea var. sintenensis Masamune

S

SELAGINELLA

keralensis *R.D. Dixit in Bull. Bot. Surv. India*, 27(1–4): 123 (1985 publ. 1987)—India.

nayarii *R.D. Dixit in Bull. Bot. Surv. India*, 27(1–4): 123 (1985 publ. 1987)—India.

SELLIGUEA

kamborangana *(C. Chr.) M.G. Price in Contrib. Univ. Mich. Herb.*, 16: 198 (1987): Polypodium kamboranganum C. Chr.

SYNGRAMMA

sect. Craspedodictyum *(Copel.) R. Tryon in Amer. Fern J.*, 76(4): 184 (1986): gen. Craspedodictyum Copel.

T

TECTARIA

sect. Sagenia *(Presl) R.E. Holttum in Gard. Bull. Singapore*, 39(2): 157 (1986 publ. 1987): gen. Sagenia Presl

subfuscipes *(Tagawa) C.M. Kuo, Stud. Ferns Ken-Ting Nation. Park (Taiwan)*: 66 (1986): Ctenitopsis subfuscipes Tagawa

THELYPTERIS

chocoensis *A.R. Smith & D.B. Lellinger in Proc. Biol. Soc. Wash.*, 98(4): 916 (1985)—Colombia.

cocleana *A.R. Smith & D.B. Lellinger in Proc. Biol. Soc. Wash.*, 98(4): 918 (1985)—Panama.

cocos *A.R. Smith & D.B. Lellinger in Proc. Biol. Soc. Wash.*, 98(4): 918 (1985)—Cocos Is. (Pacific Ocean).

crenata *A.R. Smith & D.B. Lellinger in Proc. Biol. Soc. Wash.*, 98(4): 919 (1985)—Costa Rica.

delasotae *A.R. Smith & D.B. Lellinger in Proc. Biol. Soc. Wash.*, 98(4): 919 (1985)—Costa Rica, Panama.

gomeziana *A.R. Smith & D.B. Lellinger in Proc. Biol. Soc. Wash.*, 98(4): 921 (1985)—Costa Rica, Panama.

killipii *A.R. Smith & D.B. Lellinger in Proc. Biol. Soc. Wash.*, 98(4): 923 (1985)—Colombia.

ophiorhizoma *A.R. Smith & D.B. Lellinger in Proc. Biol. Soc. Wash.*, 98(4): 925 (1985)—Colombia.

THELYPTERIS :–
proctorii *A.R. Smith & D.B. Lellinger in Proc. Biol. Soc. Wash.*, 98(4): 925 (1985)—Guatemala, Honduras.
strigillosa *A.R. Smith & D.B. Lellinger in Proc. Biol. Soc. Wash.*, 98(4): 927 (1985)—Colombia.
tapantensis *A.R. Smith & D.B. Lellinger in Proc. Biol. Soc. Wash.*, 98(4): 927 (1985)—Costa Rica.
vernicosa *A.R. Smith & D.B. Lellinger in Proc. Biol. Soc. Wash.*, 98(4): 929 (1985)—Panama.
zurquiana *A.R. Smith & D.B. Lellinger in Proc. Biol. Soc. Wash.*, 98(4): 929 (1985)—Costa Rica.

TRICHIPTERIS
bipinnata *R. Tryon in Fl. Ecuador*, 27(13): 33 (1986)—Ecuador.

V

VITTARIA
costaricensis *D.B. Lellinger in Proc. Biol. Soc. Wash.*, 98(2): 389 (1985)—Costa Rica.